The Structure of Biological Science

Also by Alexander Rosenberg

Microeconomic Laws: A Philosophical Analysis
Sociobiology and the Preemption of Social Science
Hume and the Problem of Causation (with T. L. Beauchamp)

The Structure of Biological Science

ALEXANDER ROSENBERG

Department of Philosophy, Syracuse University

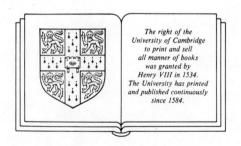

The right of the
University of Cambridge
to print and sell
all manner of books
was granted by
Henry VIII in 1534.
The University has printed
and published continuously
since 1584.

CAMBRIDGE UNIVERSITY PRESS

CAMBRIDGE

LONDON   NEW YORK   NEW ROCHELLE

MELBOURNE   SYDNEY

215794

574.01
R 813

Published by the Press Syndicate of the University of Cambridge
The Pitt Building, Trumpington Street, Cambridge CB2 1RP
32 East 57th Street, New York, NY 10022, USA
10 Stamford Road, Oakleigh, Melbourne 3166, Australia

© Cambridge University Press 1985

First published 1985

Printed in the United States of America

Library of Congress Cataloging in Publication Data

Rosenberg, Alexander, 1946–

The structure of biological science.

Bibliography: p.

Includes index.

1. Biology – Philosophy. I. Title.
QH331.R67   1985   574'.01   84–7695
ISBN 0 521 25566 X hard covers
ISBN 0 521 27561 X paperback

For Merle Kurzrock

Contents

Preface   *page* ix

Chapter 1   Biology and Its Philosophy   1
1.1   *The Rise of Logical Positivism*   2
1.2   *The Consequences for Philosophy*   4
1.3   *Problems of Falsifiability*   6
1.4   *Philosophy of Science Without Positivism*   8
1.5   *Speculation and Science*   10
      Introduction to the Literature   11

Chapter 2   Autonomy and Provincialism   13
2.1   *Philosophical Agendas versus Biological Agendas*   13
2.2   *Motives for Provincialism and Autonomy*   18
2.3   *Biological Philosophies*   21
2.4   *Tertium Datur?*   25
2.5   *The Issues in Dispute*   30
2.6   *Steps in the Argument*   34
      Introduction to the Literature   35

Chapter 3   Teleology and the Roots of Autonomy   37
3.1   *Functional Explanations in Molecular Biology*   39
3.2   *The Search for Functions*   43
3.3   *Functional Laws*   47
3.4   *Directively Organized Systems*   52
3.5   *The Autonomy of Teleological Laws*   59
3.6   *The Metaphysics and Epistemology of Functional Explanation*   62
3.7   *Functional Explanation Will Always Be with Us*   65
      Introduction to the Literature   67

Chapter 4   Reductionism and the Temptation of Provincialism   69
4.1   *Motives for Reductionism*   69
4.2   *A Triumph of Reductionism*   73
4.3   *Reductionism and Recombinant DNA*   84
4.4   *Antireductionism and Molecular Genetics*   88
4.5   *Mendel's Genes and Benzer's Cistrons*   93
4.6   *Reduction Obstructed*   97
4.7   *Qualifying Reductionism*   106

4.8   *The Supervenience of Mendelian Genetics*   111
4.9   *Levels of Organization*   117
      Introduction to the Literature   119

Chapter 5   The Structure of Evolutionary Theory   121
5.1   *Is There an Evolutionary Theory?*   122
5.2   *The Charge of Tautology*   126
5.3   *Population Genetics and Evolution*   130
5.4   *Williams's Axiomatization of Evolutionary Theory*   136
5.5   *Adequacy of the Axiomatization*   144
      Introduction to the Literature   152

Chapter 6   Fitness   154
6.1   *Fitness Is Measured by Its Effects*   154
6.2   *Fitness As a Statistical Propensity*   160
6.3   *The Supervenience of Fitness*   164
6.4   *The Evidence for Evolution*   169
6.5   *The Scientific Context of Evolutionary Theory*   174
      Introduction to the Literature   179

Chapter 7   Species   180
7.1   *Operationalism and Theory in Taxonomy*   182
7.2   *Essentialism – For and Against*   187
7.3   *The Biological Species Notion*   191
7.4   *Evolutionary and Ecological Species*   197
7.5   *Species Are Not Natural Kinds*   201
7.6   *Species As Individuals*   204
7.7   *The Theoretical Hierarchy of Biology*   212
7.8   *The Statistical Character of Evolutionary Theory*   216
7.9   *Universal Theories and Case Studies*   219
      Introduction to the Literature   225

Chapter 8   New Problems of Functionalism   226
8.1   *Functionalism in Molecular Biology*   228
8.2   *The Panglossian Paradigm*   235
8.3   *Aptations, Exaptations, and Adaptations*   243
8.4   *Information and Action Among the Macromolecules*   246
8.5   *Metaphors and Molecules*   255

Bibliography   266
Index   273

Preface

This book is an introduction to the philosophy of biology as well as an extended defense of a particular philosophy of biology. The two endeavors go together: The second will not persuade unless it offers a coherent view of the life sciences that examines and sheds light on most of their epistemological and metaphysical problems. And the first will be little more than a disconnected series of insights into isolated puzzles that do not add up to an improved understanding of the subject as a whole unless it manifests a unifying theme. Despite their many virtues, previous introductions to this subject suffered for want of such a theme, which shows how the problems of each subject are interconnected and, more important, how the solutions to each of them constrain the treatment of others. A second feature of these works is that they reflect a philosopher's agenda of problems in the philosophy of science, instead of a biologist's concerns with understanding biology. The present work aims at meeting these two needs. There is a third reason why a new introduction to this subject is required. In the decade or more since introductions to the philosophy of biology first appeared, biology itself has undergone revolutionary development, especially in its biochemical division. Meanwhile, new controversies surrounding evolutionary theory have also arisen. The details of neither of these two developments could have been anticipated, but they have materially influenced the agenda of the philosophy of biology in the years since the first introductions appeared.

In one more respect, this work differs from previous introductions to the philosophy of biology, for it presumes more biological sophistication. It does so for three reasons. First, it has had to come to terms with so much that was unknown a decade or more ago. Second, it focuses expressly on what I take to be biological concerns. Third, it constitutes a philosophy, an organized system, a definite position about the nature and extent of biological knowledge. To the extent that it offers a distinct thesis about biology it is not a presentation of all sides on current and past controversies in the discipline. To carry out the extended argument for this thesis that the work constitutes, I have had to abbreviate the exposition of certain philosophical matters. Much of the philosophical stage setting I have curtailed is available in previous introductions to the subject, and at the end of each chapter I have provided an introduction to this literature. Because the book is meant for biologists I have tried to say enough, especially about philosophical motivation, to show what the philosophical problems are and why they are biologically serious. But because the work may also be a convenient way to report to philosophers the striking achievements of the most recent period of biological research, I have tried not to stint on biological details, especially in discussing the ramifications of methods and findings in the study of macromolecules.

My aim in undertaking this project was to provide a physicalist, materialist, reductionistic account of all biology, subject only to the constraint of doing as much justice to its achievements as they deserved. That is, I have made as few concessions to the material or formal distinctness of biology from the physical sciences as is consistent with its actual character. The result has been rather different both from what I expected and from what those familiar with my previous views will predict. For it turns out that doing justice to the science of biology results in what is at best a very limited, indeed hollow, vindication of reductionism and materialism, and a refutation of antireductionists that leaves biology with as much autonomy as their view of it really requires and rather more than many a reductionist is comfortable conceding.

Among those in whose debt the pursuit of this project placed me I thank first the authors of those introductions to this subject that I now ungratefully claim to supplant: David Hull and Michael Ruse. Their books, then their stimulation, and finally their encouragement have led me both to formulate a philosophy of biology and to expound it in the present terms. I am grateful to both for detailed comments on earlier versions of many of the discussions and arguments that follow, and I am especially indebted to David Hull for reading and improving the whole of a previous draft of the work. For help of equal magnitude, and for saving me from several blunders, I must thank Philip Kitcher. I am only sorry that I could not produce the result that would fully repay his painstaking help. Additionally, portions of this book are heavily indebted to Philip Kitcher's own work. Indeed, a crucial portion of Chapter 4 reports his results, and fully half of Chapter 7 recapitulates his insights (with conclusions he would not endorse, however). Similarly, I am indebted to David Hull and Michael Ghiselin for the latter half of Chapter 7, and in the first half of that chapter I am indebted to Elliott Sober's insightful treatment of essentialism. Chapter 5 is informed by Mary Williams's approach to evolution, and I am indebted to her for encouraging me to exploit it. Parts of this chapter were also heavily influenced by published and unpublished writings of John Beatty. For using so much of the intellectual capital in which these seven philosophers have so heavily invested, I am grateful almost to the point of embarrassment.

For reading and commenting on various parts of this book, I also thank Jonathan Bennett, Daniel Hausman, Richard Burian, William Wimsatt, William Starmer, Joan Straumanis, Peter van Inwagen, Paul Teller, William Bechtel, Alan Nelson, Donald T. Campbell, Stuart Kauffman, Jaegwon Kim, and Peter Richarson. For what virtues of readability this work has I am indebted to Jonathan Bennett's encouragement and Alfred Imhoff's copy editing.

What I understand of biology and its latest accomplishments I owe to William Starmer, John Vournakis, Calvin Vary, David Sullivan, Richard Levy, Thomas Fondy, Darrel Falk, Barbara Vertel, and Samuel Chan.

This work was begun with the support of the American Council of Learned Societies, pursued under a fellowship from the John Simon Guggenheim Memorial

Foundation, and brought to completion through a grant from the National Science Foundation. I am grateful to these institutions for supporting my research.

Finally, I have learned much about heterozygote superiority from Bloomsbury and even more about Gene-regulation and Gene-expression from my son.

Syracuse, New York
August 1984

CHAPTER 1

Biology and Its Philosophy

In August of 1838, after hitting upon a mechanism for evolution, Charles Darwin confided to his notebook: "Origin of man now proved. — Metaphysics must flourish. — He who understands baboon would do more towards metaphysics than Locke" (Barrett, 1974:281). Any philosopher — and many a biologist — coming upon this prediction over the next century and more would certainly have thought it quite false. Metaphysics, the philosophical examination of the ultimate nature of reality, did not flourish during the hundred years after Darwin published *On the Origin of Species.* Indeed, it came close to vanishing. And the causes for the disappearance of philosophical and theological speculation throughout this period were to be found in the influence of Darwin's own theory.

If ever there was a theory that put an end to traditional philosophizing, it was the one Darwin expounded. By providing a single, unified scientific theory of "the origin of man" and of biological diversity generally, Darwin made scientifically irrelevant a host of questions that philosophers and scientists had taken seriously since long before the time of John Locke. The theory of natural selection has put an end to much speculation about the purpose of the universe, the meaning of life, the nature of man, and the objective grounds of morality. It has grievously undermined the theologian's most compelling grounds for the existence of God, the argument from the earth's design to the existence of a designer. Philosophers and biologists certainly recognized this effect of Darwinism, and over the course of the decades after 1859 some of them made great efforts to refute the theory as much on philosophical grounds as on biological ones. Among biologists, this work has had ever-diminishing influence, and antievolutionary philosophy has almost completely disappeared within biology. As indeed has almost all philosophy as traditionally conceived.

By making the traditional questions of philosophy biologically irrelevant, Darwin also helped make them philosophically disreputable. But when the grand questions of metaphysics were expunged from philosophy, there seemed to be nothing left to the subject but "logic chopping" and "mere semantics." Thus philosophy as a whole lost its interest for most scientists. The conclusion seems inescapable that Darwin put an end to philosophizing, at least about biological matters. By and large, Darwinians and anti-Darwinians have agreed on one thing: If Darwin was right about the origin of man, metaphysics should vanish, not flourish. For a long time, therefore, Darwin's prediction about his revolution's effects on philosophy seemed quite wrong.

But the more recent history of philosophy, and especially the philosophy of science, has vindicated Darwin after all. This chapter traces the course of the reflections that did so. This brief history of how traditional philosophical issues became respectable again in philosophy is at the same time the best argument for biologists taking the philosophical examination of their subject matter with the utmost seriousness. The history to be briefly surveyed is that of Logical Positivism – or Logical Empiricism, as some of its proponents called it. The rise and fall of this movement in the philosophy of science has revealed that the philosophy of a science is part and parcel of that science itself. The questions philosophers deal with do not differ in kind from those scientists face. Some differ in generality and in urgency, but none is a question that scientists can ignore as irrelevant to their discipline and its agenda. This means that the justification for pursuing the philosophy of science is nothing more or less than the justification for science itself.

Those who do not need to be convinced of the importance to science of philosophy and those eager to come immediately to grips with the philosophy of biology may safely leave this chapter to another occasion. Doubters may, however, profit from reviewing the argument of this chapter, for it provides the strongest basis possible for the biological relevance of the philosophy of science and does so through the examination of doctrines to be met again later in this work. Nevertheless, it is worth noting that this chapter proceeds at a level of generality much more removed from biology than the rest of the book. Indeed, the level is general enough that, if the argument to be presented is correct, this introduction can serve as the last chapter of this book instead of its first.

1.1. The Rise of Logical Positivism

Logical Positivism has certainly been the most important movement in the twentieth-century philosophy of science. Let us trace its motives, chief doctrine, and gravest difficulties. The motives were laudable, the doctrines striking, and the difficulties insurmountable. In surrendering the doctrines of Logical Positivism while honoring its motives, the philosophy of science transformed itself into something indistinguishable from science itself.

It is convenient to begin our exposition of Positivism with an important achievement of nonevolutionary biology. Throughout the latter half of the nineteenth century, embryology was at the forefront of experimental research. Among the most important of embryological experimentalists was Hans Driesch. Two striking laboratory discoveries are associated with his name. Working with sea-urchin eggs and embryos, he was able to demonstrate that the physical deformation of the egg and the subsequent rearrangement of the blastomeres – the cells produced in the first few stages of fission – had no effect on the normal development of the embryo. This experiment suggests that spatial relations among early blastomeres are irrelevant to normal development. Even more strikingly, Driesch went on to show that a single blastomere isolated from the rest at the two- or four-cell stage can give rise to a complete sea-urchin embryo normal in every respect except size.

Driesch is honored in every account of embryology for these crucial experimental discoveries. But he is ridiculed for the explanatory theory that he offered to account for them. The fact that an embryo, or indeed a single cell, can regulate its development to compensate for missing cells suggested to Driesch the operation of an organizing principle, which he dubbed an "entelechy" (after a similar notion in Aristotle's philosophy), and which he held to determine the harmonious development of living things and to distinguish them from inanimate ones. Because spatiotemporal location and physical mass seemed irrelevant to development, physics could not account for embryological phenomena. Their causes must, he thought, be sought in nonmaterial forces. Therefore he adopted the view that entelechies have a nonspatiotemporal mode of existence, although they act "into" space and time. Entelechies are elementary "whole-making" factors that have no quantitative characteristics, are unanalyzable, and, according to Driesch, are knowable to the scientist only by reflection on the orderliness of direct human experience. It was perhaps inevitable that the temptations that led this important experimentalist to adopt such speculative explanations for the startling observations he made eventually overcame his biological interests altogether. Driesch ended his days as a professor of philosophy. Contemporary works still reprint his most important experimental papers but add cautions like the following: "Most embryologists, however, have had no difficulty in explaining regulation in terms of known physiological processes, making superfluous Driesch's mystical interpretations" (Gabriel and Fogel, 1955:210).

Driesch's entelechy is just the sort of occult entity that has long bedeviled all the natural sciences. The Logical Positivist philosophers of the first half of the century expounded a philosophy of science that would eliminate such speculative metaphysics from legitimate science, that would enable us to objectively distinguish empirical claims from disguised pseudoscience like astrology and antiscience like special-creationism, and that would also determine the scope and form of intellectually respectable philosophical examinations of science. Because, according to these Logical Empiricists, knowledge is either based on observation and experiment, as in the sciences, or on formal deduction from definitions, as in mathematics, whatever transcended these limits could be safely disregarded as scientifically, or cognitively, meaningless – indeed, in the view of some, as quite literally nonsense. In the view of some of these philosophers, a claim like Driesch's that nonphysical entelechies control the development of embryos was on a par with Lewis Carroll's nonsense verse from Alice in Wonderland: "Twas Brillig and the Slithy toves did gyre and gimble in the wabe . . ."

What Logical Positivists required to eliminate metaphysical nonsense from empirical science was an objective principle or test that could be applied to statements and terms from any discipline and that would decide about the cognitive significance of the claim or concept. These philosophers searched for a principle of meaningfulness that made no demands on the specific *content* of scientifically legitimate statements but required them to have a specified relation to actual and possible empirical evidence that could test them. The history of the school of Logical Positivism is the history of attempts to find the correct formulation of such a principle. Positivists knew roughly what it had to look like, and they knew broadly what systems of statements clearly passed its standard as meaningful and what sets of statements plainly failed as meaningless. Paradigm cases, of meaninglessness like Driesch's entelechies on the one hand, and meaningfulness like Rutherford's electrons on the other, were employed to calibrate varying candidates for a satisfactory principle of "cognitive significance." Such a principle had to rule the former as meaningless and the latter as meaningful. Because the mark of science is that its claims are controlled and justified by experiment, observation, and other forms of data collection, Positivists held that, to be meaningful, expressions have to be empirically testable by observation and experiment. Those that are not have no more role to play in science than the statement that "green ideas sleep furiously." They may look respectable, and satisfy the rules of grammar of the languages they are couched in, but these pseudosentences on whose truth or falsity the empirically ascertainable facts cannot bear are literally *non*sense, or at any rate without scientific significance.

Problems arose for Positivists in formulating a manageable principle that operated along these lines and gave the right answers for the calibrating samples. Consider what is required for empirical testability. If complete verification by observations is required for testability, almost no sentences except those reporting immediate sensations are testable. Statements of physics about unobservable entities like electrons and quarks will turn out to be meaningless. Even general laws about regularities among observable phenomena will fail the test because they cannot be strictly verified, expressing as they do a claim about an indefinitely large number of events.

Accordingly, the notion of empirical testability was revised and weakened to allow for the theoretical entities of science and for the generality of its most characteristic claims, its laws and theories. Instead of strict and direct verifiability, Positivists opted for indirect confirmability: A statement is scientifically meaningful if and only if there is actual or possible empirical evidence that tends to confirm, though perhaps not completely verify, the statement. But the notion of confirmation is an unsuitably vague one, so vague that Driesch's entelechy theory might even pass its muster. Therefore many philosophers, as well as sympathetic scientists, were attracted to another formulation of cognitive meaningfulness, one due originally to Karl Popper. Its particular attraction is its ability to pass the general laws and theories characteristic of science as meaningful while excluding Driesch's entelechy theory. Verifying a law requires an indefinitely large number of positive instances, but only one negative instance seems required to falsify a law. By contrast, on Driesch's own exposition of his theory, claims about entelechies are unfalsifiable by experiments because entelechies have no quantitative properties, nor even a spatio-temporal location for that matter. Thus, it has long and widely been held, especially by scientists themselves, that the mark of a scientifically respectable proposition is that there be actual or possible empirically detectable states of affairs that could *falsify* it.

1.2. The Consequences for Philosophy

Following through on Positivist strictures on the meaningfulness of statements had the profoundest consequences for philosophy and especially for the philosophy of science: These disciplines were restricted largely to the treatment of purely "semantic" questions, in the most pejorative sense of that term. Philosophy is not an experimental science; it can claim neither a special range of facts as its subject matter nor any nonempirical mode of knowledge of the facts the "real" sciences study. It must, in the Positivist view, limit itself to the provision and examination of definitions, stipulations, and conventions about language, and to the study of their formal relations. Any other philosophical enterprise was condemned to intellectual disreputability, to the cognitive meaninglessness that characterized so much pre-twentieth-century metaphysics. It was for this reason that twentieth-century philosophy became largely the philosophy of language and that the philosophy of science became the study of the implicit and explicit definitions of the terms ubiquitous in science – like 'law,' 'theory,' and 'explanation' – and of the terms of the special sciences – like 'mass,' 'element,' and 'phenotype.' The outcome of such investigations could at most be increased clarity about usage or proposed improvements in terminology, justified by considerations of convenience and simplicity.

So circumscribed, the philosophy of science has little to offer the sciences. It may show that the way in which physicists employ the term 'law' differs from the way biologists do, or that what the latter call 'explanations' differ from what the former do. But it can hardly assess or adjudicate substantive matters within or between the sciences. According to Positivist teachings, even the linguistic differences philosophy might uncover, and the distinct patterns of reasoning it can reveal, have no factual import, for they reflect conventions utterly independent of any fact of the matter. Such linguistic differences between sciences cannot constitute or reflect anything about the nature of the sciences' subject matter.

Philosophy, along with mathematics and logic, had long been a priori disciplines, domains in which truths have always been deemed *necessary* ones. It is just because of the necessity of mathematical truths, and for that matter philosophical ones, that they had to be known a priori: Experience never reveals the necessity of any truth it communicates. This is because claims of experience are falsifiable: Things can always be conceived to be different from the way they are experienced. But now the Logical Positivists thought they knew why mathematics and philosophy were a priori and necessary. It was not because the philosopher and mathematician had a special faculty of insight into necessary truths more firmly fixed, more secure, and more important than the merely contingent findings of empirical science. The truths of mathematics, and those philosophical claims left after the banishment of metaphysics, are necessary and a priori because they are disguised or undisguised *definitions* and the logical consequences of definitions. These truths are necessary because they have no content, restrict no factual possibilities, and merely express our conventions to use words in certain ways. They are vacuous trivialities. Philosophy provides a priori knowledge because it provides linguistic knowledge, not factual knowledge. As such, it does not compete with or cooperate with the sciences in providing factual knowledge. Because its only legitimate claims are not falsifiable, philosophy was condemned to a derivative role of clarifying and reconstructing the expression of factual knowledge, but not adding anything to it.

Positivists were willing to bear the high cost of casting down philosophy from its throne as queen of the sciences mainly because in doing so they were also ending the baleful effect of metaphysical speculation and pseudoscience on the real advance of knowledge.

For all its neatness and rigor, the Positivists' program fell apart in the immediate postwar period. It did not come unstuck through the attacks of its opponents and detractors, disgruntled metaphysicians who thought that philosophy did provide an alternative route to real knowledge that science could not reveal. The Positivists' program came apart at the hands of the Positivists themselves and of their students.

They found that its fundamental distinctions could not be justified by Positivism's own standards of adequacy. The collapse of Logical Positivism is best illustrated for our purposes by examining more closely the claim that scientific knowledge must be falsifiable. More than any other slogan, this one has become the outstanding shibboleth of contemporary biological methodology.

1.3. Problems of Falsifiability

A proposition is scientific if and only if it is falsifiable. This is the criterion or principle of falsifiability. Falsifiability must be distinguished from falsity, of course. To *falsify* a proposition, that is, to show it is false, it is sufficient to infer from it some implication that is in fact not borne out by observation or experiment. For a proposition to be *falsifiable* it must only be logically possible to do this, not actually, physically possible; otherwise we should have to say that a true empirical law is unfalsifiable because it cannot in fact be shown to be false.

Consider such an expression as, say, Ohm's law, which states the relation between resistance, voltage, and amperage: $R = E/I$. To test the simple claim that, for a potential-difference of E volts, and a current of I amperes, the resistance, R, in ohms, is equal to E/I, we require an ammeter, a voltmeter, an ohmmeter, a conductor, a resistor, and a source of electrical potential. Testing Ohm's law by setting up the appropriate circuit and observing the deflection of the point on the ohmmeter while varying the voltage and amperage requires a host of subsidiary, auxiliary hypotheses be true: not just assumptions about the presence of an electrical potential, or that the meters are functioning properly. What is assumed when Ohm's law is put to the test is the whole body of physical and electrical theory that, first, underwrites the construction and reliability of the meters; second, enables us to alter the amperage and voltage; third, assures us we can ignore certain forces acting on the circuit; and, fourth, adjusts for other forces. In particular, trusting the voltmeter involves embracing Maxwell's equations, which describe how the electric field generates a magnetic field, which twists the needle on the meter's dial. Additionally, we must implicitly appeal to Newtonian mechanics, which governs the needle's resistance to a spring and its deflection of a pointer. Accordingly, all these assumptions, hypotheses, and background theories meet the test in a body, together with the law we set out to test. Science meets experience not sentence by sentence, but in large blocks of theories and laws, blocks that are themselves divided from others by only constraints of practical manageability. Adopting these constraints constitutes substantial contingent theoretical commitments.

Suppose, now, that in our test of Ohm's law the meters do not read as the law predicts. Where does the fault lie: What proposition is falsified? Ohm's law? The assumptions about the construction and reliability of the meters? The assumption that there are no relevant intervening forces, or that they can be neglected? Are Newton's laws or Maxwell's equations at fault, or is the special theory of relativity that lies behind them? Of course it will be replied that none of these wider theories is thrown into doubt by such a test. Good sense directs that we check the wiring, the conductance of the metal it is made of, the springs in the meters, etc. So far as practical matters are concerned, once defects at this level are excluded, it is Ohm's law that would be suspect. But so far as matters of *strict* falsifiability are concerned, we see there is no such thing. For a disconfirmation does not point the finger at one particular statement under test; there is no one statement under test, for the entire conjunction of propositions is required for the prediction that fails. We are free to give up any one of the conjuncts and preserve all the rest. And this is not a mere matter of logic; the actual practice of scientists interpreting their data often reflects this freedom. Indeed, the most radical of scientific revolutions results from a scientist finding the fault to which an experimental anomaly points deeply in the center of a research program, instead of at its peripheral assumptions about the accuracy of measuring instruments.

How deeply can the falsification of a test, or of several of them, point? In the history of science it has certainly pointed at least as far as the falsity of Newtonian physics and its "philosophical" assumption of causal determinism. The discovery of the irreducibly random phenomena of radioactivity in effect falsified the belief behind Newtonian mechanics that every event has a cause that produces it in accordance with strict and exceptionless laws. Quantum mechanics rests on the rejection of a Newtonian principle that physicists and philosophers spent two hundred years attempting to prove as a necessary truth of metaphysics. In fact, difficulties in reconciling quantum mechanics with the most fundamental aspects of physical theory and its mathematical structure have led to the questioning of even more central and more "metaphysical" assumptions. In particular, some philosophers and physicists view the Heisenberg uncertainty principle of quantum mechanics as good reason to surrender the logical principle of bivalence, that every meaningful proposition is either true or false. Even more radically, responsible physicists have held that recent experiments require either the surrender of quantum mechanics or the "metaphysical" thesis that there is a world of enduring physical objects that exist independently of our knowledge of them. If these two proposals are coherent, then the experimental evidence that tests quantum mechanics can lead us to surrender, for factual reasons, principles of logic and mathematics we supposed to be necessarily true, and metaph 'cal theses Positivists supposed to be without empirical significance.

If any proposition can be surrendered as a result of a falsifying experiment, and if in the actual history of science the most central and firmly held of our beliefs have sometimes been surrendered, then we cannot identify propositions as necessarily true – as propositions we embrace *come what may* – that are known a priori. We cannot draw a contrast between such statements and contingent factual propositions – statements that may or may not survive attempts at falsifications – and so have scientifically significant empirical content. Similarly, any proposition, no matter how apparently factual, no matter how apparently vulnerable to falsification, can be preserved in the face of any possible falsifying experiment. We may in all consistency maintain that the earth is flat, attributing all apparent evidence against this belief to the falsity of one or another of the auxiliary assumptions that, together with it, are jointly falsified in photographs of the earth taken by an astronaut. Similarly, claims that Positivists stigmatized as pure metaphysics may also be surrendered in the aftermath of a falsifying experiment. Is the thesis of thoroughgoing universal determinism one of metaphysics? Is it scientifically empty speculation to assert that every event has a cause? It has certainly been a traditional thesis of philosophy, and yet it is one that has certainly come to be doubted as a result of the discovery of quantum-mechanical phenomena.

If we are to conclude that quantum phenomena have falsified metaphysical determinism, then we must conclude that metaphysical principles are testable after all and therefore cognitively significant. The only way to deny this power to experiment and observation is to deny that they ever falsify any single proposition at all. Either way, falsifiability no longer distinguishes between meaningless metaphysics and factual science.

Testing Ohm's law involves adopting Maxwell's equations for electromagnetism, and adopting these involves buying into the relativistic electrodynamics that accounts for them. And behind this theory stands the post-Newtonian "world picture," the research program that has animated modern science since the seventeenth century. It would of course be fatuous to hold that all this is at risk when an experiment does not corroborate Ohm's law. Any concern that would give an experimentalist real pause must be livelier than this abstract possibility. Even a theorist need not lose any sleep over the furthest mathematical, conceptual, and logical foundations of modern science. But the theorist cannot hold them logically irrelevant to his or the experimentalists' day-to-day concerns, and he has assuredly taken sides on their truth. What is more, at least sometimes in the history of science, and the lives of scientists, these broadest theoretical concerns do take a serious turn — either because they are called into question or because they suggest a direction for research.

These conclusions provide cognitive legitimacy to the speculative philosophy from which the Logical Positivists thought themselves to have freed "real" science. The justification for eliminating or embracing such notions as Driesch's entelechy is no different in kind from that employed to assess claims about the existence of electrons, magnets, or virons. It differs from them by degree, and very great degree at that. But ridding biology of such notions is not after all a matter of applying some rule against useless metaphysics. For deciding on the existence or nonexistence of entelechies is nothing less than questioning the adequacy of competing embryological theories altogether. But because this question is surely not an excursion into cognitively meaningless speculation, it follows that disputes about entelechies are not scientifically idle after all. Driesch's vitalism or the mechanism it opposed are indeed metaphysical theories, but they do not stand apart from "real" science. For better or worse, they stand on a continuum from sheer speculation through research programs and grand unifying theory to general theory and special models, all the way across to particular empirical findings. Unpalatable as this conclusion may be for empiricist philosophers and empirical scientists, to deny it without providing a workable distinction somewhere along the continuum would be unprincipled dogmatism — a dogmatism that the Positivists and their students would not accept.

1.4. Philosophy of Science Without Positivism

The end of Positivism means an end to philosophy's proscriptions against either treading on the subject matter of the empirical sciences or engaging in empty metaphysics. For metaphysics can no longer be distinguished from theoretical science. And neither can be distinguished from logic, linguistic conventions, or their analysis. For the necessity and unrevisability that was supposed to mark these subjects also fails to distinguish them from science or metaphysics. Although the fall of Positivism frees philosophers (and scientists for that matter) to turn their attention to more exciting activities than the study of language, it also transforms the significance of the very study. It turns the linguistic and logical analysis Positivists produced into the kind of metaphysical and epistemological exploration of the foundations of science to which philosophy has traditionally attended. It reveals that the analysis of concepts is just metaphysics carried out under a different name.

This change is well illustrated in the philosophical problems generated by the apparent goal-directedness, or purposiveness, of living things. The *teleology* (from the Greek words for "ends" or "goals" and their study) of the animate world has always been a focus of philosophical debate. Vitalists held that the purposiveness of things could only be the result of special forces, like Driesch's entelechies; mechanists insisted that teleology was only a special and complex form of mechanical causality, ultimately to be understood through the application of physics and chemistry alone. Materialism is of course just as metaphysical a thesis as vitalism. So Positivism invoked a plague on both these houses and enjoined philosophers to turn their attention to the purely linguistic question of giving the *meaning* of characteristic teleological expressions of biology. A cottage industry sprang up, in which philosophers provided definitions of terms like 'goal,' 'purpose,' and especially 'function'; these definitions were in turn rejected by other philosophers on the strength of counterexamples – clear cases of teleology that did not satisfy the definition or, still worse, nonteleological phenomena that did; the result was a cycle of revisions, qualifications, and reformulations that elicited another round of counterexamples, and so on.

With hindsight, however, philosophers came to see that the question of whether teleological expressions are definable in nonteleological physical terms is really just the ancient debate between vitalists and materialists carried out under the guise of linguistic analysis. If teleological statements can be translated into nonteleological, causal ones, then teleological processes are causal ones. If there is no difference between the formal claim about translation and materialists' allegedly factual one that living systems are just physical systems, then the linguistic question is identical to the metaphysical question of whether vitalism or materialism is correct.

In fact, the distinction between linguistic, metaphysical, and methodological problems and empirical issues is groundless. Biologists' attempts to uncover the purely causal mechanism of an apparently goal-directed activity like photosynthesis may or may not succeed. If it does, then this may strengthen a materialist metaphysical view. It will certainly encourage the continued exploitation of a methodology of searching for causal mechanisms to explain teleological behavior. But, of course, success in any one area of investigation cannot establish the general claim that all purposive phenomena are really causal. Nor does it establish the universal propriety of the methodology of searching for such mechanisms. What would? Well, nothing can ever be *established* in science. Nevertheless, a cogent explanation of why this method works will certainly strengthen the confidence of one biologist's particular account of photosynthesis.

On the other hand, suppose no causal mechanism for some goal-directed phenomena is detected, despite great effort. Under such conditions, biologists would be within their rights to insist that, nevertheless, further industry – better experimental materials and techniques – will eventually reveal such nonteleological mechanisms. Their conviction in this case may even be stigmatized as unfalsifiable. Nothing will convince them that the phenomena are irreducibly teleological. To this extent, the biologists' convictions about the facts and the appropriateness of their methods are in effect metaphysical. What if they succeed? Shall we withdraw the charge that the biologists' convictions were empty metaphysics, or shall we recognize that in the last analysis metaphysics is either unavoidable or indistinguishable from empirical science?

By providing and criticizing various definitions of teleological phenomena in terms of causal mechanisms, philosophers are in effect taking part in biologists' searches for such mechanisms. If biologists succeed, then the explanation for their success will appeal to the adequacy of philosophers' definitions of the phenomena. If biologists fail, then the explanation of the failure may appeal to philosophical criticisms of the definition under which the phenomena are described. Failure must in the long run support vitalists' alternative theories as much as success sustains materialists' views.

1.5. Speculation and Science

The decline of Positivism has thus led to a recognition of alternative philosophies of biology as explanatory theories of extremely high levels of generality. Philosophies of science both explain the most general characteristics of the objects of scientific investigation and also explain the successes and failures, the limits and prospects of alternative theories and methods in the sciences. Grand speculations in the philosophy of biology motivate and justify research programs. Their influence extends right through the hierarchy of theories, models, and experimental designs all the way down to findings of particular fact. Such commitments and convictions can retard as well as foster scientific progress by wrongly excluding lines of research, potential explanatory hypotheses, or improvements in intertheoretical depth and unity. Whether they foster or retard advance, they cannot be written off.

Nevertheless, because they are so far removed from the day-to-day work of biologists, it would be a mistake to think that all the matters broached in such speculations need be settled before biological progress can be made, recognized, or certified. It would be equally silly to suppose that the day-to-day progress of the field can be expected to settle these questions once and for all. To a large extent, therefore, the biologist is right not to keep one eye firmly focused on controversies in philosophy, and the philosopher should not expect that metaphysics and epistemology require the minutest attention to the latest biological results. If the importance of the philosophy of science for biology rested only on the distant relations between speculation and observation, biologists would probably do better to err on the side of neglecting philosophy rather than the side of absorption in it to the neglect of narrower matters.

But, in fact, specific biological and philosophical results are in much closer contact than it might be supposed, as the rest of this work will show. The area of contact is in fact so wide that biologists and philosophers find themselves stumbling into one another's field, sometimes inadvertently and sometimes intentionally. At these points of contact, what counts as philosophy and what counts as biology become matters of arbitrary labeling at best. Because it is impossible to say where one field stops and the other starts, decisions about the significance of an inquiry in the no-man's-land between these two subjects must be made on their merits and not by appeal to jurisdictional determinations or dicta about the irrelevance of these subjects to one another. In fact, at least some of what contemporary philosophers have accomplished in their study of biology is not recognizably philosophy and so may slip by unnoticed in any discussion of the bearing of their subject and its methods on biology.

The remainder of this book will focus on a connected set of specific problems at the intersection of biology and philosophy, in a domain that cannot be conveniently labeled philosophy or biology exclusively. Because the problems are as much biological as philosophical, the justification for examining them is nearly the same as the one biologists advance for their work: Their examination can be expected to increase our knowledge of the world and our ability to improve on the knowledge we have already acquired.

The broad questions of metaphysics for whose legitimacy and importance the present chapter has argued will be largely left aside in what follows. This is chiefly because the broadest theories in the philosophy of biology can only be assessed by answering the narrower, technical problems of the discipline. In this respect, the philosophy of biology is no different from biology itself: The most theoretical issues in evolution are to be settled only through detailed investigation of model systems in the *Drosophila* lab or by sifting stratigraphic records in the paleontologist's dig.

In the final analysis, however, the justification for pursuing the philosophy of biology rests on the fact that biologists cannot avoid the great questions that transcend their day-to-day concerns. For if there are correct answers to the questions faced every day in the lab and the field, and if the theories biologists propound are definitely true or false as a matter of the objective facts about the way the world works, then there must also be correct answers to the great questions of metaphysics and epistemology as well. If there is objective knowledge in biology, there is objective knowledge in its philosophy as well, for the two subjects are indistinguishable and inseparable. This is why Darwin was correct after all. For no theory has had greater impact on providing biological knowledge than his, and this expansion of knowledge must result in the flourishing of metaphysics.

Introduction to the Literature

The classical English-language introduction to the philosophy of Logical Positivism is A. J. Ayer, *Language, Truth and Logic* (New York, Dover Books, 1961). Many important papers by Positivists are brought together in A. J. Ayer, *Logical Positivism* (New York, Macmillan, 1959), which also contains an extensive bibliography.

A well-known Positivist attack on metaphysics in biology, and specifically on Driesch, is to be found in Moritz Schlick, "The Philosophy of Organic Life," in H. Feigl and M. Brodbeck (eds.), *Readings in the Philosophy of Science* (New York, Appleton, Century, Crofts, 1953, pp. 523–36). J. H. Woodger, *The Axiomatic Method in Biology* (London, Routledge and Kegan Paul, 1937) and *Biology and Language* (Cambridge, Cambridge University Press, 1952), are the works of a biologist strongly influenced by Logical Positivism.

Popper's criterion of falsifiability was first advanced in Karl Popper, *The Logic of Scientific Discovery*, originally published in German in 1935, and in English in 1958 (London, Hutchinson; New York, Harper & Row, 1963). The strongest arguments against a falsifiability criterion of scientific legitimacy have their earliest sources in a work that antedates Popper by over twenty years: Pierre Duhem, *The Aim and Structure of Physical Theory*, originally published in French in 1914, and in English in 1954 (Princeton, N.J., Princeton University Press). The importance of Duhem's thoughts only became widely recognized after arguments similar to them were mounted by Willard Van Orman Quine. In the most influential essay of twentieth-century American philosophy, "Two Dogmas of Empiricism," Quine attacked the view, shared by Positivists and Popperians, that a distinction can be drawn between necessary truths and contingent ones, or between empirically testable and untestable statements. This essay more than any other led to the abandonment of the Positivist program in the philosophy of science and to the liberation of philosophy from Positivism's linguistic limitations. The argument of this chapter is heavily influenced by "Two Dogmas of Empiricism." This essay and other influential ones are reprinted in W. V. O. Quine, *From a Logical Point of View* (New York, Harper & Row, 1961).

W. V. O. Quine, *Word and Object* (Cambridge, Mass., MIT Press, 1960) is an impressive illustration of philosophy that honors Positivism's motives while rejecting its restrictions. Thomas Kuhn, *The Structure of Scientific Revolutions* (Chicago, University of Chicago Press, 1970) is an extremely influential introduction to the interplay of philosophy and science. Its debts to Quine are manifest and acknowledged. Quine would not, however, endorse all Kuhn's claims on their most popular interpretation.

CHAPTER 2

Autonomy and Provincialism

Biology is not a physical science. Is this statement merely a truism, reflecting a bit of nomenclature? Does it mark merely an administrative boundary between scientific disciplines? Or are the life sciences different from physics and chemistry in respects important enough to turn the truism into an important conclusion about the nature of the different subjects of these disciplines and about different means appropriate for studying them? This is the central question of the philosophy of biology. Whether and how biology differs from the other natural sciences is not the only question addressed in the philosophy of this subject, but it is the most prominent, obvious, frequently posed, and controversial issue the philosophy of biology faces. Moreover, the way in which it is answered, and the ramifications of these answers, reflect the alternative answers to almost all the other narrower questions that preoccupy the philosophy of biology. The more specific questions philosophers and biologists have posed about the logic, epistemology, metaphysics, and methodology of science are all organized around this question. What is more, alternative answers to it reflect important differences in the direction that biologists advocate their discipline should be moving.

This chapter will canvass four main lines of the most important alternative answers to this question of whether and how biology differs from physical science. Two of these answers are motivated by philosophical concerns and reflect an agenda of controversies largely limited in their interest to philosophers. Two of them reflect issues faced by biologists. It is on these latter two answers that the rest of this work shall focus.

2.1. Philosophical Agendas versus Biological Agendas

In the last few decades many philosophers have turned their attention to biology to assess the adequacy of a philosophy of science that has been drawn from an almost exclusive examination and reconstruction of physics. That physics should have been the chief source of inspiration for theories of the nature of science is natural and obvious. That biology should provide the next target of philosophical scrutiny is equally natural if less obvious. Once having in hand a philosophy of physics, an account of its logic and methodology, its epistemological foundations and metaphysical implications, it is natural to apply this account to another scientific discipline, in particular to one that appears to bear important differences from physics. If biology fails to measure up to the structure and standards of scientific adequacy that this philosophy has claimed to detect in physics, then there may be something seriously wrong with the philosophy of science in question. If, on the other hand, biology satisfies the strictures on scientific respectability that this philosophy propounds, then its adequacy as an account of all natural science, physical and biological, is vindicated. The philosophical theory in question has its origins in the work of the Logical Positivists. Although many of the details of their account of the nature of science have been surrendered by their postpositivist followers, much of its spirit continues to have wide currency in contemporary philosophy of science.

The postpositivist doctrine holds that science proceeds by the formulation, testing, and explanatory employment of empirical generalizations of greater and greater generality, organized together into theories that broaden and deepen the explanatory unification and predictive precision of these generalizations. Postpositivists insist on the need for laws or at least improvable generalizations in every science just because scientific explanation consists in the subsumption of what is to be explained under laws. They insisted on such laws because the practical prediction and control required for the confirmation of claims to scientific knowledge can only proceed by the employment of laws. Furthermore, these philosophers expected that the findings, laws, and theories of the various sciences will ultimately constitute a coherent hierarchy of theory, and not simply a body of independent, merely compatible – but autonomous and isolated – disciplines. There is much more to this doctrine, but enough has been said to make it clear that biology may not immediately and obviously satisfy this description. It has no laws with the simplicity, precision, interconnection, and explanatory or predictive power of, say, those of Newtonian mechanics. Many of its findings, and its descriptive vocabulary, are insulated from those of physics and chemistry; it seems to be restricted to the study of model systems with only limited generality. All these features make it a strong test case for the adequacy of postpositivist philosophy of science.

Among philosophers, the question whether biology differs significantly from physics has taken the form of the question whether it differs significantly from the postpositivist picture of physics. Those who answer this question affirmatively go on to infer that the postpositivist picture is seriously inadequate, either as an account of science in general or perhaps even as an account of physics. In effect, these philosophers employ more or less isolated components of biological science as data to test a philosopher's account of the nature of science. Finding that account disconfirmed, their program requires them to go on to provide a better account of science in general, an analysis that will accommodate both biology and physics.

Of course, identifying components of biological descriptions, explanations, theories, and methods that test a philosopher's conception of science involves considerable discernment and interpretation: One must find significant and characteristic examples from biological practice and then give the correct explanation of their salient features. Many disputes among philosophers of science focus on these matters of selection and interpretation of the data that test accounts of science. This is not surprising, for those who hold that the postpositivist philosophy of science does adequately characterize biology must challenge the examples offered by philosophers denying its adequacy or generality. Thus, defenders of postpositivism focus mainly on showing how its opponents have misconstrued the biological examples they employ against the theory. And defenders try to show how these examples should be properly analyzed to reveal their compatibility with the view of science constructed by attention to physics.

Thus, among philosophers of biology there are two principal research agendas: Both reflect answers to the question of whether biology differs significantly from an account of physical science inspired by the empiricist philosopher's study of the growth of physics. We may call these two agendas the postpositivist and antipositivist ones. Between them, much has been disputed about the nature of science and the nature of biology. Important questions and domains of dispute have been isolated, patterns of argumentation clarified, priorities among alternative issues ironed out. To a considerable extent, each of these research agendas has provided the grist for one another's mill. Theses, examples, and counterexamples offered by each have been the principal input determining the output of the other. As in any academic discipline, the philosophers on each side of this dispute have been influenced far more by their philosophical opponents, and their agendas have been set more fully by philosophical imperatives, than they have been by biologists and biological needs or problems. This is not only inevitable, it is even desirable. Philosophy cannot be productive when it follows a course plotted by nonphilosophers. But these agendas clearly reflect philosophers' preoccupations. They do not directly answer the questions about science posed by nonphilosophers concerned with something more than the adequacy of postpositivist philosophy of science. In particular, biologists concerned with the broad question of whether and how biology differs from physical science are not interested in refuting or tidying up any particular account of science. They are interested in the ramifications of answers to this question for the course of future work in their own discipline.

Biologists (and nonbiologists, for that matter) who have reflected on this question for its implications about the future course of their science have generally begun with the assumption that much contemporary biology does differ in very important respects from physics and chemistry. Differences have arisen about exactly how it differs and whether this difference is a good thing, given the aims of the subject as a natural science. Of course, some biologists hold that there are no important differences between biology and physics – that there are at most divergences in local tactics of inquiry and experimentation, dictated by differences of degree between the subject matters of biology and chemistry, for example. These biologists detect no serious philosophical problem about the differences between these disciplines. As we shall see throughout the chapters to follow, this is not a view that any one who has examined the matter can endorse as initially very plausible. Indeed, the assertion that irreconcilable differences do exist between biological and physical science seems more reasonable at first blush. For the most impressive achievements of biology do not seem to be naturally linked to physical science. Self-conscious attempts to reconstruct biology along lines motivated by physics have been largely sterile and uninteresting. As we shall see, even molecular biology differs strikingly from its neighbors just across the border in organic chemistry.

Among philosophically interested biologists, there has been general agreement that the subject does differ crucially and ineliminably from physical science. The disagreement surrounds the question of whether this difference needs to be preserved or must be obliterated. In effect, the disagreement is over whether biology should retain its distinctive character or needs to be radically reconstructed to bring it into harmony with the physical sciences. On one side of this dispute stand biologists and philosophers, who advocate the autonomy of biological science. These *autonomists,* as we shall call them, hold that the really important aims of biology and the appropriate methods for attaining these aims are so different from those of the other sciences that biological theory and practice must remain permanently insulated from the distinctive methods and theories of physical science.

By contrast with the autonomists, there are biologists and philosophers whom we shall call *provincialists.* The provincialist holds that at best biology is a province of physical science, a dependency that can advance only by applying the methods of physical science and, nowadays especially, the methods of physical and organic chemistry. According to this doctrine, biological findings and theories must be not merely compatible with those of physics but must actively cohere with its theoretical achievements. To do this it must, at a minimum, surrender some of it aims and methods, adapt others in the light of what we know about how physics proceeds, and import from the physical sciences its research programs, theories, laws, and concepts. Provincialists hold that the differences between biology and physics are either inconsequential matters of investigative tactics or reflect deficiencies in biology that must be extirpated. Their answer to the question of whether biology confirms the conventional philosopher's picture of science drawn from physics is that much of it does not and that where it does not it needs changing, because the philosopher's picture of physics is, in most essentials, correct.

Some of contemporary biology, particularly the latest achievements of molecular biology, provincialists will leave untouched. Indeed, for them, this area of biology is a paradigm of what the rest of biology must come to. The key notion for understanding provincialists is 'coherence.' The mark of scientific fertility and depth is the degree to which a given discipline's theory extends, applies, and is explained by those theories that have provided the most detailed and precise account of nature over the vast dimensions of contemporary physics. For instance, chemistry comes into its own as a predictive and explanatory science through its interconnection with atomic theory and thermodynamics. For it was this achievement that converted the subject from the unsystematic, anecdotal cookbook character of alchemy and its successors, even up to Lavoisier. So, too, biology must acquire more than a passive consistency with the findings of physics and chemistry; it must actively articulate the leading ideas of these subjects. It is not enough to show that the existence of living systems is compatible with, for example, the second law of thermodynamics. Their existence must be shown to be a consequence of the operation of such principles. Thus provincialists, following Logical Positivists, hold up the unity of science, both in method and subject matter, not just as a long-term goal or ideal but as a pervasive criterion of current scientific respectability.

By contrast with provincialists, autonomists hold that biology is a natural science clearly enough, but one that is and ought to be independent in its methods, aims, and results from physical science. In their view, the only constraint physics can provide biology is that of consistency: Biological claims cannot violate the second law of thermodynamics, for instance. Biology, in the autonomist's view, seeks answers to questions that physical science cannot answer and that therefore must be approached by means that physics cannot provide. In the autonomist's view, the life sciences are free to borrow from physics, but they cannot proceed simply by adopting its methods or results. They must forge their own tools, and neither can nor should they search for a justification of these tools in the tactics and strategy of physical theory or experimentation. The autonomist holds that the conventional picture of science drawn from physics is, despite all its virtues, simply wrong for biology. For this subject is undoubtedly a natural science, even though it answers to only a few features of the philosopher's account of science, constructed as it is from a narrow fixation on physics. Autonomists admit that it is logically possible to pursue a study of living systems by means limited to those suitable in physics. But the result, in the autonomist's view, would be an impoverished, restricted, powerless discipline bearing no results comparable to the wonderful achievements of biology, achievements attained by modes of thought entirely autonomous from those of the physical sciences. Autonomists reject the priority demanded for scientific unification on the ground that it blinds us to biological insights and blocks the advance of our understanding of living systems.

It is important to keep in mind that, for all their disagreements, most provincialists and many autonomists share a conception of the nature of physical science; indeed, both often write as if the Logical Positivist or postpositivist conception of the nature of science is largely adequate for physics. They also agree that this conception cannot capture the actual character of modern biology. Where they disagree is on the upshot of this failure for the future of biology.

Some autonomists, casting about for arguments in behalf of the autonomy of biology from the other sciences, have taken comfort in philosophers' arguments against the adequacy of postpositivism as an account of physics. They have found support for their views in antipositivism. Opponents of Logical Positivism have attempted to show that the methods and nature of the concepts, laws, and theories prescribed by postpositivism are honored in the breach by much of the history of physical science. Autonomists reason that if physics need not show the features philosophers have demanded of it, then the claim that biology must show these features too is weakened. But this argument is a two-way street, one that autonomists tread at their own risk. For if the postpositivist conception of science is wrong even for physics, then perhaps the correct philosophical account of science will show that biology and physics do not differ significantly. In that case, arguments for autonomy may turn out to be gravely compromised.

Provincialists need not be so closely tied to a particular conception of the nature of physical science. All their doctrine requires is that whatever physical science is like, biological science must share that character because of the fundamental unity of the methods and subject matter of the sciences. Any philosophy of science that embraces this unity will suit the needs of the provincialist. However, because the thesis of the unity of science is perhaps the most distinctive of postpositivism's claims, provincialists are inclined to accept it in at least its broadest outlines.

There are thus four broad agendas in the philosophy of biology: the two preoccupied by predominately philosophical concerns, postpositivism and antipositivism, whose aims are to improve philosophy's understanding of the nature of science in general; and the two whose aims are of more narrowly biological interest, autonomy and provincialism, which dispute the nature and significance of biology's differences from physics with a view to entrenching or reconstructing its character and future direction. The present work will attend much more fully to the dispute between provincialists and autonomists than to the problems that face the philosopher. We shall track the biologist's dispute through the main problems of the philosophy of biology, reflecting on how each side attempts to solve these problems and on what the ramifications of these alternative solutions are. Although we shall occasionally come face to face with points on the agendas of postpositivism and antipositivism, this will be largely because they intersect with disputes between provincialists and autonomists. In this respect, the present work is a departure from previous studies in the philosophy of biology, for they have largely focused on the philosophical agendas.

Although the philosophical dispute is important and may have the broadest consequences for our understanding of science as a whole, its payoff must of necessity be less immediately significant than an examination and adjudication of the biological dispute. For if the provincialist is correct, much of biology must be recast, if not cast down. If the autonomist is right, then we cannot hope to see the revolutionary assimilation of biology into the physical sciences that, for instance, recent breakthroughs in molecular biology have encouraged some to hope for: We cannot hope for a science of living systems that has all the generality and precision of, say, physics. We shall have to be satisfied with a different kind of science. If neither of these two views is correct, then the future course of biology is even less clear. For between them autonomy and provincialism seem to exhaust the possible consequences of biology's apparently irreducible differences from physics.

2.2. Motives for Provincialism and Autonomy

In the examination of general arguments lodged by provincialists and autonomists on behalf of their conclusions, we must distinguish *reasons* offered in favor of the separation of biology from physics or the assimilation of biology into it from *motives* that impel philosophers and biologists to take the sides they do. The motives are only rarely revealed in the controversies, but their existence explains much of the vigor and heat with which the debate is conducted. They reveal the stakes biologists have in the controversy, and exploring them may help us understand the broader commitments of provincialists and autonomists.

There are two chief motives for autonomism: One is professional, the other ideological.

Autonomism is a natural reaction among scientists trained in traditions that antedate the revolution in molecular biology. The theories, laboratory techniques, applications, and implications of this extremely fast-moving subject are naturally perceived as threats to the traditional preserve of evolutionary biologists, naturalists and ethologists, ecologists, and others trained to examine hypotheses and employ methods appropriate to whole organisms, and especially to large groups of organisms. By attracting attention away from older foci of biological research, molecular biology represented one kind of professional threat. By addressing itself to traditional problems, or to the reformulation of these problems in new and incomprehensible terms, and by solving some of them with apparently greater power and precision than traditional strategies of biological thought, molecular biology represented another kind of threat.

Two responses to this predicament present themselves to the nonmolecular biologist. First, one can retrain oneself to accommodate and assess the advances and pretensions of the new field; second, one can offer philosophical, methodological, and substantive arguments to show that some large part of biology will remain impervious to the molecular approach, for the foreseeable future or forever. Typically, autonomists have done both these things, accommodating themselves to and welcoming advances in biochemical approaches to some problems but also arguing that there are definite limits to the encroachment of physically inspired methods of biology, thus preserving much of its autonomy.

The second ideological motive for autonomy is a belief that the independence of biology from physical science reflects something special about organic phenomena, and particularly human life, and that provincialism would devalue or dehumanize man as a moral agent or social being. Autonomists often rail at 'reductionism,' as though it were a term with the moral connotations of 'racism,' or 'fascism.' Indeed, they describe the pretensions of provincialists as 'imperialism.' They hold that attempts to make biology a compartment of physical science impoverish the picture of our place in nature and relegate the human being to the status of just another physical object, bereft of agency, responsibility, or rights to humane treatment. In addition, some autonomists attack provincialism as reflecting a hidden political agenda of Western capitalism (see for instance Lewontin, 1983).

The motives for provincialism are, so to speak, equal and opposite to those of autonomism. There is first of all the desire to capitalize on the successes of molecular biology and provide in other areas of biology results of the power and precision that it has provided. The theoretical investment and technological cost of the revolution in biology will thereby be more efficiently amortized, and the strengths and special skills of its pioneers will become more valued within biology as a whole. The notion that traditional biological problems might not succumb to methods inspired by physical science's success in molecular biology simply strikes the provincialist as contrary to the best available evidence and as a hollow rationale for those biologists who have been left behind in the recent revolution of the subject.

At the same time as these institutional and professional motivations clash, normative and ideological differences also abound. The provincialist tends to hold that there is nothing of *scientific* relevance that distinguishes humans, and other living things, from inanimate nature. In their view, the rejection of the world picture of the physicist is a species of medieval or dialectical obscurantism. It is an interesting irony of twentieth-century intellectual history that provincialists, and the Positivist philosophers of science with whom they sympathize, long viewed themselves as a radical minority battling established doctrine. They considered that their views, to the extent that the views had any normative consequences, were liberating, enlightened doctrines that undermine claims of special authority or privilege associated with the traditional philosophical and ideological status quo. The fact that, nowadays, their views are held up to attack by autonomists as fostering conservative attitudes – such as defending reactionary and antisocial standards of scientific authority – is both surprising to such philosophers and biologists and disarming as well. Having come to be viewed as the established, orthodox, received view about the nature of science, provincialism now finds itself attacked from the same point on the political and social spectrum that gave it birth. It may well be in part because of the continued sympathy of many biologists and philosophers toward the perspective of the alienated intellectual critic that they are reluctant to attack autonomism as vigorously as they attacked the antimaterialist, religiously inspired anti-Darwinian philosophy of biology they originally reacted against.

This earlier anti-Darwinian version of autonomism, however, bears a crucial difference from contemporary autonomism: The former was founded on a theologically inspired commitment to vitalism and was undermined by an attack on this commitment. Contemporary autonomism rejects any association with this doctrine and so protects itself from a natural point of attack. By wrapping itself in a mantle of antiauthoritarian, humanizing, methodological tolerance, it disarms potential critics who share the moral agenda of the autonomists and are unwilling to jeopardize it or be identified as opposing it. Thus many biologists are reluctant to oppose autonomist claims about the independence of biology from physics. The notion that 'reductionism' might just possibly play into the hands of 'racists' or 'fascists,' as some autonomists claim, gives many provincialists serious pause. It makes them reluctant to press their insistence that what cannot in principle be rendered actively coherent (and not just passively consistent) with physical science must be excluded from the corpus of scientific biology.

For our purposes, the motives of provincialists and autonomists are worth identifying so that we may put them aside. Were the present volume an exercise in the sociology of science, or an examination of alleged ideological presuppositions or implications of competing biological philosophies, these topics would be among our principal concerns. If it were true that the *motives* for embracing alternative views in this controversy were inseparable from the *reasons* offered for why these views are right or wrong, then a *purely* epistemological, metaphysical, methodological examination of biology would be likewise impossible. Indeed, it is the fervent claim of some autonomists that no such separation can be made between the assessment of ideas on their cognitive merits and the causal explanation of whose interests are served by the adoption of these ideas. This book will not address the conviction that social, psychological, political, and indeed economic forces determine the currency and acceptance of philosophical and scientific theories. It shall simply be assumed to be false. If this conviction is correct, then the project of assessing provincialism and autonomism on their cognitive merits is not just puerile and jejune – it is pointless. Pointless because cognitive merit will have little or no bearing on which if either of these views comes to be adopted. Anyone reading the present work to help decide which philosophy of biology is right is ipso facto committed to the falsity of the claim that the adoption of such a philosophy is determined by noncognitive considerations alone. Otherwise, you wouldn't be reading this book. If, after reading it, the reader considered its conclusions entirely false and set about to refute them, the reader would still implicitly have rejected the claim that the decision between these views of biology is caused by social, psychological, political, and other nonrational forces. Whatever influence these external forces may have on the adoption or repudiation of belief, in what follows it shall be assumed that such influence can be distinguished from rational considerations. It will be further assumed that these forces can be kept at arm's length while the merits of the cases are weighed and that their influences on these disputes can even be moderated.

2.3. Biological Philosophies

Although provincialists and autonomists have motives for their views, they also have reasons for them — considerations that they offer as confirming or justifying their claims, as opposed to factors that induce them to advance their respective theses about biology. Between the autonomist's reasons and the provincialist's there is a fundamental asymmetry, quite unlike the symmetry of motives behind them. In brief: The fundamental philosophical argument for autonomism is epistemological; it bears on the formulation, extent, and justification of biological *knowledge*. The philosophical basis for provincialism is metaphysical; it rests on convictions about the ultimate *kinds* or nature of the *objects* biology deals with.

This is a surprising development in view of the intellectual origins of provincialism and autonomism. For autonomism has traditionally had a metaphysical basis, whereas provincialism has an equally old epistemological pedigree.

The notion that biology is an autonomous science, which differs in crucial and irreconcilable ways from physics, is a doctrine long opposed to empiricism and its twentieth-century descendant, Logical Positivism. The reason is that it long rested on convictions that Positivism stigmatized as paradigmatically nonscientific nonsense: vitalism and organicism. The autonomy of biology was traditionally justified by the allegation that living systems are different in kind from inanimate ones, that their behavior is the consequence of nonmechanistic, vital forces. Or, again, it was held that, in consequence of organizational principles that accord properties to a whole organism, these properties are not explicable by appeal to the behavior of its component parts. This force and/or its organization make the organism's behavior independent of its physical properties and are responsible for the evident purposiveness of life. The metaphysical difference between sentient creatures, like humans, and purely physical systems was held to be obvious from introspection. And the equally unbridgeable metaphysical difference between living creatures and dead nature was held to be obvious from observation.

The fact that the secrets of organic nature did not succumb to methods that had so admirably succeeded in the study of the merely physical was *explained* by appeal to this metaphysical difference of substance and organization. The difference did indeed provide an appealing metaphysical foundation for an epistemological double standard. It led to the conclusion that our knowledge of living systems was neither as general or as precise as our knowledge of mere matter in motion. Yet it also reconciled us to this conclusion by showing that our limited knowledge was not a scientific deficiency but a reflection of facts about nature. Whatever it is that imparts life to living things, it cannot be captured in notions like position or momentum, nor is life a mere aggregation of nonorganismic physical or chemical properties of the ultimate constituents of living things. Accordingly, the vitalist held, if we are to acquire scientific knowledge of living things, it will have to be through means that are autonomous from those of physics. Thus, the foundation of autonomism's epistemological claims was traditionally metaphysical.

Biological provincialism, on the other hand, is a product of the sustained epistemological attack on metaphysical conceptions, an attack associated with modern empiricism. Vitalism and organicism, for all their appeal as explanations of the difference between biology and physics, were notoriously subject to epistemological objections. Like other theoretical notions, vital forces or emergent organismic properties are not directly observable. But, unlike other unobservable notions, they seemed incapable of lending any additional strength to biological prediction. It was easy for Positivists to stigmatize these notions as empirically empty and cognitively meaningless. Their only explanatory role was negative, to explain why physical science could not accommodate biological phenomena. The expansion of physics and chemistry into new domains further undermined the plausibility of organicism and vitalism. The end result was surrender by almost all biologists, including autonomists, of biological speculation of the vitalist and organicist sort and of any attachment to metaphysics at all. But with its metaphysical underpinnings gone, autonomism loses much of its rationale.

The epistemological doctrines of Positivism strongly encouraged the assimilation of biology into the physical sciences. It especially encouraged attempts to show that cognitively respectable biology can and should be pursued by methods common to physical science and expressed in concepts that can in principle be based on notions drawn from chemistry and physics. In the heyday of Positivism, mechanism or materialism was treated as a metaphysical doctrine no less otiose than vitalism or organicism. No Positivist who embraced the thesis of the unity of science took the claim as one about the metaphysical assimilation of the biological to the physical, for this thesis was treated as just as empty of empirical content as its vitalistic denial.

As the Positivist straitjacket on metaphysics was loosened, however, philosophers and biologists came increasingly to find in metaphysical doctrines the ultimate justification of their epistemological claims for the unity of scientific method across the disciplines. The methods had to be broadly the same, and the theories, laws, and concepts had to be interconnected. For the subject matter of all the sciences is broadly the same. The subject of them all is matter in motion – admittedly, matter aggregated at varying levels of organization, some levels even capable of description autonomously from others, but all ultimately the product of and explainable by appeal to the variables of chemistry and physics. Metaphysical materialism, however, faces an epistemological embarrassment: the very same limitations on biological knowledge that vitalism and organicism purported to explain. These the materialist has to explain *away*, has to show to be strictly temporary, practical obstacles, or at least to be in principle surmountable as physics or chemistry surmounted its obstacles. But despite the great successes of molecular biology, these obstacles have remained unsurmounted in many areas of biology; indeed, in the ones to which biologists have long attached the greatest importance. What exactly these limitations and obstacles are we shall shortly examine.

The epistemological embarrassment of materialism has rekindled interest in autonomy as a respectable philosophy of biology. Contemporary autonomism is buttressed now not by doubtful metaphysics, but by epistemological arguments. It is now the autonomists' turn to stigmatize their opponents' view as empty speculation, likely to distract scientific advance by imposing on it an unsuitable positivist epistemology grounded in a gratuitous metaphysics. The irony of this denouement is considerable. Provincialism is a response to the epistemological embarrassment of the Positivists: It holds that where biological theory cannot be, in principle at least, connected to physical theory the biology should be rejected as unscientific. But the grounds for this claim are sheer metaphysics: Nature is nothing but what physics tells us it is, so any account of nature, including animate nature, that is not reducible to physics must be wrong. Contemporary autonomists, whose predecessors grounded autonomism in metaphysics, evince a profound contempt for metaphysical *dictat*. They insist on the integrity of biological methods and theory and search for an epistemological justification of them.

But because the failure of Positivism has returned metaphysics to a position of respectability, if not preeminence, the autonomist's position is seriously unstable. For influential autonomists no longer feel themselves free to adopt any version of vitalism. As the most visible proponent of the autonomy of biology, Ernst Mayr, has written: "All biologists are thorough-going 'materialists' in the sense that they recognize no supernatural or immaterial forces, but only such that are physico-chemical" (Mayr, 1982:52). Mayr embraces a thesis he calls "constitutive reductionism" and attributes it to "virtually all biologists":

Constitutive reductionism . . . asserts that the material composition of organisms is exactly the same as found in the inorganic world. Furthermore, it posits that none of the events and processes encountered in the world of living organisms is in any conflict with the physico-chemical phenomena at the level of atoms and molecules. These claims are accepted by modern biologists. The difference between inorganic matter and living organisms does not consist in the substance of which they are composed but in the organization of biological systems. Constitutive reductionism is therefore not controversial. Virtually all biologists accept the assertations [*sic*] of constitutive reductionism, and have done so (except the vitalists) for the last two hundred years or more. Authors who accept constitutive reductionism but reject other forms of reductionism are *not* vitalists, the claims of some philosophers notwithstanding. (Mayr, 1982:60)

Mayr contrasts the constitutive reductionism that he and other autonomists accept with what he calls "explanatory" and "theoretical" reductionism. The first is the thesis that everything biological can be explained without residue or remainder as molecular interactions. The second holds that the laws and concepts of biological theories can be translated into, or deduced from, those of physics. Both these theses Mayr rejects.

But this combination of acceptances and rejections of varying kinds of reductionism leaves the autonomist bearing an onus of proof. Because no appeal is made to a metaphysical difference to justify methodological and epistemological claims of autonomy, these claims are ungrounded and unexplained. The possibility is thus left open that obstacles to explanatory or theoretical reduction are merely practical and can be overcome by some new biochemical breakthrough. The bare possibility of such an outcome seems enough to refute interesting and strong versions of biological autonomism. For the mere theoretical possibility of reduction invites biologists to attempt to turn it into an actuality that substantiates the denial of biological autonomy. Accordingly, an autonomist who repudiates vitalism must find a replacement for it. One must find an argument that will show that the epistemological differences between physics and biology are unbridgeable in principle, even though there are no important metaphysical differences between them. It is hard to see how an acceptable argument of the required strength could be constructed, for it would have to rest on some conceptual limitation of the human mind – a limitation that empowers it to discover physical theories of overwhelming power and complexity about simple objects while precluding it from discovering such theories about complex ones.

For all their denials of vitalism, some autonomists recognize that the difference they find between biology and physics must be based on an unbridgeable difference between their subject matters. As Mayr suggests, this difference is usually described as the irreducible organizational properties of whole organisms. No one doubts that biological entities betray a more complicated organization than physical ones. But for organization to make the sort of difference autonomism requires, the organizational properties and relations of living things must be qualitatively different from those of merely physical things. Such a difference is required if no explanatory or theoretical reduction of the biological to the physical is even theoretically in the offing. But this qualitative organizational difference would be no less metaphysical, and no weaker, than outright vitalism. It will differ from vitalism only in postulating irreducible, inexplicable *properties* – instead of irreducible and inexplicable *entities* or *forces* lurking in living matter.

Contemporary autonomists are not likely to be any more comfortable with this strong a version of organicism than they are with unabashed vitalism. But it remains to be seen whether, without a rationale like this, they can substantiate epistemological differences between physical and biological science.

By contrast with the autonomist, provincialists ground their demands that our knowledge of biological phenomena be epistemologically continuous with physical knowledge on strong metaphysical commitments. Starting from what Mayr calls "constitutive reductionism," they go on to embrace explanatory and theoretical reduction as demands on biology that follow inevitably from this materialist conviction. Francis Crick has expressed the epistemological upshot of this attitude clearly:

> The ultimate aim of the modern movement in biology is in fact to explain *all* biology in terms of physics and chemistry. There is very good reason for this. Since the chemistry and the relevant parts of physics [i.e.] . . . quantum mechanics, together with our empirical knowledge of chemistry, appears to provide us with a "foundation of certainty" on which to build biology. In just the same way Newtonian mechanics . . . provides a foundation for, say, mechanical engineering.

Crick goes on to say:

> I do not mean that biology must be studied *only* from the atomic viewpoint. The professional scientist will attack wherever he can, whenever he can see that the subject matter may yield and produce those wonderful patterns of organized knowledge which we call "science." The point of attack is always a matter of tactics.
>
> [But] eventually one may hope to have the whole of biology "explained" in terms of the level below it, and so on right down to the atomic level. (Crick, 1966:10–12)

Physics and chemistry provide a "foundation of certainty" for biology just because the world is ultimately composed only of the stuff of physical theory. And this fact ensures the success of a purely physical explanation of all biological phenomena. Those parts of biological theory *permanently* immune to such assimilation are to be surrendered. The rest will prove to be derivative from physical theory and will retain its biological identity only for reasons of convenience and economy of expression. J. J. C. Smart, one of the earliest philosophical advocates of provincialism, expressed this attitude toward biology aptly, echoing Crick's analogy between biology and applied science:

If it is asked whether biology can be made an exact science the answer is 'No more and no less than technology.' If by 'exact science' is meant one with strict laws and unitary theories of its own, then the search for an exact biological science is a wild goose chase. We do not have laws and theories of electronics or chemical engineering, and engineers do not worry about this lack. They see that their subjects get scientific exactness from the application of the sciences of physics and chemistry. . . . There are no real laws of biology for the same reason that there are no special "laws of engineering." (Smart, 1963:58)

The trouble with this attitude is that, for all its vast claims about what is in principle possible and what must be foregone, it offers no positive explanation of the recalcitrance of many biological phenomena to physical explanation and the apparent incommensurability of important achievements in biological science to the physical-chemical foundations provincialism stakes out. Provincialists reject the epistemological double standard of autonomy. But this leaves them with no convincing explanation of why biology has developed and continues to be pursued in ways independent of those through which physical science attained such imposing success. Provincialists' attitude toward those achievements of biology not assimilable to physics is simply to reject them as scientifically unacceptable. In refusing to do justice to a great deal of biology, provincialism raises as many questions about itself as autonomism does.

Provincialism can of course be interpreted as leaving biology pretty much untouched. It is easy to treat its claims as prescriptions about the very long run, meanwhile allowing all the practical autonomy that ecology, paleobiology, evolutionary theory, and other divisions of nonmolecular biology require. As such, provincialism would have no more upshot for the actual direction of research in biology than a commitment to the merely temporary or practical autonomy of biology from physical science. An attenuated, practical, or temporary autonomy constitutes an invitation to biologists to remove the practical impediments to a physically inspired biology. Similarly, a provincialism requiring the assimilation of biology into physics only in the long run gives carte blanche to biologists to carry on exactly as they have hitherto for the foreseeable future. If it is to have real bite, provincialism must be a strong claim about how biology is to be pursued over the near future. And it must have a convincing case for turning away from those lines of enquiry that are resistant to translation into its terms.

Autonomists and provincialists both agree that much biology differs from physics in irreconcilable ways. From these unbridgeable differences they draw opposite conclusions about the status of biology and its best future prospects. Both doctrines are in a sense pessimistic ones, holding that much biology cannot be linked to physical science in a way that will enable us to transmit the power and precision of the latter to the former. Autonomists must therefore be satisfied with a science of life significantly less powerful than physics. Provincialism must jettison much of what this science has already achieved because these achievements lack the power of physical theory.

2.4. *Tertium Datur?*

Is there a third alternative to autonomism and provincialism? Doubtless many readers are certain that there is. Others, reflecting on the unpalatable consequences of these two doctrines, must surely hope so. Meanwhile, many will refuse to grant the very premise that seems to force upon us a choice between these views. Provincialism and autonomism are both based on the conviction that irreconcilable differences exist between physical science and biology. But no such differences have been enumerated so far. Yet without a firm conviction that such differences in kind exist, a choice between autonomism and provincialism is not required. Most biologists who have no time for philosophy hold this view. To them it is evident that biology is a science like any other; that philosophical reflections on its differences from physics make heavy weather out of inconsequential differences in the tactics of research. The grand strategy of biological research seems to these biologists identical to that of chemistry and physics.

What is more, the indifference of such biologists to philosophy may extend further. For even if philosophers and biologists together show that there are no differences in kind here, only differences of degree, what will the result be? At most, biology will have been shown to be a science alongside physics and chemistry, each different from the others in some respects and similar to the others in other respects. So what? Isn't this something we know already?

In an important respect, these philosophically indifferent biologists are correct. If providing a third alternative to provincialism and autonomism, or cutting their common ground from beneath them, has only the consequence of showing that biology is biology, a science among the other sciences, then the game is not worth the candle. Or, at any rate, the result will have little claim on the attention of biologists. To convince these biologists of the importance of the dispute between autonomism and provincialism, and of the need to resolve or dissolve it, two things must be done. First, they must be shown that the differences between biology and physical science are much more substantial than widely supposed, so that the choice between autonomism and provincialism becomes forced, momentous, and alive. Second, it must be shown that correctly understanding these real differences is crucial to refuting both these views and makes a real difference to the actual future course of biology itself. Each of the chapters to follow is devoted to these two tasks. If they can be accomplished, the biologist indifferent to philosophy will turn out to be arguing from ignorance.

Autonomism and provincialism are extreme but nevertheless intelligible responses to an undeniable fact: the real differences between biology and physical science. The middle ground between them may seem attractive, but it is untenable without a great deal of argument about and reexamination of this undeniable fact. This reexamination proceeds most efficiently by tracking the arguments between the extreme views, seeing where they go wrong and why. In effect, they set the agenda of the philosophy of biology. Whereas autonomists and provincialists both locate a difference of fundamental strategy between physics and biology, what actually obtains is a difference of local tactics. But it is a difference large enough to make autonomism plausible and yet difficult enough to be unbridgeable in simple and direct ways, making provincialism tempting. Furthermore, these differences will continue to have important ramifications for biological practice. The catalogue of these differences makes no gulf between the sciences, but it explains why biology is the way it is and suggests the directions in which it should move in the future. No biologist can forswear an interest in these questions.

What exactly is the contrast between *global* strategy and *local* tactics? We may illustrate the difference with two extended examples. Consider a difference in tactics:

the scope of laboratory experimentation in physics and in the life sciences. Here the differences seem very much ones of degree – in our ability to observe, replicate, and control phenomena.

The physical sciences are preeminently observational and experimental sciences. Although theoretical advance is greatly prized in these disciplines, theorists persistently face experimenters' demands for means to collect data that test theory. Some divisions of physical science, notably cosmology and astronomy, are devoted to the study of systems we cannot experiment on. But these subjects are continually assessed against experimental findings on laboratory systems that can be manipulated. Now no one doubts that biology is also an experimental science, but there are substantial constraints on experimentation and its influence in biology as opposed to physical science. These constraints do not reveal themselves everywhere in biology. There is little difference between the methods of a molecular biologist and an organic, or even a physical, chemist working on macromolecules; similarly, some studies of enzyme kinetics or the thermodynamics of metabolic processes are experimentally indistinguishable from parallel studies of chemical catalysis and energetics. But ethological phenomena discovered in the environment are well known for their recalcitrance to laboratory replication. Evolutionary hypotheses are difficult to test in the field because of the relatively slow generation time for many species compared to the generation time for the human experimenter. On the other hand, laboratory tests of these hypotheses are plagued by qualms about their relevance to nature, qualms that are unknown in chemistry or physics. For our ignorance of the degree to which these experiments isolate relevant variables and are closed to interfering forces is much greater in biology.

Such differences in amenability to experimentation are widely and correctly viewed as differences *of degree,* not differences *of kind,* between the testability of theory in life versus physical science. The reasons these are differences of degree are, first, that there are clear cases in physics where an experiment is subject to the same constraints or qualifications as in biology, and in biology there are cases where the differences with physics disappear. Second, we can grade successive steps of practical and material impediments to experimental manipulation or replication along a continuum in which the differences fall in a natural order. Thus, astronomy and cosmology, like some biological subdisciplines, do not lend themselves to experimental manipulation. By the same token, there are components of biological theories as easy to substantiate in a laboratory as Galileo's principle of the pendulum. What is more, just as theories in nonexperimental disciplines like astronomy are tested by experiments in, for instance, high-energy physics, claims in behavioral ecology must at least be consistent with the animal psychologist's laboratory findings about, say, stimulus generalization.

Some differences of degree between biology and physics may have profound effects on their relationship without raising any philosophical questions whatever. Thus, consider the nature and role of units of time in physical theory and evolutionary theory. In the former, the unit of time has long been the revolution of the earth on its axis, a process repeated thousands of times in the life of a human observer. In some evolutionary enquiries, the unit of time varies from species to species, depending on the time between birth and end of reproductive period. These units cannot be synchronized with physical units of time; some are greater than the lifetime of the average research grant, and some are longer than the professional life of a single investigator. They limit the replication of observations or experimentation in biology and obstruct the linkage of temporal findings in biology to physical phenomena described in the fixed units of physical duration.

Despite their magnitude, such differences of degree will not sustain arguments to the effect that biology differs from physics in ways that have epistemological or metaphysical ramifications. Such differences are real, and they are significant impediments to scientific advance. But they do not undermine biology's claims to be a science or its obligations to meet the kinds of standards demanded of physics. These problems require biology to adopt novel experimental techniques to circumvent the practical problems its research objects pose. But these creative variations on traditional procedures of observation are differences in *tactics,* not strategy.

By contrast, there are differences between physics and biology that look like differences of *kind,* differences that cannot be scaled, like size, or weighed, like simplicity versus complexity. There are differences in the means, methods, and standards of research between biology and physics. These differences appear to reflect the divergent goals and incommensurable conceptions of these disciplines. The most famous example of such a difference in the basic strategies of biology and physics is the claim that, whereas the explanatory scheme of the physical sciences is mechanistic, that of biology is purposive, teleological, or functional – and not mechanistic.

Mechanism is, broadly, the view that the behavior of a system is dependent exclusively on the Newtonian properties of its components: position and momentum (or their quantum-mechanical surrogates). The behavior of a mechanical system is a simple mathematical function of the values of the position and momentum of the components of the system. Physics has managed to explain a vast domain of phenomena by the exploitation of this mechanistic conception. Perhaps its most famous success in the exploitation of this conception is the kinetic theory of gases – which explains the relations between the pressure, temperature, and volume of a gas by attributing position and momentum to its unobservable component molecules.

However, it has long and widely been held that the fundamental explanatory variables of biology cannot be those of Newtonian mechanics and that the behavior of biological systems is not to be explained by decomposition into the mechanical behavior of the physical components of these systems. Rather, biology explains a system's behavior by uncovering the needs, purposes, goals, or functions that the whole system and its parts serve. It is the correct identification of the goals served by a biological system that explains its behavior. But there is no need, space, or scope within physical theory for goals, purposes, and functions. Accordingly, there really does seem to be a difference in kind between the two disciplines. The grand aims of the physical and biological sciences must diverge: One explains by securing a mechanistic disaggregation of phenomena into its component parts, whereas the other explains by identifying a functional network within which a given phenomenon is shown to be a part. In this case, the fundamental strategies of the two domains must differ. At most, biologists and physicists need to be concerned with whether their conceptions of nature and their detailed explanations are logically consistent with one another.

The claim that these two disciplines are animated by incommensurable conceptions does not do justice to every compartment of physical and biological science.

Parts of biochemistry are as "mechanistic" in their approaches as those of their nearest neighbors in divisions of chemistry. There are, as well, physical phenomena, electromagnetism for instance, that do not involve mechanistic explanation. But, as a general description of the texture of these two fields, the nomination of one as mechanistic and the other as teleological is a good first approximation and reflects a serious difference between them. And the difference is qualitative, like one between apples and oranges, not quantitative, like one between batting averages.

Any attempt to slip between the horns of provincialism and autonomism must show that what appears to be a fundamental strategic difference is a tactical concession to the practical difficulties of dealing with living as opposed to physical systems. It will have to show that the difference in kind alleged by both autonomists and provincialists is really composed of one or more differences in degree, differences that are commensurable once they are uncovered and that do justice to the practical autonomy of biology while making clear its theoretical integration with physical science.

Two things need to be kept in mind about any attempt to show that alleged differences in kind, qualitative differences, are really differences in degree, quantitative ones. Such a conclusion will be cold comfort if these quantitative differences turn out to be too great to overcome. For then, correctly identifying these differences as "merely" differences in kind will have no value in the improvement of theoretical explanation and practical control. As Karl Marx's collaborator Friedrich Engels long ago pointed out, sometimes a difference in degree is so great that it becomes indistinguishable from a difference in kind. Taking $1,000 from a millionaire has a far different kind of effect on him than taking the same amount from someone at the poverty line. Similarly, if a difference in degree between two disciplines has exactly the same consequences for their scientific prospects as a difference in kind would have, then substituting one for the other becomes a purely philosophical exercise. To conclude that all living systems are merely macromolecules in motion would be a merely spiritual consolation to materialist biologists if there are practically insuperable obstacles to explaining any part of their behavior with all the power and precision that physics brings to bear on nonliving things. It would not underwrite their research program to the exclusion of its competitors. Undermining the choice between provincialism and autonomism must be more than an exercise in pure philosophy showing the existence of an abstract possibility.

The second thing to keep in mind is that some differences in global aims are perfectly compatible with complete uniformity in method and metaphysics. Autonomists and provincialists may turn out to be right about the fundamental differences that separate the aims of biology and physics while utterly mistaken about their means of attaining these aims. It may be differences in aims between these two fields that dictate differences in the frequency with which common, shared means and methods are employed and dictate differences in the degree of generality and precision deemed satisfactory (as opposed to possible). If so, then there may after all be qualitative distinctions between these disciplines that make a real difference in the results they reach. But these differences in aim will not underwrite either the dead end of autonomism's or provincialism's repudiation of the achievements of biology. For they will be achieved by concepts and methods commensurable with those of physics.

2.5. The Issues in Dispute

Answers to the grand question of whether and how biology differs from physical science reflect the shape of answers to a host of narrower questions about biology's differences from physics and chemistry. It is answers to these narrower questions that should determine our answer to the broader one. Any other procedure substitutes preconception for analysis by letting a conviction about the broad question govern answers to the narrow ones.

Even among these narrower questions there are grades of generality. Most prominent among them is of course the issue of whether the research programs of mechanism and teleology are reconcilable. This is a debate in which postpositivist and antipositivist philosophers have long taken part, spawning a series of analyses of teleological description and explanation. The sequence of analyses, counterexamples, and reformulations was directly animated by the question of whether teleology could be formally assimilated into mechanism. Biologists have taken little interest in this cottage industry of definitions for terms like 'goal' or 'function.' Nevertheless, as we shall see, reflection on the results of this largely philosophical debate can help us take the first and most crucial step toward constructing an alternative to provincialism and autonomism.

Only slightly less general than disputes about research programs are issues surrounding differences in the nature, number, and relations among theories in biological versus physical science. Chemistry and physics between them count a score or more of theories independently discovered and developed, each of which has spawned subdisciplines, that reveal an apparently natural division of nature. These theories reflect levels of interaction that can be studied separately and yet can be physically and mathematically integrated. Mechanics; optics; thermodynamics; electromagnetism; and theories of the chemical bond, of chemical kinetics, and of equilibrium can be independently expounded and extended. Yet they are so interconnected that we may rank them from the more fundamental to the more derivative and explain the more derivative in terms of the more fundamental. And we may trace the influence of new advances in one theory on improvements in other theories over relatively brief periods of time.

By contrast, biology does not reflect so rich and articulated a theoretical hierarchy. There are many bodies of general statements in this discipline called theories and even more called models. But it is sometimes claimed that there is only one theory in the field that really deserves the name: Darwin's theory of evolution by natural selection (and perhaps its modern successor, the so-called new synthesis or synthetic theory of evolution). There are, of course, theories and models of greater and lesser power in molecular biology: the "central dogma" governing DNA transcription and translation, the theory of the operon, and the theory of regulatory genes generally. At the threshold of physiology there are theories of the relation of primary chemical structure to higher structure and function of polypeptides and proteins, accounts of enzyme kinetics, and theories of the regulation of metabolic cycles and biosynthesis. But many of these theories seem to reflect the incursion of physical science into biology and are not distinctively biological theories.

There are, of course, the Mendelian theory of genetics and its successors, and a whole host of models in ecology, ethology, paleontology, embryology and development, and physiology generally. The theoretical accomplishments of biology, especially in the last thirty years, cannot be gainsaid. But, for all their excitement and advance, they do not yet present the picture of rich articulation manifested in physical sciences. This is a difference that needs explaining or explaining away.

The theory of evolution itself is riven with controversy, and agreement cannot even be claimed on the canonical expression of its central ideas. Indeed, there remain members of the biological community who deny its warrant and even reject its claim to cognitive legitimacy. Leaving aside special creationists with fundamentalist Christian objections to the theory, there are serious biologists who hold it to be a vacuous and circular triviality. Others insist that, though a respectable theory, it does not provide even in principle the explanatory and predictive results physics has led us to expect of a theory. Moreover, there is no consensus on how this theory is to be related to the rest of biology.

This is in stark contrast to the place of, say, Newtonian theory in physics, a theory no less central to physics than Darwinian theory is to biology. Many of the theories of physical science enumerated above bear clear, often deductive, relations to the leading principles of Newtonian mechanics. But no such clarity and formality accompany the relation of the theory of natural selection to such disciplines as systematics, paleontology, morphology, embryology, ethology, and genetics – even though it is said to illuminate, unify, and explain them.

Consider an example of these tangled intertheoretical relations. The "synthetic theory of natural selection" is the name given to the result of harnessing Mendelian laws of genetics to Darwinian selection. Together, they explain gradual evolutionary change by natural selection over small heritable variations. On the strength of this view, evolution and Mendelism are said to bear the sort of intertheoretical relations characteristic of such physical theories as chemical bonding and stoichiometry. But the relations could not be as simple and clear-cut as this, for Mendelian segregation and assortment in sexual species are not just among the *causes* of evolutionary change, they are the *effects* of it as well. An explanation of why the genes of sexual individuals assort independently and segregate randomly must itself appeal to the adaptive advantage of this mechanism of variation. That is, it must appeal to evolutionary forces that were in operation long before the appearance of Mendelian genes. This circle of intertheoretical relations suggests that, even if we allow that biology embodies more theories than just Darwin's, their relations cannot be as hierarchically straightforward as those of physics and chemistry.

The sheer number of general theories produced in biology differs considerably from the numbers produced in physics and chemistry. Added to this difference are the difference in formulation and interconnection of these theories to one another and differences in the amount of controversy that surrounds even their most informal and general exposition. The disparity with physics and chemistry is particularly striking when we consider that it is reflected in the most central and most imposing of biological accomplishments, the theory of natural selection. These differences have led to radical claims, endorsed by both autonomists and provincialists, that biological theory – its structure, the evidence that tests it, its relation to other theories – cannot be understood along the same lines as physical theory.

Theories are themselves composed of laws, or so it has been in physics. And yet the question whether biology does or should contain any laws, in the physicist's sense of this term, has long persisted. There are those who doubt or deny the existence or the need for laws in the life sciences altogether. But, assuming they are mistaken, there has been a persistent dispute about whether biological laws will be of the sort familiar from physical science. This dispute is a continuation of ones surrounding mechanistic versus nonmechanistic science. In this context, the broad dispute about research programs is changed into a narrow one about the logical form of explanations and explanatory laws in biology and physics. For it is widely held that the laws governing the behavior of living systems are not causal laws or mechanistic ones. They differ in their form, in the connections they express between events and objects, in their predictive content, and in their insulation from the laws of other theories and disciplines.

As noted above, physical law reflects the operation of familiar push–pull causal mechanisms, instanced in the changes of position and momentum of balls on a billiard table (or molecules in a gas). But biological laws describe relations between goals, ends, purposes, and systems that function so as to attain them. The relationship between a goal and the behavior that it explains cannot be causal in the same way that physical relations are causal, if only because later goals cannot cause the earlier events whose occurrence they explain. The explanatory laws of biology, accordingly, are not causal laws, or so it is said. This difference is said to be due to the fact that biological systems are not merely physical systems, but differ from them in being goal-directed. Their goal-directedness is to be reflected and explained in so-called teleological or functional laws. And because the processes reported in such laws are not causal, they cannot be linked to the nonteleological, causal laws of physical theory.

Philosophers opposed to this view have long attempted to analyze the meaning of such laws and their terms (like function, purpose, and goal) to show that they can also be expressed in nonteleological generalizations. Thus biological laws would be shown to be either convenient abbreviations for such physical laws, or else dispensable way stations to the more accurate description and absorption of biological phenomena within the corpus of mechanistic science.

Much of the grand debate about the nature of biology and its differences from physical science has come down to disputes about the eliminability or translatability of such distinctive biological generalizations. It is a particularly salient debate because the characteristic concern with teleology is common to biological thinking right down to the level of the macromolecule and its behavior, as we shall see in detail in the next chapter.

Presumably the difference between, say, molecular biology and organic chemistry will be one of degree only if, by the dint of inconvenient but feasible translation, the teleological claims of biology can be replaced by the purely causal mechanical descriptions. Even assuming that this translation is feasible, can we conclude that this difference between biology and physics is therefore theoretically as well as philosophically inconsequential? Some biologists and philosophers hold that even though the elimination of teleology is in principle possible, nevertheless its practical infeasibility has great consequences for the aims and expectations of biology. In particular, the best possible systemization of biological phenomena will never have the technological application or predictive precision of physical science. This conclusion of course affords scope for both provincialists and autonomists. The former infer from it that biology needs to be superseded by a more thoroughgoing physical approach to living systems that will provide the precision characteristic of physics; the latter infer that the penetration of physical science and its methods into biology cannot proceed much further than it has hitherto.

There are of course autonomists and provincialists who argue that teleological claims are not even in principle translatable into, still less eliminated in favor of, causal descriptions. These biologists and philosophers treat the presence of such claims in biology and their absence from physical science as the clearest mark of the difference in kind between them. But this difference needs to be explained. The traditional autonomist's explanation has been that the inevitability of teleological description in biology betokens a crucial *fact* about biological systems: namely, that they are not merely physical systems, that they are not just complex aggregations of matter in motion. Therefore they cannot be adequately understood by methods appropriate only to the study of inanimate objects. Accordingly, biology is metaphysically *autonomous* from physical science, because its subject matter is.

Among provincialists, however, this same ineliminability gives rise to a suspicion with a strongly contrary conclusion. Provincialists begin with the presumption that biological phenomena are just a complex species of physical phenomena. From this presumption, they infer that biology's dependence on teleological description and functional explanation shows there is something seriously wrong with it. Its credentials as a natural science are not in order. Its differences from physical science are due to defects and deficiencies that cannot be corrected while leaving biology in its present character. In short, we may well conclude that biology is autonomous because it is inadequate to its tasks and that it should make way for a physically inspired study of living phenomena.

Thus there appear to be differences between the global research programs of biology and physical science, between the nature and relations of theories in these disciplines, and between the character of laws and their explanatory roles. Additionally, there are important issues in dispute about individual concepts and terms that figure in biology, but not in physics or chemistry. Not only do broad disputes about teleology haunt arguments about individual terms like 'fitness,' and 'adaptation,' as well as more restricted notions like 'mimicry,' 'predation,' and 'competition'; but the immunologist's willingness to employ terms like 'recognition' to describe cell-to-cell interactions and the molecular biologist's appeal to 'codes,' 'information,' 'error,' and a host of other intentional terms to describe DNA transcription and translation are remarkable. For they betray a commitment to the use of notions utterly beyond assimilation by current physical science.

Even more daunting are disputes about the meaning and propriety of the term 'species.' It should be a matter of considerable embarrassment to biology that one of its central notions – indeed, one of its oldest concepts – should remain to this day the subject of heated intratheoretical controversy. No definition of the term has found universal adoption; indeed, some responsible biologists dispute its intelligibility altogether. Despite its apparently crucial role in the most imposing of biological achievements, the theory of *On the Origin of Species,* the term is still without clear meaning. Some biologists hold it names a class of types of organisms, unified by interbreeding and separated from others by reproductive isolation; others hold it names a class delimited by evolutionary or ecological forces. Still another group looks to boundaries on gene flow or demographic bottlenecks to define the notion of a species. Others have rejected the very notion that species name classes; they hold instead that particular species are spatiotemporally scattered particular objects. Among this welter of divergent opinions about a single term there are vast consequences for the nature of evolutionary theory and therefore for biology as a whole. For consider, if species are general kinds of organisms like other natural kinds of things, say chemical elements, they cannot coherently be said to evolve and change. This makes the theory of evolution potentially incoherent as an account of speciation. On the other hand, if species name individual lines of descent, finite collections of particular objects, we can expect no universal laws about these particulars, because there can be no laws about particular objects. Moreover, we can expect no systematic linkup between statements about species and general laws, whether biological or nonbiological. This is a conclusion both autonomists and provincialists are likely to endorse and exploit. If the costs of an alternative to provincialism or autonomism are thorough reformulation of the theory of natural selection, then these costs may be prohibitive.

Thus, between biological and physical science there are apparently important differences at the level of the global research program, the explanatory strategy, the form and content of theories, the nature of laws and regularities, and the character and status of individual concepts. This is an imposing budget of problems for anyone who denies the common ground of autonomy and provincialism.

2.6. Steps in the Argument

Although the general proportions of the domains of conflict in philosophy of biology seem to reduce in size as one moves from differences in research programs all the way down to differences among individual terms, it will eventually be seen that all these controversies can be unified around a few central problems. Thus, although the chapters to follow appear to follow roughly the sequence of issues sketched, the same problems will recur in different guises. The final resolution of each of them will be seen to turn on the adequacy of a general philosophy of biology, whose lineaments will be clear only by the end of the present work.

Thus, Chapter 3 takes up the problem of teleology – the goal-directedness and purposiveness of biological systems, the naturalness of, and the need to find, functions for the objects of biological explanation. It then turns to the question of how descriptions and explanations couched in such terms can be successfully grounded in a theoretical system within which such notions are literally anathema – that of physical science.

But although the problem of teleology is among the broadest problems to be dealt with, it is simultaneously a special case of the general issue of the reduction of biological theory to physical theory. So Chapter 4's discussion of this more general problem of the relation between molecular biology and the rest of the discipline should be expected to shed further light on the results of Chapter 3.

Chapter 5 turns to the structure of evolutionary theory. Because this theory is the most imposing of biological achievements, whether it reflects the significant characteristics and the intertheoretical relations of theories in the other sciences is crucial. What we discover about its character and relations must influence the findings of Chapter 4 about reductionism. This is because reduction is a relation between theories, and if reflected anywhere in biology it must be manifested by the theory of natural selection. In particular, it must be satisfied by the individual laws and generalizations that embody this theory.

The question of whether there are such distinctive evolutionary laws is in part a question faced in Chapter 5, but also one to which Chapter 3's discussion of teleological laws is crucial. For such evolutionary laws as we can isolate will turn out to be teleological themselves. Moreover if, as many biologists and philosophers hold, the functional and purposive character of biological laws and explanations rests on their implicit or explicit appeal to evolutionary theory, then this theory is the real locus of the problem of teleology and not after all the place of its solution.

The teleological character of the fundamental laws of evolutionary biology hinges on evolutionary theory's exploitation of the notion of fitness. But this notion has long been identified as the Achilles' heel of the theory. Chapter 6 is devoted to a full account of its theoretical status and to the ramifications of this status for evolutionary theory, for the prospects of reduction, and for the assimilation of teleological into nonteleological theory. Accordingly, its results bear directly on all three of its predecessors. Much the same must be said of Chapter 7, "Species," which is devoted to understanding the scope and significance of taxonomy and systematics. But more importantly, this chapter brings the rest of the book's arguments to bear, finally settling the provincialist–autonomist problem. This is done by showing that the differences between the natural sciences reflect differences among the scientific and technological interests but not differences among the tools and methods that biologists and physicists bring to bear on the empirical data available to assess their theories.

Finally, in the last chapter, the results of our enquiry are employed to illuminate some recent controversies about the direction in which both evolutionary and molecular biology should proceed. Those that cannot be settled will at least be left more fully understood. This illumination is the final test of any philosophy of biology.

Introduction to the Literature

The best brief defense of the postpositivist analysis of science is found in Carl Hempel, *The Philosophy of Natural Science* (Englewood Cliffs, N.J., Prentice-Hall, 1965). An equally excellent, though more extended, treatment of many of the same issues is found in Ernest Nagel, *The Structure of Science* (New York, Harcourt, Brace, and World, 1961; and Indianapolis, Hackett, 1979). This classic defense of postpositivist philosophy of science includes influential chapters on several issues treated in the present work: reduction, teleology, and organicism, as well as lucid treatments of mechanism and the nature of physical explanation.

Antipositivists have taken their best inspiration from Thomas Kuhn, *The Structure of Scientific Revolutions* (Chicago, University of Chicago Press, 1961; 2d ed., 1970). Other influential antipositivists have been N. R. Hanson, *Patterns of Discovery* (Cambridge, Cambridge University Press, 1958), and P. K. Feyerabend, "Explanation, Reduction and Empiricism," in *Minnesota Studies in the Philosophy of Science,* vol. 3 (Minneapolis, University of Minnesota Press, 1962).

The first two important introductions to the philosophy of biology were Michael Ruse, *The Philosophy of Biology* (London, Hutchinson University Library, 1973), and David Hull, *The Philosophy of Biological Sciences* (Englewood Cliffs, N.J., Prentice-

Hall, 1974). Although both have long served as extremely useful texts, both were dedicated to exploring the philosophical agendas of postpositivism and antipositivism respectively. Because Hull's book is self-consciously designed as a textbook, it tends to be more balanced in its assessment of these two views. A more straightforwardly antipositivist work is Thomas Goudge, *The Ascent of Life* (Toronto, University of Toronto Press, 1961). This was the very first work by a philosopher devoted wholly to contemporary biology. In other works of the period, this subject is usually an afterthought to discussions of physics.

Among contemporary biological autonomists, the most distinguished and most vocal is Ernst Mayr, *The Growth of Biological Thought* (Cambridge, Mass., Harvard University Press, 1982). Richard Levins, *Evolution in Changing Environments* (Princeton, N.J., Princeton University Press, 1968), chap. 1, expounds autonomy from a Marxist point of view. Quite different is F. Ayala, "Biology As an Autonomous Science," *American Scientist*, 56(1968):207–21. Explicit endorsement of provincialism is rare in biology, for reasons given in Section 2.2, as well as because it involves sacrificing so much of the discipline. Its leading proponents are the Nobel laureates in molecular biology, Jacques Monod, *Chance and Necessity* (New York, Knopf, 1971), and Francis Crick, *Of Molecules and Men* (Seattle, University of Washington Press, 1966). A well-known philosopher's defense of provincialism is J. J. C. Smart, *Philosophy and Scientific Realism* (London, Routledge and Kegan Paul, 1963).

Motives, as opposed to reasons, for embracing provincialism are recounted in Horace Judson, *The Eighth Day of Creation* (New York, Simon and Schuster, 1979). This work is a nonscholarly but accurate history of the revolution in molecular biology. R. Levins and R. Lewontin, "Dialectics and Reductionism in Ecology," *Synthèse*, 43(1980):47–78, reflects the ideological motivations behind much contemporary autonomy. These motives are also plain in R. Lewontin and R. Levins, "The Problem of Lysenkoism," in H. Rose and S. Rose, eds., *The Radicalization of Science* (London, Allison and Busby, 1976), as well as in Lewontin's review of S. Rose (ed.), *Against Biological Determinism* (London, Allison and Busby, 1982); see R. Lewontin, "The Ghost in the Elevator," *New York Review of Books*, 29(1983):34–7.

Claims that sociological, political, economic, and psychological forces cannot be separated from considerations that justify scientific views are exemplified in D. Bloor, *Knowledge and Social Imagery* (London, Routledge and Kegan Paul, 1977). B. Barnes and S. Shapin, *The Natural Order* (Oxford, Blackwell, 1979), includes chapters applying this doctrine to biology in some detail. Among biologists, this view is vocally defended by S. J. Gould; see *The Mismeasure of Man* (New York, Norton, 1981) for an extended example. This thesis is strongly opposed in L. Laudan, "The Pseudo-Science of Science," *Philosophy of Social Science*, 11(1981):173–98.

Among nineteenth-century metaphysical autonomists, the most prominent was Hans Driesch, *The History and Theory of Vitalism* (London, Macmillan, 1914). Provincialism, and in particular mechanism, was long defended by Jacques Loeb, *The Mechanistic Conception of Life* (Chicago, University of Chicago Press, 1912). Positivist attacks on both these doctrines as unintelligible metaphysics are exemplified in Felix Mainx, *Foundations of Biology*, vol. 1, no. 9 of the Positivist *International Encyclopedia of Unified Science* (Chicago, University of Chicago Press, 1939). Contemporary autonomy without metaphysical vitalism is espoused in Mayr, *Growth of Biological Thought*.

CHAPTER 3

Teleology and the Roots of Autonomy

Nothing is more striking in biology than the apparently goal-directed phenomena of embryology and development. And the more we know about the details of development in simple systems and complex ones, the more striking the phenomenon seems. The reason is that not only do whole organs and limbs develop in accordance with a plan, with a goal that is reached even in the face of interference and obstacles, but their component tissues and cells also differentiate and develop, and sometimes even regenerate, in an apparently goal-directed way.

The stages through which a chick limb bud develops, for instance, are narrowly fixed, easily identified, uniform in their result, and plastic in their capacity to compensate for variations and manipulations of the tissues. Of course, some manipulation of the chick limb bud will result in deformations of various sorts, but these oddities simply reinforce the goal-directed appearance of embryological development. For instance, if the top of the limb bud is grafted to its side, the result is either two wings or one wing with additional, "supernumerary" digits. When we attempt to explain this goal-directed, teleological process by appeal to the behavior of the cells out of which the limb bud is composed, we are struck by further goal-directed activities. The limb bud reaches its goal, a fully formed wing, because its component cells also have goals or states that they reach in a regular pattern, in spite of vicissitudes and obstacles. Research has shown that cellular pattern formation and differentiation seem sometimes to reflect the operation of a program of development that is internal to the cell and switched on independently of its environment. In other cases cells develop or even change their goals depending on "signals" they receive from their environment. The goal-directed process of wing development is explained by the goal-directed processes that its constituent tissues and individual cells undergo.

The goal-directedness of complex biological systems seems beyond doubt. What has been disputed for at least three hundred years is whether this feature of them makes living things different enough from nonliving ones to make the biological study of them an autonomous science or a provincial one. For goal-directedness was long ago read out of respectable physical science. If its persistence in biology cannot be explained or explained away, the choice between autonomism and provincialism is forced upon us. It has been long assumed that descriptions and explanations of goal-directed systems were ultimately to be cashed in for nonteleological, theoretical explanation at the level of molecular biology. Developmental biologists do not seem

3'—U U A A U C C C A C A G—5'   Messenger RNA

5'—A A T T A G G G T G T C—3'   Complementary DNA template ⎫

                                                          ⎬ Double helix

3'—T T A A T C C C A C A G—5'   Homologous DNA   ⎭

Figure 3.1. A small portion of tryptophan-operon gene. Note presence of uracil on messenger RNA at places where homologous DNA contains thymine. Like thymine, uracil base-pairs with adenine on complementary DNA chains. 3', 5' denote orientation of nucleic-acid chain. The nucleotides are labeled: A = adenine, C = cytosine, G = guanine, T = thymine, U = uracil.

at present to be very close to such molecular explanations of cellular development, still less to such explanations of the emergence of whole organs like the chick's wing. But, it will be said, surely the obstacles to such explanations are only temporary. Thus many biologists and philosophers have been led to expect that description and explanation in terms of goal-directedness is only a way station on the path to teleology-free molecular biology. But such descriptions and explanations are in fact as central to biochemistry as they are to embryology. And this makes the character and significance of teleology a problem of even greater importance now than it was when we knew nothing of the mechanisms that underlie the teleology of nature.

To see this, consider what has been discovered about the structure of the genetic material, DNA and RNA. DNA is composed of four bases, four different molecules: adenine, guanine, cytosine, and thymine. The double helix of DNA is the result of pairing off between complementary pairs of these four bases. Adenine bonds to thymine, and guanine to cytosine. The order of the bases on a strand of DNA conveys the genetic code. RNA is produced on the template of a DNA strand by the attachment of bases in an order determined by the DNA's sequence of bases. Proteins are subsequently produced out of amino acids linked together under the RNA's direction in an order determined by the initial sequence of adenine, cytosine, guanine, and thymine in a DNA molecule. But RNA differs from DNA. The bases out of which it is constructed include adenine, guanine, and cytosine. But instead of thymine, RNA contains uracil. (The substitution of uracil in RNA for thymine in DNA is illustrated in Figure 3.1.)

Why does DNA contain thymine, whereas RNA contains uracil? This is a characteristic question of contemporary molecular biology, one that has only been provisionally answered within the last few years. The explanation of why this difference obtains between DNA and RNA reflects the most glaring difference between biology and physical science. Moreover, it reflects this difference at so basic a level and in such an advanced domain of biological research that the difference between molecular biology and its nearest neighbor, organic chemistry, cannot be ignored or expected to disappear. And if this difference cannot be expected to disappear at the level of biochemistry, its persistence throughout the rest of biology is even more secure. As we shall see, the explanation of why DNA contains thymine whereas RNA contains uracil raises in new terms the oldest problem of the philosophy of biology, the problem of the apparent purposiveness of living systems and its appropriate explanation.

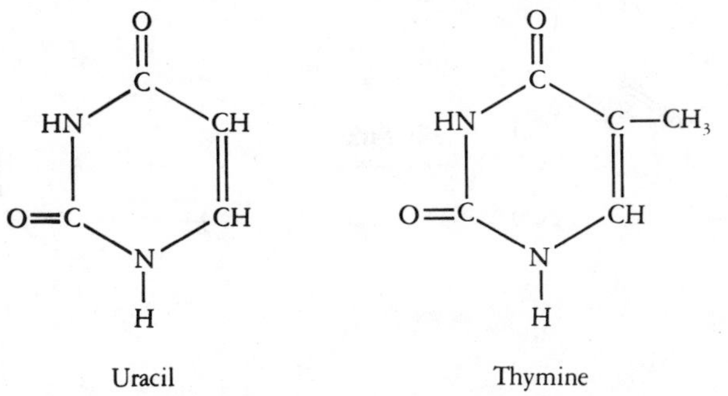

Figure 3.2. Only difference between uracil and thymine is substitution of a methyl molecule (CH$_3$) for hydrogen (H) at number-5 carbon atom of thymine.

3.1. Functional Explanations in Molecular Biology

The difference between DNA and RNA is universal. Throughout the plant and animal kingdoms, DNA contains thymine and RNA uracil. This difference between the two molecules is surprising for several reasons. For one thing, the difference between thymine and uracil is chemically very small. Both are identical ring-structures, except that there is a methyl group on thymine where uracil contains only a hydrogen atom. (This difference is illustrated in Figure 3.2.) In fact, thymine is synthesized from a uracil deoxyribonucleotide, and this synthesis is energetically costly to the biological systems that must undertake it. The extraenergetic cost does not seem worth it, because the two obvious crucial effects of the presence of thymine in DNA can both be accomplished by uracil. Indeed, in RNA uracil does both these jobs: First, thymine's role in expressing the genetic code in DNA is faithfully discharged by uracil in RNA. Second, thymine's structural role in maintaining the lateral stability of the double helix (by pairing with adenine) is also accomplished by uracil during the transcription of RNA on the DNA template. Accordingly, the presence of thymine in DNA instead of uracil was long a biological puzzle.

Because the extraenergetic costs are not repaid by these two capacities of thymine, we must conclude either that at this basic level nature is not minimizing energetic costs of life or that thymine has other *effects,* which uracil cannot have, that explain its presence in DNA and absence in RNA. It must be that thymine does something in DNA that uracil cannot do but that it can do in RNA or that need not be done in RNA. Now, both thymine and uracil are ring-structures, and the only chemical difference between them is that thymine's bulkier methyl group at the place on the ring where uracil has a hydrogen atom. It is the substitution of this methyl group for the hydrogen that is energetically costly. What effects on DNA might it have?

To answer this question, we must turn our attention to another of the bases, cytosine, which figures in both DNA and RNA. Cytosine is also very close to uracil and thymine in structure, and in fact it may turn into uracil by losing an amide group. This process, called deamination, is illustrated in Figure 3.3. When this happens in DNA, the result is potentially mutagenic. For cytosine pairs with guanine, and uracil (and thymine) pair with adenine. On replication, one of the

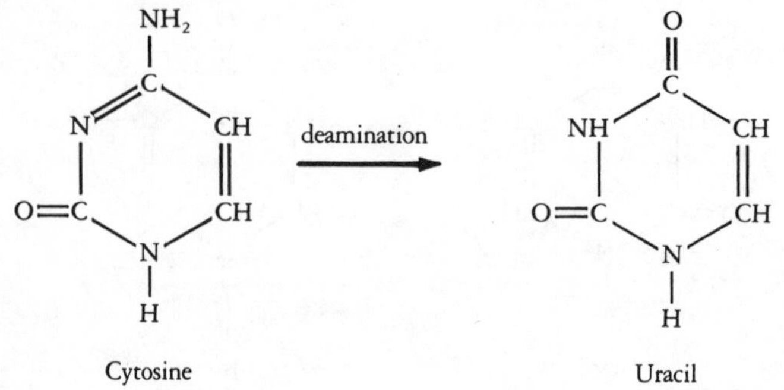

Figure 3.3. These closely similar structures differ in presence of amide group (NH_2) in cytosine and of oxygen atom (O) in uracil. Cytosine sometimes deaminates, loses its amide group, and is subsequently transformed into uracil.

DNA daughter helicies will therefore contain an adenine–uracil base pair instead of a guanine–cytosine base pair. The result would be a change in the coding instructions for RNAs constructed in the daughter, carrying the mutation. The steps in this sequence of events are illustrated in Figure 3.4.

This sort of point mutation is regularly prevented by a DNA-repair mechanism that employs three or more enzymes to remove uracil (i.e., deaminated cytosine) whenever it appears in DNA and replaces it with cytosine (The operation of this repair mechanism is illustrated in Figure 3.5.). The first of these enzymes, uracil–DNA glycosidase, recognizes the uracil to be foreign to the DNA and removes it. However, it has no effect on thymine, because the latter has a methyl group that blocks the enzyme's attachment to the thymine ring and thus prevents it from removing the correctly placed thymine molecule. If uracil were one of the constitutive bases of DNA, then this cytosine-repair mechanism would be unable to distinguish correctly placed uracil bases from those that result from cytosine deamination. Because thymine is just uracil plus a methyl group, it has the same consequences as the presence of uracil while being impervious to the action of the repair mechanism. Despite its extra cost of production, the thymine is worth it, for DNA requires the highest fidelity. The needs for faithful replication free from mutation are so great in DNA that there are several different verification and repair systems for DNA sequences. But the need for fidelity of the RNA transcribed from DNA is much less critical. For a gene produces many messenger RNAs, and, if a few are mistranscribed or deaminate before translation to protein, little damage will be done. Accordingly, no similar checking and repair mechanism is needed for RNA; thus uracil, which can do everything thymine does except protect the code against inadvertent correction, replaces it in RNA, at lower cost. DNA contains thymine and RNA contains uracil because thymine does something crucial for the proper functioning of DNA that uracil cannot do, whereas RNA does not perform this function (i.e., replication) and so uracil suffices.

This explanation has several remarkable features, which distinguish it from the sorts of explanations offered just across the border from molecular biology, in organic chemistry. One distinguishing feature is the appeal to capacities of thymine and uracil to fulfill functions of DNA and RNA. Explanations in physical science do not

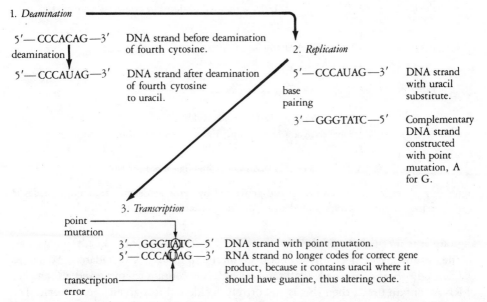

1. *Deamination*

5′—CCCACAG—3'   DNA strand before deamination
deamination   of fourth cytosine.

5′—CCCAUAG—3'   DNA strand after deamination
                             of fourth cytosine
                             to uracil.

2. *Replication*

5′—CCCAUAG—3'   DNA strand
base   with uracil
pairing   substitute.

3′—GGGTATC—5'   Complementary
                             DNA strand
                             constructed
                             with point
                             mutation, A
                             for G.

3. *Transcription*

point
mutation

3′—GGGTATC—5'   DNA strand with point mutation.
5′—CCCAUAG—3'   RNA strand no longer codes for correct gene
                             product, because it contains uracil where it
transcription   should have guanine, thus altering code.
error

Figure 3.4. Uncorrected deamination of cytosine can lead to point mutations and failure to produce correct gene product.

assign functions to the phenomena they explain. They assign causes. But the molecular biologist's explanation of the difference between RNA and DNA proceeds by citing the *effects* of the difference and not its *causes*. This in fact is what a function seems to be, *a certain kind* of effect.

Compare an organic chemist's treatment of the question why DNA contains thymine, whereas RNA contains uracil. His initial response would be that from the chemist's point of view the question contains a false presupposition: Both RNA and DNA can be constructed to contain uracil and/or thymine. Both thymine and uracil are pyrimidines, and the thymine-base is produced from the uracil-base by the donation of a methyl group from a tetrahydofolate-derivative in a reduction-reaction. The synthesis of DNA or of RNA is a polymerization reaction. The difference between DNA and RNA is caused by the absence of uracil-sugar, deoxyuridylate, and the availability of the thymine-sugar, deoxythymidylate, at the point of DNA polymerization, and vice versa for RNA. Organic chemistry can supply all the chemical reactions involved. It can also explain the effects of this difference: all the details of the reactions involved in the deamination of cytosine to uracil, the cleavage of the polynucleotide – the DNA strand – at the site of deamination, the re-polymerization of cytosine, the religation of the DNA, and the steric hindrance of thymine's methyl group that blocks these reactions. In one sense, organic chemistry can explain why DNA contains thymine and RNA uracil and what the upshot of this difference is.

But there is another sense of the question that it cannot answer and about which chemistry and physics have always been suspicious. This is the sense in which the question seems to ask what the difference between DNA and RNA *is there for*. In the chemist's view of the matter, it is not there for any further reason whatsoever, or at

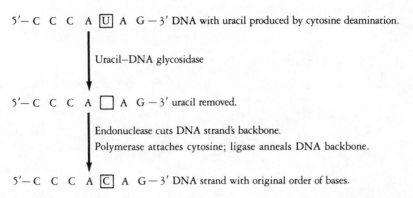

5′— C   C   C   A   [U]   A   G — 3′ DNA with uracil produced by cytosine deamination.

Uracil–DNA glycosidase

5′— C   C   C   A   [ ]   A   G — 3′ uracil removed.

Endonuclease cuts DNA strand's backbone.
Polymerase attaches cytosine; ligase anneals DNA backbone.

5′— C   C   C   A   [C]   A   G — 3′ DNA strand with original order of bases.

Figure 3.5. Repair system removes uracil generated by cytosine deamination and replaces it with cytosine, restoring DNA strand to its original order of base pairs.

any rate it is no business of chemistry to find a further cause not already mentioned in the complex causal chain of chemical interactions it has elucidated. So far as chemical theory is concerned, if all samples of DNA contain thymine and all samples of RNA contain uracil then this is an entirely accidental fact about the universe. For biology, it is very far from mere happenstance.

Of course what biologists want is not just a prior cause, or a more fundamental explanation of the difference between DNA and RNA, say at the quantum-mechanical level instead of the level of the chemical bond. They want an explanation that appeals to the *upshot* of this difference, to its *effects* on the organisms that contain DNA and RNA. But such effects of this difference are chemically irrelevant to its immediate chemical causes, and so irrelevant to its chemical explanation: In chemistry, effects do not explain their causes; rather, the reverse is true. Chemists can indeed explain how, in the circumstances, the difference between DNA and RNA leads to the prevention of point mutations in DNA and the increase in transcriptional errors in RNA, but, again, for them this is no part of the chemical explanation of the difference.

Because the biological explanation does involve effects, it is *prima facie* different from the chemical explanation. And this is a difference even clearer in the explanation of phenomena at higher levels of biological organization than the DNA polynucleotide. Cell organelles, like mitochondria, are accorded functions, and these functions are cited in the explanation of organelle physiology; cells, tissues, organs, and behavioral traits are assigned functions; sometimes whole organisms are said to have functions with respect to populations of which they are members; properties of whole populations are accorded functions. The identification of phenomena by appeal to their functions, the search for functions of items whose structure is understood, and the appeal to functions to explain findings is probably the most pervasive characteristic of the biological sciences.

For a variety of reasons, not the least of which is the absence of such concerns from physical science, it was long held or hoped that this pervasive character might disappear from the life sciences at the level of molecular biology. Moreover, the dispensability of the notion of function or goal at this level would be a harbinger of change throughout modern biology. Our example shows this expectation to be moot at best. The ubiquity of such characterization in biology and its absence from physics suggest to the autonomist that life science must be signficantly different at least in explanatory aims and standards from physical science. It convinces the provincialist that the entire discipline is vitiated by a practice without scientifically respectable credentials. In this chapter we shall explore both these views, as well as the possibility that both are mistaken, that biologists' appeal to functions can be legitimated on standards drawn from physical science, so that it need not be forgone and yet generates no obstacle to the articulation and coherence of biological and physical science.

3.2. The Search for Functions

Why do biologists search for functions? Because biological systems have them. Nothing could be more obvious. Organisms have aims and purposes, which their behavior serves; their component parts serve to fulfill these purposes and have functions in meeting the needs of cells, tissues, organs, whole biological organisms, and systems like ant colonies made up of large numbers of individual organisms. Because the objects of biological interest have goals, ends, purposes, and functions, it is incumbent on biology to identify, describe, and explain these teleological – end-directed – features. And because biological systems and their components have these teleological features, it is perfectly appropriate to cite and appeal to them in the explanation of biological phenomena. The search for functions seems dictated by the biological facts alone, right down to the level of biochemistry.

And yet this conclusion is problematical. The objects with which physical science deals do not have goals, ends, purposes, or functions (except as they serve explicit human purposes). And in physical science the appeal to teleology to explain phenomena has been ruled out as illegitimate since the early seventeenth century. Not since Galileo overthrew post-Aristotelian mechanics has physical science tolerated teleological explanation – explanation of events and processes in terms of the ends they serve. The work of Galileo, Kepler, and Newton showed that physical displacement, motion, could be explained and predicted solely by appeal to its prior causes in position and momentum. Accordingly, there is no scientific need to suppose that the motion of a physical object answers to any purpose, either its own or something else's. In the four centuries of physical discovery since Galileo this conclusion has been enormously strengthened by the assimilation of so much of physical and chemical phenomena into Newtonian mechanics. Its relativistic and quantum-mechanical successors have as little need for goals, purposes, and functions as Newtonian theory has. If anything, they have reinforced the conviction that the behavior of physical systems is to be explained and predicted by a small number of purely physical variables. In fact, this commitment long ago hardened into the complete proscription from respectable physical explanation of any appeals to nonmechanistic causation.

Physical science will not countenance goals or ends in the explanation of physical processes just because they are attained *after* or at the end of the processes they are supposed to explain. For, first of all, causation of the sort physics deals with cannot operate backward in time; indeed, the very idea seems incoherent to the physicist, for it involves something that has not even happened yet, the end, bringing about an a priori state, the process that leads to it. Even worse, processes that aim at a goal or end often fail to achieve it because they are interfered with. In these cases, the explanation of processes by appeal to their goals involves attributing a causal power to something that never happens at all, before, during, or after the process it is alleged to explain. In short, explaining something by citing its goal, purpose, end, or function cannot be a *causal* explanation in the physicist's sense of the term.

Even the explanation of human action in terms of its goals or purposes seems to substantiate this conclusion about the explanatory impotence of teleology. It is certainly true that human actions, the choices we make and means we employ, are explained by the ends they aim at, but the ends are not the causes of the actions and only indirectly explain them. It is our prior desires and beliefs, expectations and preferences, in which these ends are represented, that cause our actions. Thus, the explanation of purposive human action seems only apparently teleological. And if it is not yet mechanistic, at least it is causal in the physicist's sense: It finds the determinants of human action in prior beliefs and desires, not future or even unattained ends and goals.

Of course, when biologists attribute a goal or function to a process or explain it by citing its goal or function, they make no equivalent overt or covert appeal to prior mental states that somehow register, record, or represent a goal or purpose in the way beliefs and desires do. Thus it is said that plants exhibit heliotropism, that is, turn in varying directions *in order to* maximize their exposure to the sun. But no biologist is prepared to find the causes for this behavior in a plant's beliefs about the role of photons in photosynthesis and in its desire to maximize the number of photons landing on its leaves. Of course, for a long time it was held by responsible biologists that this sort of behavior, and all other teleological processes in nature, did reflect the intentions of an agent: God, who designed biological systems to serve His purposes. But this understanding of teleological explanations is no longer widely accepted in biological science. The absence of desires, and beliefs about how to attain them, from nature leaves a vacuum that threatens to undermine functional or purposive descriptions and explanations in biology.

Reflecting on the history of science as the progressive elimination of purpose and ends from the scientific picture of nature, many commentators on biology have stigmatized teleological appeals as the dead metaphors and vestigial remains of an anthropomorphic view of nature, one that should have disappeared with the advent of Darwin's theory of natural selection. This theory offers an alternative in variation and selection for the theological foundations traditionally provided the teleology of nature. Its vindication is held to have revealed the attribution of goals and ends, purposes and functions as an illegitimate "overlay" on purely physical processes that are ultimately to be explained in mechanistic, nonteleological terms. Opponents of teleology hold that although it may just barely be permissible to *describe* biological processes in terms that attribute functions to them, such descriptions must be treated as purely metaphorical; that we must ever be on our guard against converting such picturesque metaphors into literal claims; that despite the naturalness and convenience of such descriptions, we must on no account suppose that teleological attributions have any real explanatory power.

But this attitude runs up against the fact that biological systems do have functions and do serve purposes. This is an observational conclusion so obvious that it is difficult to seriously and sincerely deny. Similarly, the proscription on the explanation of biological phenomena by appeal to their functions is hard to take seriously after the close examination of such teleological explanations as those at the heart of contemporary molecular biology.

The autonomist's conclusion is that the presence of functional descriptions throughout biology, and its acceptance of teleological explanations, betoken a crucial difference between physical science and life science. What others see as a problematical divergence from a four-hundred-year-long tradition in physical science, autonomists have always viewed as the cornerstone of their argument for the autonomy of biology. For a long time the focus of their analysis of this difference was upon the notion of causation. Traditionally, autonomists accepted that mechanical, push-pull causation, of the type we are familiar with from mechanics, is the sole appropriate connection between the events and processes studied by physics. But they believed that a different sort of causation held among events in the biological sphere. Since Aristotle, philosophy has distinguished so-called efficient-causes, prior determinants of events, from final causes, ends for the sake of which the events occur. Appealing to this distinction, autonomists held that biology is the domain of final causes and that efficient causes are either absent or inconsequential so far as explaining biological phenomena are concerned. This view of the matter involves appeal to as strong a metaphysical supposition as the traditional autonomist's commitment to vitalism. To deny the writ of physical causation beyond a certain domain, to hold that future states (or even never-realized ones) can bring about prior ones, not only contravenes established physical theories but violates the fundamental physicalist assumptions of modern natural science. For biologists committed to vital forces that did not operate in accordance with physical law, this consequence was no difficulty.

For contemporary autonomists, the proscription of physical causation, or the assertion of a different kind of nonphysical causation, represents as unattractive a foundation for the autonomy of biology as out-and-out vitalism; indeed, it is nothing more than a disguised version of this now-repudiated thesis. Accordingly, contemporary autonomists hold two distinctive theses: that teleological description and explanations are perfectly compatible with the operation of purely physical, efficient, mechanical causation in the biological as well as the physical realm; but that such descriptions cannot be reduced to or replaced without loss by non-teleological descriptions characteristic of the rest of science. The irreplaceability of the molecular biologist's functional explanation of the difference between DNA and RNA by the organic chemist's explanation is a perfect illustration of this thesis. The autonomist's problem is to reconcile these two theses, one metaphysical and one epistemological.

Failure to effect this reconciliation paves the way for the provincialist view that teleological description and explanation is an illegitimate overlay on purely causal phenomena that should be treated with the resources of physics and chemistry alone. The provincialist's problem is that this attitude in effect throws out the baby with the bath water. For surrendering all teleology renders utterly unintelligible much of the history of advances and discoveries in biology and leaves unexplained the undeniable fact that biological systems are teleological and do have functions.

To solve their problems, both provincialists and autonomists have had recourse to Darwin's theory of natural selection, the latter to underwrite the legitimacy of teleological thinking in biology and the former to explain it away. Many autonomists argue that the causal connection back – from the goal or end to the process that attains or attempts to attain it – is established by variation and selection in the past. Thus, for instance, Ernst Mayr holds that biological processes have "proximate," efficient causes, as do any purely physical phenomena, but also "ultimate" evolutionary ones that in a distant past selected biological processes because of their adaptive effects (Mayr, 1982:67–9). It is this evolutionary winnowing and shaping that give biological processes their teleological character and underwrite the explanations and descriptions that distinguish biology from the sciences of "proximate" causation only.

Provincialists by contrast hold that it was Darwin's achievement to show how variation and selection can give the *appearance* of teleology, can mislead us into seeking the explanation of biological processes in terms of their functions and goals, even when their persistence and presence have a purely physical cause. As both sides admit, functional attributions and explanations have a far more limited prospect of explanatory and predictive improvement than physical descriptions, explanations, and predictions. We shall see why below (in Sections 3.5 and 3.6); meanwhile, part of the provincialist's argument against teleology is that Darwin's discoveries undermine it, as opposed to underwriting it, just because of teleology's predictive and explanatory weakness. By showing that processes incompletely described and weakly explained in a teleological manner are after all purely the products of physically caused variation and selection, Darwin provides us the grounds for seeing their dispensability.

Neither of these alternative appeals to the theory of natural selection is in the end decisive. But their defects can be made clear if we recall one of the signal discoveries of pre-Darwinian biology.

In 1628, Harvey published *On the Motion of the Heart and Blood.* It had previously been held that food was converted to blood in the liver and that some of the blood was driven to the heart to receive a quantity of "vital spirit." The heart's function was deemed to be dilatation, and it was not held to be a muscle. Harvey's achievement was to show that, to the contrary, the heart is a muscle, and that its most important function was to pump the blood into the blood vessels. Harvey first showed the theory of food–blood conversion wrong by arithmetic considerations; if the heart pumps two ounces of blood sixty-five times a second, it would pump more than ten pounds of new blood into the body per minute. But this amount of blood cannot arise from the amount of food consumed. Thus, he inferred, the blood must circulate within the body. His experiments on animal aortas and vena cavas and his observations of human wounds led Harvey to the conclusion that the arteries convey blood from the heart with nourishment and air for the body and that the veins return impure blood. Harvey lacked the apparatus and technique for discovering the capillaries, nor could he explain how food was converted into blood. This of course does not diminish Harvey's importance to the history of biology. In fact, his importance is not solely based on his actual discovery that the heart is a muscle whose function is to circulate a fixed amount of blood. His importance rests equally on the details of his anatomical discoveries and on his methods. For his seem to have been the earliest important empirical research in biology consciously motivated by the Galilean revolution in physical methods, especially the demand "to make measurable what cannot be measured." Harvey summarizes his experimental results as follows:

The blood passes through the lungs and heart by the force of the ventricles, and is sent for distribution to all parts of the body, where it makes its way into the veins and porosities of the flesh, and then flows by the veins from the circumference on every side to the center, from the lesser to the greater veins, and is by them finally discharged into the vena cava and right auricle of the heart, and thus in such a quantity or in such a flux and reflux hither by the arteries, hither by the veins, as cannot possibly be supplied by the ingesta, and is much greater than can be required for the mere purpose of nutrition; it is absolutely necessary to conclude that the blood in the animal body is impelled in a circle, and is in a state of ceaseless motion; that this is the act or *function* which the heart performs by means of its pulse; and that it is the sole and only *end* of the motion and contraction of the heart. (Harvey, 1962; chap. 14:114; emphasis added)

Is it plausible to say that these claims are either undermined or underwritten by Darwinian evolution, a theory unknown for a further 225 years after the publication of Harvey's results? Harvey's discoveries constituted a definite and substantial improvement on biological knowledge. They were offered with the support of a wealth of experimental, observational evidence and theoretical reasoning. They were well justified, responsible scientific conclusions. It would be fatuous to say that Harvey's claims were ungrounded because he knew nothing of a theory that could explain how natural selection for blood circulators led to the persistence and configuration of the heart. Even today we can provide only the most schematic sketch of such an explanation. It would be only slightly less implausible to claim that the theory of natural selection now in hand adds anything *directly* to Harvey's evidence for the function of the heart. Nor does it increase our detailed understanding of the heart. This is not just the claim that the *meaning* of Harvey's functional explanation of the "ceaseless motion" of the heart involves no conceptual connection to natural selection. Rather, the latter provides no further, needed, evidential warrant for it and so cannot be cited in any autonomist's rationale for the legitimacy of functional modes of thought in biology. The same considerations militate against the provincialist's conclusion. It seems undeniable that Harvey made an important scientific discovery, an improvement on previous beliefs about the matter, and that he provided excellent experimental and theoretical grounds for his functional claim. To stigmatize his achievement as an inappropriate and illegitimate "overlay" on purely physical processes, shown as such by Darwinian considerations, is a gross misunderstanding of an outstanding scientific achievement.

3.3. Functional Laws

It is true that functional claims in biology, including ones made before 1859, are all ultimately explained, though not immediately justified, by the mechanism of natural selection. But merely adverting to such a mechanism does not deprive functional attributions, and the explanations in which they figure, of unusual features absent in physical descriptions and attributions. Harvey's claim about the function of the heart is frequently analyzed into the following terms (adapted from Ruse, 1973:186):

1. The circulation of the blood is an effect of the heart's pumping.
2. The circulation of the blood is an adaptation.

This analysis can be applied to any process or system and the function it serves. Here, of course, the controversial term is 'adaptation,' and many accounts have been offered of this term to accommodate the vast range of levels and units for which a biological phenomenon may be adaptive. A function may be adaptive for the organism that manifests it and therefore may enable it to survive and produce viable offspring, or it may foster the survival of the organism's kin group, its local population, or more controversially its species or ecosystem.

But even if we could nail down the level of organization at which the functional phenomenon provides adaptative advantage, the analysis would still betray a fundamental difference from that of nonfunctional claims. There are many ways to skin a cat, and nature has at one time or another employed most of them. Adaptational claims do not explain why in a given case nature has "chosen" the actual adaptive process over all the other available alternative ways of meeting the same need. Circulation is an adaptation because of the metabolic needs of the organism. These needs could have been met by any of a large number of mechanisms, and indeed in other organisms are met by many other mechanisms that do not involve circulation. What is more, circulation could have been provided for by pumps of a different arrangement than the heart or by the circulation of a substance with different characteristics from the blood. No doubt the actual course of evolution, which resulted in hearts and blood and its circulation, is the product of *constraints* that made some alternatives either less adaptive or altogether unavailable. But the functional attribution, understood, in 1 and 2 above, as a disguised appeal to evolutionary adaptation, does not give even a hint of the causes that restricted or encouraged the particular sequence that led to the actual manner in which evolutionary needs were met. In this respect a functional attribution is radically unlike a straightforward causal one. The causal claim about phenomena traces the actual *route* that nature takes to the effects it explains. It tells us not only that an effect came about, but how it did so. If functions are just adaptations, then a functional claim does not do this. At most it provides a generic assurance that the functionally described phenomenon is the product of variation and selection: that there is some route or other. But if Harvey's functional claims are given the credit they deserve, then in general functional explanations must have more content than merely this guarantee of a pedigree in evolutionary theory.

Provincialists will agree with this conclusion but will diagnose the problem in terms of an absence of laws or empirical generalizations connecting functions and adaptations. Not only is the existence of such laws required for causal relations, their absence deprives teleological accounts of explanatory and predictive power. This last point is the real argument for the provincialist's appeal to causes and of their rejection of functions as an "overlay," a relic from anthropomorphic conceptions of the world. It is not just causal description and explanation that they demand, it is "nomological" description and explanation: Biological phenomena, like the data of other sciences, must be systematized under general laws of the form and force such laws take in physics and chemistry. Because functional claims cannot be subsumed under the nonteleological laws of physical science, they float in a scientific void.

To this claim some autonomists respond that there are *special* functional laws, which such phenomena reflect; that of course these functional laws are distinctive, undiscoverable, and inexpressible by the methods and concepts of physics; and that it is upon this incommensurability that the autonomy of biology rests. Once we recognize these laws and how they underwrite particular functional attributions and explanations, the suspicions that surround biological teleology should disappear.

Laws that underwrite the biologist's claims, all the way from Harvey's to those of molecular biology, take a distinctive form: Given a purposive, goal-directed system, S, say, an animal, heart, or replicating polynucleotide, in a particular environment, E, with a goal, end, purpose, or need, G, that can be brought about by behavior B, there is an empirical, contingent, testable law of the form

(T) Whenever a system of S's type in an environment of E's type has a goal of G's type, behavior B occurs, because it brings about (or tends to bring about) goal G.

A particularly clear example of such a law is the so-called rule of intercalation, which systematizes a wide range of limb-regenerative phenomena in cockroaches, amphibians, fruit flies, and crustaceans. Each cell in the limb of such a creature is said to internalize a "positional value," roughly a set of its coordinates on a grid covering the whole limb surface. Contiguous cells normally have positional values that differ by only one unit on each axis of the grid. When a limb is severed and the wound heals, cells with widely differing positional values are brought together. The rule of intercalation holds: "Discontinuities of positional value between adjacent cells result in the growth of new cells with intermediate values *so as to* restore continuity of positional values" (Alberts et al., 1983:866).

Restoring continuity in these values and regenerating the lost limb are one and the same phenomenon. Here the system, S, is the limb stump; the environment, E, is not explicitly mentioned but includes the absence of forces developmental biologists have identified as interfering with regeneration; the goal, G, is regeneration; and it is brought about by behavior B, intercalation of positional values of the component cells.

The autonomist insists that the teleological and functional laws of the form of (T) that biology uncovers are just as testable and just as strongly confirmed as those of physics, but such laws cannot be brought within the theoretical network of physical laws. Teleological laws are autonomous, we cannot do without them in biology, and we cannot explain them by appeal to the nonteleological laws. We cannot deduce them from causal statements, nor can we substitute causal descriptions for them if we are to explain most biologically significant phenomena.

These are strong claims, and they have evoked a long history of attempts to overturn them, attempts motivated by the desire to retain teleological and functional language in biology while allowing for its coherence with the laws of physics. For if the claims of irreconcilability and incommensurability cannot be overturned, we will have to choose between provincialism and autonomism.

The claim that teleological laws of the form of (T) are testable and have been empirically confirmed is important because one traditional argument against teleology is its alleged empirical vacuousness and its tendency to erect metaphysical forces where only causal forces operate. That organisms in environments emit characteristic behaviors is obvious; equally, that organs, organelles, or their molecular components behave in repeatable ways in similar circumstances is a well-confirmed fact. What seems less open to test is the claim that they literally *have goals* and that the behavior emitted occurs *for the sake of* these goals. To see that such claims are quite testable, it is urged, we need only expound the meaning of an expression like S does B for the sake of goal G. All instances of this scheme, like "DNA is composed of thymine in order to minimize mutation," or "the heart pumps for the sake of circulating the blood," mean nothing more than statements of the following form:

1. Behavior of type B invariably, or frequently, brings about G.
2. Behavior B occurs because it tends to bring about goal G.

Thus, in our DNA example:

1. The presence of thymine tends to bring about the avoidance of mutation.
2. Thymine is present in DNA just because it has this tendency.

That functional claims can be shown empirically to meet condition 1 seems uncontroversial. Condition 1 is just a standard kind of causal claim, familiar in physical science. To establish claims of type 2 is harder, but still entirely possible: One simply varies the circumstances in which S may emit behavior B; if it does so only when its emission subsequently (tends) to bring about G, then 2 is confirmed. In some cases, like the DNA–thymine case, this will be difficult. Varying the conditions and waiting to see if the minimization of mutation is brought about by the presence of thymine in DNA might take as much time as the evolution of the system actually required. In other cases, it will be easy; for example, animals in laboratories are easily shown to exhibit behavior of certain kinds when and only when the behavior tends to bring about some goal or end, meet some need, or serve some purpose. No doubt claims of the form of 2 are mysterious and have the appearance of backward causation – the effect, the goal, the function determines the behavior that enables the system to attain it. But, for all their mystery, they seem unimpeachable in respect of testability.

Moreover, it is often held, in given cases their mysteriousness may be dispelled by Darwinian considerations. The attribution of a function or a goal need have no implications for backward causation or the bringing about of causes by their effects. In fact, such attributions are sometimes taken to be implicit claims about the origins of the processes they describe (see, for instance, Wright, 1976). For example, when it is said that behavior B occurs because it tends to bring about goal G, we are not talking about *particular* bits of behavior and their *particular* effects. We are talking about *kinds* of behavior and their *kinds* of effects. The beating of my heart is not brought about by its current or subsequent circulation of my blood through my circulatory system but by the fact that in the past the beating of hearts has had this effect of circulating the blood; it is the existence of hearts in general that is explained by the fact that organisms with hearts that circulate blood have been and are more likely to survive and reproduce than organisms with hearts that fail to circulate blood, or organisms that employ other means to provide oxygen and remove CO_2 (under the same constraints that we face). A similar selectionist story can be told for other functional attributions. This is the ploy to which autonomists like Mayr appeal when they distinguish "proximate," physical causes from "ultimate," evolutionary ones.

But of course this explication of the notion that functions are types of effects that occur because in the past their ancestors brought about a certain goal leaves it open to the objections against explicating function in terms of adaptation. In fact, if we insist on this as the mechanism that underlies all functional claims, the present analysis of function is tantamount to the adaptational one.

Both are difficult to accept because of the great "distance" they seem to put between a function and its causal determinants. In seeking and providing a causal explanation, we reflect a widely held scientific conviction that a phenomenon and the forces that determine its occurrence cannot be decoupled, that there cannot be action-at-a-distance. A scientific explanation must cite the chain of causes behind the event to be explained. Selectionist accounts that explain a function in terms of the adaptiveness of paleontologically venerable ancestors fulfill this demand only in the most limited sense. The evolutionary account we can give for almost any function will be identical in detail to the account available for many other functions. All we can say about them is that they are phenotypes thrown up by variation in the genome and that they provided selective advantage in the past. We cannot fill in very many of the details of how and why natural selection eventuated in hearts or kidneys, mitochondria and contractile vacuoles, or even the structure of the hemoglobin molecule, for that matter, as opposed to their many alternatives. We cannot trace backward from the current presence of four-chambered hearts in mammals, through a genetic code that programs their appearance in particular organisms, to the crucial circumstances in which differential fitness of such hearts as opposed to other designs led to their being fixed in the mammalian population.

This gives the adaptational gloss on functional attribution its uninformative, generic flavor. It appears to be an interpretation generalized – from some circumstances in which it truly does underwrite functional attributions – to cover all claims of biological teleology with the express aim of providing them a scientifically respectable pedigree. But the grip of functional attributions in biology reflects a presumption that there is something in the immediate causal vicinity of the functional phenomenon that accords it a teleological character, rather than in some event in the Mesozoic era 150 million years ago.

One attractive suggestion is that we can find in the individual genome a compromise between global evolutionary explications of function and the conviction that such statements rest on a more intimate, immediate, causal basis. Individuals have hearts that circulate blood because of the operation of the genomes with which they were endowed at conception. The DNA of which these genomes consist directs the production and regulation of polypeptides and proteins that interact together in development to form cellular assemblies that eventually constitute the heart and cause it to circulate the blood. Thus, it has been argued, functional attributions are to be understood as reflecting the operation of genetic programs. According to Mayr,

a physiological process or a behavior . . . owes its goal-directedness to the operation of a program. . . . All the processes of individual development (ontogeny) as well as all seemingly goal-directed behaviors of individuals fall into this category, and are characterized by two components: they are guided by a program, and they depend on the existence of some endpoint or goal which is foreseen in the program regulating the behavior. The endpoint might be a structure, a physiological function or steady state . . . (Mayr, 1982:48).

Exploiting this suggestion, we may understand the claim that some function B is undertaken because it tends to bring about a goal G as an elliptical claim about the developmental program of the genome: The nucleotide sequence contains instructions for generating a process that leads toward goal G. The instructions generate behavior B, and that is why B may be said to occur because it leads to G.

Although it is certainly true that the genome does code for the appearance of many functionally characterized phenomena, this suggestion does not really advance our understanding of such claims very far. By appealing to the idea that an end is (in Mayr's word) *foreseen* in the genome, it attempts to take advantage of parallels with human, sentient teleology that seem to be free of biological controversy. But the notion of foresight at work is clearly metaphorical, and itself functional. It stands in need of the same kind of clarification the original functional attribution requires. If genetic programs foresee the ends they ensure the attainment of, just because they have been shaped by selective forces to do so, then the appeal to programs is no better than a direct appeal to adaptation to explain functional attributions.

Even more seriously, there are many effects of genetic transcription, translation, and expression that are clearly not functions of the organism or its parts; hereditary disorders, like diabetes or sickle-cell anemia, are as fully programmed as any of the body's normal functions but are dysfunctions. So being the product of a genetic program cannot be the whole story of what functions consist in. What is more, much animal behavior is clearly goal-directed, yet obviously too plastic, too persistent, too irregularly emitted to be usefully explained by appeal to the animal's genome. Surely the animal's original genome sets limits on plasticity and persistence. But, like natural selection, it is also too far removed in space and time from the chimpanzee's nit-picking behavior to explain it as having the function of reducing an itch.

The attribution of functions to the genome is part of the problem of teleology, not its solution. That is one of the chief lessons of the functional explanation of the difference between DNA and RNA.

The attempt to remove the aura of backward causation, or some sort of occult noncausal determination, from teleological claims by linking them with natural selection is too costly. It works only by treating all such claims as equally generic statements about the adaptive course of evolution over the eons. The attempt to find a biologically distinctive but more proximate prior cause in the genome leaves ungrounded nonhereditary, purposive behavior. Still worse, it leaves unanswered the same questions about the functional features of the genetic material that it hoped to answer for the functional phenotypes that these genes code. Between the Scylla of unimprovable vacuity and the Charybdis of backward causation, autonomists must search for an alternative. Otherwise their claims that teleological descriptions, explanations, and now "laws" are all autonomous will be accepted, but to them will be attached the stigma of unscientific mystery mongering or triviality, the very charges lodged by the provincialist.

3.4. Directively Organized Systems

The assertion that laws of the form (*T*) are autonomous — not deducible from or explainable by appeal to nonteleological laws of organic chemistry, physical chemistry, or physics — is hard to establish. The reason is the logical form of the claim: It is what philosophers call a negative existential claim, the denial that something can be done. As such, no number of failed attempts to do that something can rule out the possibility that the next attempt will succeed. On the other hand, one positive instance of a teleological or functional phenomenon explained by causal laws seems all that is necessary to undermine the notion that teleology makes biology autonomous. It might also undermine the claim that biology is a provincial science, from which teleology should be expunged, for the success would show that biological teleology can be coherently articulated within a purely physical perspective. For this reason philosophers and biologists have long sought such positive instances of a purely physical explanation of a teleological regularity. What is needed is a set of causal laws in which components of system S, environment E, functional process B, and goal G are mentioned, but in which no goals are attributed to S or functions to B, and in which nonteleological properties of S, together with properties of E, lead to phenomena like behavior B and through it to the occurrence or satisfaction of goal G. If from such laws the truth of (T), or rather its instances, can be deduced, then teleology will be given a physical explanation. Most attempts to show that such laws exist and that they demystify teleology turn on notions introduced in the development of cybernetics and general systems theory: *feedback loops* and *feed-forward loops*.

The idea is that systems that appear to be moving toward a goal are controlled by ordinary causal interactions that are often large in number, that limit one another over time, that are sometimes redundant, and that interact to produce the *appearance* of purpose in the eye of the observer. Such systems have been described as "directively organized" ones. Well-known examples of such directively organized systems and their goals studied in physiology include the body and its goals of maintaining an internal temperature of $37°$ C and of maintaining water balance. The general form of a directively organized system is specified in the following conditions. Given the factors mentioned in a teleological law of the form (T) – the environment E, the goal-directed system S, its behavior B, and goal G – we can specify a set of component subsystems of S that generate its goal-seeking behavior B in environment E because they are causally related in the following five ways (due to Nagel, 1979):

1. The state of each of the component subsystems of B at a given time, t, together with the states of the others at that time, causes, in a purely physical way, the attainment of goal G at a later time $t + d$.
2. The state of each of these subsystems at t is instantaneously independent of the states of the others at t.
3. Each subsystem has only a restricted range of states.
4. If the state of one of the subsystems changes greatly enough at t, then, in the absence of changes in the other subsystems, at $t + d$, the whole system S will be caused not to attain goal G.

But

5. The subsystems are so causally linked that whenever such a great change occurs in one of them at t, this change causes changes in the other subsystems at a later time $t + e$, which together with the initial great change at t causes the whole system to attain its goal at $t + d$.

How meeting these conditions generates goal-directedness in a purely causal way can be illustrated by the liver. A major function of the liver is to maintain the level of glucose within the blood between certain fixed limits. It does this by converting glucose taken from the blood into glycogen, which can be stored in liver cells and broken down back into glucose by them. When the blood-glucose level approaches the upper limits, the glycogen-producing machinery of the liver is switched on, and the glycogen-degrading mechanisms are turned off, thus drawing glucose away from the blood and lowering its level. As blood-glucose levels decline toward their lower limits, the reverse happens, and glycogen stored in the liver is converted to glucose and restored to the blood, thus shifting the blood-glucose level above its lower limits. This regulation of blood-glucose levels involves both a complex series of chemical reactions that convert glucose to glycogen and rearrange the glycogen molecule to increase its solubility for storage and also a series of entirely different reactions that convert glycogen back into glucose. The steps in these reactions are all effected by catalysts, enzymes found in the liver cells. The liver maintains blood-glucose levels through the control of these enzymes. Several of these enzymes are "allosteric," that is, their molecular shapes can be changed by the molecules whose reactions they catalyze or by other enzymes. These changes in molecular configuration activate or deactivate their catalytic powers by uncovering or covering their "active sites," the atoms where the reactions they catalyze take place.

The major subsystems through which the liver keeps blood-glucose levels within certain limits include two hormones produced in the pancreas, insulin and glucagon, and epinephrine, a hormone secreted by the adrenal medulla. These hormones' effects on glucose–glycogen conversion are equal and opposite, so to speak. Thus, for instance, an increase in the blood-glucose level causes an increase in the secretion of insulin. When it arrives at liver-cell membranes the insulin suppresses glycogen degradation and stimulates glycogen synthesis. A decrease in blood-glucose levels stimulates the production of glucagon. Its arrival at liver-cell membranes stimulates the breakdown of glycogen and suppresses its synthesis. Epinephrine operates the same way, though its effects are more concentrated in muscle tissue than in liver cells. Moreover, epinephrine also directly stimulates the production of glucagon and the suppression of insulin. Finally, the glucose level of the blood within the liver itself allosterically activates and deactivates enzymes that control glucose–glycogen conversion. Thus there are at least four interacting subsystems that together keep glucose levels within certain limits by causing changes in one another's concentrations and configurations. And they do so because they are causally connected to each other in the five ways specified above.

We may demonstrate the causal influence of each subsystem on the others by following out the cycle illustrated in Figure 3.6, starting at any point in the whole system. Thus an increase in insulin – produced, say, by direct injection – will lead to net glycogen synthesis until the blood glucose level is drawn down toward its lower limits. This will set off the secretion of epinephrine, which suppresses the body's own production of insulin. The epinephrine also works together with reduced blood-glucose levels to cause the secretion of glucagon and thus eventually the degradation of glycogen back into glucose, thus raising the blood-glucose levels.

One of the most important results of recent work in molecular biology has been the detailed chemical account of crucial links in the causal chains interconnecting these subsystems. That is, biochemists can now describe much of the molecular mechanism that underlies the directively organized system described in Figure 3.6. The mechanism it describes is but an intermediary along the way to a full tracing out of how a vast number of enzymes and the substrates (like glucose and glycogen) on which they operate themselves constitute a more finely grained molecular system that operates to generate the nonmolecular feedback and feed-forward loops. A diagram of this directively organized system would require several spatial dimensions to accommodate all the causal connections among enzymes and substrates,

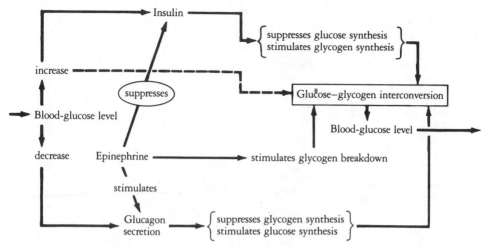

Figure 3.6. Directively organized system of blood-glucose level maintenance.

their interdependence and redundancy, the way in which changes in the molecular structure of each enzyme cause compensating changes in the molecular structure, their catalytic activity, and the amounts of other enzymes.

For example, glucose itself, epinephrine, and two enzymes that degrade and synthesize glycogen form a directively organized system whose function is to regulate glycogen production. The components of this system operate roughly as follows: Epinephrine catalyzes a cascade of at least four separate reactions, which result in the glycogen-synthesizing enzyme, glycogen synthetase, shifting its molecular configuration from one that makes its site of catalytic activity active to one that makes it inactive. Glucose in the liver cells binds to two other enzymes, enabling the first, a phosphatase, to remove a phosphate molecule from the second, a phosphoralase, thus making the second inactive in glycogen degradation. This second enzyme is the very one that epinephrine makes inactive. Moreover, the phosphate-removing enzyme, when free from the glycogen-degrading enzyme, can combine with glycogen synthetase, removing a phosphate group from it and making it active in glycogen synthesis. Matters are complicated by the fact that the enzyme that glucose activates and epinephrine deactivates, glycogen synthetase, has two distinct control sites for catalytic activity: one controlled by glycogen via the phosphatase, the other controlled by epinephrine via a cascade of three other enzymes and macromolecules. The molecular interactions of glycogen production are schematically illustrated in Figure 3.7.

Of course, glycogen and epinephrine levels are linked by causal chains of dependence beyond this directively organized subsystem. But the subsystem has exactly the same features as the larger one that contains it and moreover is part of a fuller explanation of how the components of the larger directively organized system interact and fulfill the five features of such systems noted above:

1. The state of each of the enzymes, the hormone epinephrine, and the glucose molecules at a given instant together determine net glycogen production at a slightly later time.

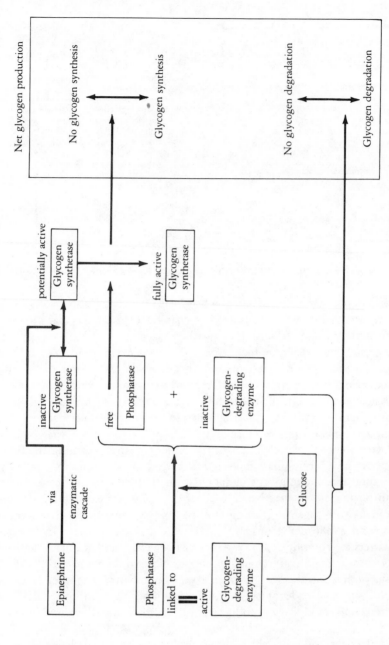

Figure 3.7. A molecular directively organized system underlying directively organized system of Figure 3.6. Small boxes surround components of this system. Arrows indicating causal connections between glucose and epinephrine are left out for legibility.

2. The concentration of glucose at any given time is instantaneously independent of the concentration of epinephrine or any of the other enzymes at that time.
3. There are narrow limits to the organically possible concentrations and the molecularly possible configurations of the enzymes and substrates.
4. Uncompensated, a great change in any one of them will cause a breakdown in net glycogen production or degradation even though the change is within permissible limits specified by condition (3).

But

5. The causal links among the enzymes and substrates are such as to cause compensating changes that prevent the breakdown of the system and prevent goal failure.

It is important to recognize that the compensations generated are simply the result of causal chains and not themselves harbingers of further teleology. This is particularly apparent in the subsystem just sketched. We may trace a purely chemical path from great changes in concentrations or molecular configurations to other changes that have the compensating effect without any appeal to enzymes or other molecules that have the function of compensating for changes or disturbances.

It is worth noting what distinguishes a directively organized system at the level of chemical and physical interactions from a purely physical system to which no function can be attributed. Consider a substance whose state is maintained as a gas through the interaction of its pressure, temperature, and volume. Remaining in the gaseous state is not a goal, purpose, or function of the system, and yet its "subsystems," pressure, temperature, and volume seem to satisfy most of the five criteria cited above for directive organization and thus for constituting a teleological system as a whole – whose aim is maintenance of the gaseous state. But one thing is missing in this case. Here the states and values that the subsystems, pressure, temperature, and volume take on at any instant are simultaneously dependent on the states of the other subsystems. It is a law of nature that the pressure, temperature, and volume of a gas – independent of other conditions – are simultaneously fixed by one another. In the case of a goal-directed system, there are no such exceptionless laws simultaneously linking components independent of the conditions of the rest of the system.

It is this independence of subsystems and the plasticity of the goal attainment they permit that gives rise to the appearance of goal-directedness. Each of the enzymatic subsystems can be influenced by biochemical forces outside the liver, and the others will respond causally to changes in it produced by these forces; the compensating changes give the liver the appearance of flexibility and persistence in its quest for its goal and the appearance of autonomy in achieving its goal. In addition, the account explains why such goal-directed systems sometimes fail to attain or maintain their goals: If outside, or for that matter internal, influences are so large that they shift one or another of the subsystems to states that will causally preclude goal attainment, then – no matter what states the other subsystems can or do attain – goal failure will ensue. Thus if one of the enzymes involved in blood-glucose maintenance is defective or deficient, then – no matter what states the other enzymes can or will attain – the liver will not be able to maintain the blood-glucose level required for life. Such defects have been clinically diagnosed under half a dozen different names.

The account appears to be complicated, but then the phenomenon of blood-glucose regulation is complicated and involves a large number of biochemical interactions. For our purposes, the important thing is that there is nothing teleological in the detailed biochemical description of the interaction of the subsystems. Yet the appearance of goal-directedness of the whole system, of behavior in accordance with a law like (T), is a natural consequence of the organization. This shows that because we can find the relevant causal laws relating the subsystems in question, we can explain the teleology of the whole phenomenon in a purely non-teleological way — a way that moreover makes no reference to evolutionary notions like adaptation.

Moreover, we can now see why explanations of processes that cite their functions are really explanatory. If functional explanation were nothing more than the implicit suggestion that the phenomenon to be explained has an evolutionary origin, such explanations could not survive the provincialist's charges of uninformativeness and unimprovability. But if a functional explanation is the implicit suggestion that the phenomenon to be explained is the purely causal consequence of the operation of a directively organized system of ultimately nonteleological subsystems, then not only will such explanation be initially informative, but we can expect further details about the operation of the subsystems to enrich and deepen it as well.

The explanatory role of the statement that the function of the liver is to regulate blood-glucose levels is not after all that of citing one of its *effects* as the determinant of the liver's presence and operation. Rather, the explanatory role of such a statement is played in a wider context (as Cummins, 1975, points out). The claim is explanatory *relative* to a broader account of the body's needs and how it meets them, in this case its capacity to meet the energy needs of the muscles. The function of the liver is to maintain blood-glucose levels because (1) the liver is capable of doing so, and (2) the body that contains it meets its needs to supply energy to the muscles, in part *through* the capacity of the liver. In its turn, the liver's capacity to keep blood-glucose levels within certain limits is explained by attributing functions to its component parts, in this case its cells, which have the capacity to interconvert glucose and glycogen. Again, the liver cells have this function because (1) they have the capacity to do it, and (2) the liver's capacity is explained through their capacities. We may pursue the cycle further: explaining the liver cell's capacity to interconvert glucose and glycogen by appeal to the functions of the enzymes to be found within. It is below this point that functional language drops out, for the explanation of how the enzymes have their capacities is no longer functional, but purely physical and chemical. The fact that we can, so to speak, discharge functional attributions at this lowest level is what gives the functional explanations at higher levels their content and interest. We appeal to specialized and simpler functions to explain more general capacities, and the significance of such explanations is dependent on how much simpler and how different the explaining functions are from the explained capacities, and how complex is the organization of components whose causal interconnection produces the function. The attribution of a function to the liver, relative to the body's capacity to meet energy needs in its muscles, has explanatory power because it is the first step in the analysis of this general capacity of the body into the much simpler capacities of its smallest subsystems.

This account of the nature and significance of functional explanations is nicely illustrated in the example with which this chapter began: the difference between

RNA and DNA. Among its several effects, minimizing mutations is a capacity of thymine. For DNA's function of faithfully replicating and transcribing itself is in part a result of thymine's blocking the action of uracil–DNA glycosidase. This capacity of thymine, together with the capacity of the enzyme to remove uracil bases, and the capacity of endonucleases, polymerases, and ligases to replace deaminated cytosines (i.e., uracil bases), altogether provide an important part of the purely causal mechanism behind DNA's continuing fidelity. It is against this background that thymine's presence in DNA is functionally explained. It is part of a directively organized system that generates the continued fidelity of the DNA that contains it. Thymine is not itself a directively organized system. Its components do not exhibit the complex interconnections characteristic of such systems. It just sits there in the DNA, so to speak. But this is just what we should expect in a functional explanation at the border of molecular biology and organic chemistry. It is at this point that functions are, so to speak, discharged.

But there is another sense in which thymine can be accorded a function. This is the sense of function that stands behind the notion (examined in Section 3.2) that a function is an effect with an evolutionary or adaptive pedigree. Because of its effects in the preservation of DNA's fidelity, it is a fair supposition that the presence of thymine in DNA is not an evolutionary accident, a by-product of the process that resulted in DNA's being the vehicle of heredity. The presence of thymine in DNA is an adaptation, whose advantages over a limited number of known available alternatives (like uracil) can be specified in detail and not just announced or merely speculated about. Because the presence of thymine in DNA can be explained in the ultimate context of evolution, as well as in the immediate proximate context of DNA's fidelity, it has an evolutionary function. But it has this functional role with respect to evolution *because* it has a crucial part in the subsystems that provide the capacities of DNA to remain faithful. Thus, the sense in which adaptations are functions is parasitic on the sense in which they play a role in the capacities of larger directively organized systems that contain them. (We shall explore this matter further in section 3.7 below and in Chapter 8.)

3.5. The Autonomy of Teleological Laws

The upshot is of course that functional explanations and descriptions are justifiable because they rest on purely nonteleological foundations. There are, however, two questions about this strategy for justifying teleology. The first is the question of whether the features of directively organized systems maintaining a preferred state can be generalized to all the contexts in which biologists make functional and teleological characterizations. The second question is of course whether even if there were a directively organized system underlying each and every functional one, this would undermine the claims of provincialism or autonomism.

This strategy for analyzing teleological systems involves the discovery of internal components of functional systems that conspire together, so to speak, to produce behavior that has the external appearance of teleology. If for many functional systems we cannot discover such internal components, then we cannot provide a nonteleological explanation for the external teleological regularities of (*T*)'s character that they manifest. One of the appealing features of this demand that we search for internal components that produce external teleology is of course the fact that a great deal of biological research is devoted to this very task. The autonomist can take comfort in this fact, for the task would itself be pointless unless we took seriously the teleological regularities this strategy aims to explain. These teleological laws organize the search for internal components that work together to explain the laws. But, the autonomist continues, not only will success in discovering such internal components and their relations presuppose teleology, but (1) there is no guarantee that success will always be achieved, and (2) even when achieved it will not permit us to forgo teleological characterization. The reasons are that we have no guarantee a priori that we can discover an internal characterization for every goal-directed system S. Such a characterization must show that the internal components of S invariably operate in environment E to produce behavior B when goal G is available but do not generate B when G is not available or when E makes G inappropriate as a goal. Otherwise we will not have given a nonteleological explanation of the crucial characteristics of teleology. In fact this demand may not even be met by the example sketched above. For it is easy to arrange circumstances in which the glycogen-synthesizing and -degrading enzymes can all operate within the range that could maintain a blood-glucose concentration of 100 mg/100 ml, and yet that concentration would not be maintained. What is required is unusual intervention, say, adding an inhibitor for one or more of these enzymes to the diet. What this shows is that the stipulated internal components are not causally sufficient for — do not guarantee — the attainment of the goal in question. Similarly, they can be shown not to be causally necessary for it either. For example, by some external intervention we can keep the blood-glucose level at required levels while the liver is rendered inoperative because of damage or surgical intervention, etc. Again, this is an abnormal situation, but it shows that our theory of internal components is at least formally incomplete. As it stands, it does not explain some cases where the goal is attained and some cases where the goal is not.

The natural reply to these complaints is that our sketch of this directively organized system is indeed incomplete; we need to mention more of the internal components to give the whole story, and we need to add ceteris paribus — other things being equal — clauses to rule out unintended irrelevant circumstances. But, the autonomist replies, we have no guarantee that we can tell the whole story, that there is a *finite* number of components, and a well-motivated way of excluding irrelevant conditions, that will enable us to avoid this objection.

Moreover, as we have noted above, nature has more than one way to build a mousetrap, and it usually uses lots of them. Given types of goal-directed systems within different species or higher taxa — whether they be blood circulators, body-temperature regulators, prey capturers, or mutation minimizers — are not likely to employ the same internal components to get the same job done, to reach the same goals. For example, both reptiles and mammals are directively organized body-temperature regulators, yet the internal components for doing this vary greatly between them; some blood circulators have hearts with three chambers whereas others have four; similarly, at the biochemical level, thymine is not the only way DNA minimizes its mutations, etc. Not only do different teleological systems attain the *same* goals by use of different internal subsystems, but the same goals are often attained by similar organisms employing different means, even when the available internal determinants of goal attainment are the same: Two animals of the same species faced by the same predator may both avoid predation by different means, say, flight and camouflage. The walking-stick mantis and the chameleon avoid predation by the same behavior, mimicry. But the underlying mechanisms by which they attain the same goal, survival, through that behavior are vastly different. At a lower level of organization, the function of the mitochondrion is to produce ATP molecules, and it does this by employing two different means: Some ATP is produced in the citric-acid-cycle process, and more is produced by oxidative phosphorylation, a different process with the same goal; moreover, this latter process generates ATP molecules by the operation of any or all of at least four different independent internal subsystems.

A full nonteleological account of a teleological law about the behavior appropriate to a goal will have to be *disjunctive*. It will not be enough to stipulate one set of subsystems directively organized to attain a preferred state. Rather, a collection of alternative sets of such systems will be required, each one of which can attain the state. But no one of them will be essential or causally necessary for its attainment, because of backups and redundancies. Moreover, many of these alternatives or "disjuncts" will be absent from most teleological systems that do attain the goals they are organized to attain. So, even if we could completely itemize all the components of one set of subsystems and all the laws relating them to one another and to the preferred state, we would still have to identify all the other sets of subsystems that are or could be employed to attain a given end. At least we must do this if we are to provide anything like a complete nonteleological substitute for a law like (*T*).

The difficulty of meeting this requirement will vary with the degree of precision employed to describe the goals of a system and the means appropriate to it. Thus, in the case of animal behavior, if we describe goals as avoiding predators and means as fleeing or camouflage, the number of different sets of subsystems that generate such behavior, like running, jumping, taking wing, diving, remaining motionless, and changing shape or color, are vastly beyond our powers to practicably enumerate. Even if we describe the goal as finely as producing ATP molecules in the body, we shall have to enumerate at least three different types of behavior that results in their production: glycolysis, the Krebs cycle, and oxidative phosphorylation. These three processes are themselves carried out by three different, independent subsystems in the cell-cytosol and the mitochondria, each one of which itself produces ATP through the operation of different enzymes at different sites. Similarly, the minimization of mutation in DNA replication is attained by several different means, only one of which involves the presence of thymine. When we get down to the level of primary molecular structure, the complete specification of directively organized subsystems seems a manageable undertaking. At this lowest level, the only thing missing from the chemist's explanation of why DNA contains thymine, whereas RNA contains uracil, is the recognition that cytosine, uracil, thymine, and the various DNA ligases, polymerases, and glycosidases are directively organized to minimize RNA-production costs while maximizing DNA-replication fidelity. (Here the notion of "directive organization" has a completely nonteleological reading in terms of the causal interactions between cytosine-to-uracil deamination, uracil–DNA glycosidase activation, thymine's methyl-group-blockage, base pairing in replication, etc.).

As we move up the levels of biological complexity and organization, our ability to tell the full story of directive organization becomes weaker and weaker. For the number, size, and redundancy of alternative sets of subsystems that attain preferred states increase astronomically. A full nonteleological account of the functions of the liver is complex beyond all expectation. After a certain point, the enumeration of these sets of subsystems and their laws of operation becomes practically useless, even when we know the relevant facts. What is more, in most cases we do not know the relevant facts. Consider a complex teleological regularity, like "plovers lay four eggs in order to maximize inclusive fitness." Any attempt to provide an account of the directively organized subsystems of plovers to *fully* explain this regularity would be unmanageably long, baroque, and without further theoretical and predictive payoff beyond the functional explanation.

This means that for most of biology above the level of molecular genetics, the provision of complete nonteleological explanations of teleological phenomena are practically unattainable. We must admit the practical autonomy of such constructions: We cannot completely explain them, still less could we substitute the baroque, disjunctive nonteleological descriptions of directively organized systems for them, even if we could identify fully such systems.

But autonomists want a stronger conclusion than mere practical autonomy for biological teleology. They argue that it is an open, empirical question whether we can ever in theory provide these nonteleological descriptions. What is more, some autonomists go on to claim that for really interesting phenomena, like animal learning, the failure to provide such theoretical accounts hitherto is good reason to think that they will never be provided. They complain that in the absence of a full empirically confirmed theory of the directively organized subsystems that have the appearance of teleology, the notion that such systems are the source of teleology is sheer dogmatism:

The belief that the regularities cited in teleological type explanations must also be accountable in terms of non-teleological laws is a manifestation of the tendency on the part of those who are opposed to the view that organisms have a special status to assume the problems away and to close an empirical question with a logical clasp. (Taylor, 1964:14)

3.6. The Metaphysics and Epistemology of Functional Explanation

In the absence of the complete nonteleological theories demanded, the dispute about the nature and eliminability of teleology does indeed become philosophical. For it ultimately hinges on the biologist's attitude toward two grand philosophical theses: determinism and the finitude of nature. To maintain autonomism in its strongest form, one must reject one or both of the claims that natural phenomena behave in accordance with laws at least as deterministic as those of quantum mechanics and that there are only finitely many kinds of causal routes to the same kind of effect — that the number of ways of skinning a cat or building a mousetrap is finite.

Some autonomists hold that teleological laws are basic to the explanations they figure in — that they cannot even in principle be replaced by any number of nonteleological laws, no matter how large, describing a vast disjunction of mechanisms that jointly exhaust the means whereby a given goal can be accomplished. As a leading defender of autonomy has written:

Now this claim to have reached the rock bottom of explanation is not one which is usually made in scientific theory, the possibility always being left open, however unlikely it may seem, that another set of laws will be discovered which are more basic. In this way, therefore, teleological explanation represents a deviation from the modern norm and a throw-back, to an earlier type of explanation. (Taylor, 1964:21)

A throw back to Aristotelian science, but not, in the autonomist's view, an error. There are two possible grounds for this claim that another set of laws cannot be discovered to explain or explain away teleological ones: Any set offered will be incomplete because it does not cover all the alternative means to the end it explains; the means it describes are not always sufficient to attain the end; or the set of laws is complete but is indefinitely, indeed, infinitely large, just because there is an indefinitely large number of means to a given end. If the set is infinitely large, then it cannot be discovered and employed as a more basic explanation of teleology than teleological laws.

But suppose that the number of different things in the universe that can happen is finite, because the number of rearrangements of the basic material out of which the universe is made is finite. If the laws about the relations among things that happen are deterministic, then the existence of the required set of laws more basic than any teleological law is guaranteed: In principle, there will always be a nonteleological description of the internal components of a system that together with causal laws about these components determine the persistence and plasticity of its goal-directed behavior. Why? Because if the number of different things that can happen is finite, then the number of different ways a teleological process is realized must be correspondingly finite. If every event is deterministically linked to finite classes of joint causes and joint effects, then a nonteleological mechanism that suffices in one case to produce goal-directed behavior must do so in other cases where the same initial conditions are realized. Together these two principles guarantee the existence of directively organized systems for any teleological law. To assert the strong, substantive autonomy of teleology one must deny these two strong metaphysical principles. It may come as an unwelcome surprise to autonomists that they are committed to such strong principles as the denial of determinism and of the finitude of nature. On the other hand, those who reject autonomism are committed to the assertion of these principles.

It would, of course, be a mistake to attempt to adjudicate them, for any final decision about their truth will hinge on the final character of the best scientific theories available. But the direction in which modern science is proceeding does seem to put the burden of disproof upon the autonomist. For the most fundamental account of the universe, quantum mechanics, holds out no prospects either for infinitude or a biologically important indeterminism. Although the theory is technically indeterministic on its standard interpretation, the indeterminism it describes is not what autonomism requires. Quantum mechanics describes a world that has asymptotically approached determinism by the time we get to the level of aggregation that concerns molecular biologists. Besides, below that level the statistical regularities are identical in form with the fundamental deterministic regularities of prequantum physics.

Because autonomists are eager to put distance between themselves and doctrines like vitalism, they may equally be loath to base their claims for biology on metaphysical principles so discordant with contemporary physical science. Moreover, determinism and the finitude of nature seem required for biological theories that no autonomist wants to forgo, especially the theory of natural selection. For instance, if there are in fact an infinite number of different ways a particular selective force could be successfully adapted to, then the one actually established in the course of evolution would not be explainable as *the most* optimal strategy given actual constraints, for the simple reason that there may be no *one* such strategy but an indenumerable number of them. In some views, it would not be explainable at all, for no reason could be given to expect it, as opposed to any of the infinite number of its alternatives.

The autonomists' argument does not sustain their strong conclusion. But it does underwrite a weaker claim, one with much of the practical force for the tactics of biological research that the strong thesis of autonomy provides. If the number and complexity of underlying subsystems that constitute directively organized entities are not infinite, but very large, then it is only to be expected that these internal mechanisms will be hard to describe and unwieldy at best in the useful explanation and prediction of the behavior of teleological phenomena. Furthermore, we are no better off with only a partial or incomplete description of these underlying causal mechanisms, one that does not include all alternative available subsystems or stipulate the regularities that connect them with much detail. We are no better off, for our explanations and predictions will be no more detailed or powerful than those provided by careful teleological generalizations about the same phenomena. At the level of molecular biology, we can expect to uncover alternative mechanisms simple enough and small enough in number to effect a substitution for teleological descriptions in particular cases. But even at the very next level of organization, that of cellular assemblies and organelles, the number of alternative molecular mechanisms becomes too large for compendious description and predictions that actually take advantage of everything we know about the molecular level.

This heterogeneity of alternative mechanisms, coupled with the difficulty of discovering precise laws governing their interactions in the attainment of the subcellular functions they jointly realize, makes teleological description inevitable as a practical matter at even relatively low levels of organization and high levels of detailed molecular knowledge. Naturally, as our knowledge of molecular details improves, and as our computational devices become more powerful and useful, the range of detailed nonteleological treatment of biological phenomena will grow.

But beyond a certain, as yet unknown, level, we cannot really expect to find such alternatives to teleological treatment useful. Beyond a certain level of organizational complexity the sheer amount of time required to amass the data about the values of nonteleological parameters and variables might exceed the amount of time it takes for the predicted event to occur. If real-time prediction requires teleology, then its persistence in biology will be ensured by the most practical of considerations.

Similarly, unless we crave explanations for only the narrowest class of teleological phenomena, and are willing to be satisfied by explanations for these phenomena that are simultaneously schematic and minutely detailed beyond all but the specialist's comprehension, we shall have to allow the attribution of a wide variety of functional ascriptions in biology on a permanent basis. Probably by the time we reach the level of complexity of the opening and closing of the eye's iris, explanation in terms of all the alternative redundant cellular systems directively organized to optimize illumination of the retina would be unmanageably long and too detailed to highlight the crucial facts about the mechanism. Even such an account would make functional attributions to cellular assemblies, which must themselves be cashed in for directive organization at the molecular level. Accordingly, the prospects for eliminating teleology become even more daunting. So long as biology, like the other sciences, embraces the aims of *intelligible* explanation and useful prediction, teleological attributions will always have a place. They will be less evident the closer we come to the intersection of biology and chemistry, but even here they will be conveniences we can forgo only at considerable cost.

The many but finite number of ways in which nature can attain a goal undermine both the claims of autonomists and provincialists. They show autonomists to be wrong, for, short of very strong metaphysical convictions, they cannot establish a difference in kind between biology and its phenomena as opposed to physics and its domain. They cannot secure autonomy on the basis of our ignorance of detailed underlying internal mechanisms for the plethora of functions realized in nature. But the vast heterogeneity and redundancy of nonteleological processes that conspire to give nature its goal-directed character are no comfort to provincialists. It is true that they win the metaphysical and epistemological battle by securing admission that teleological matters are nothing more than the operation of nonteleological forces and can in principle be known as such. But they lose the argument at the level of research tactics: For the purposes that they wish to see biology serve, scientific explanation and prediction, are only attainable in practice by the employment of teleological language. Thus, biology must continue to remain different from physics, if it is to provide us with manageable scientific knowledge. Provincialists' recommendations that teleological ascriptions be expunged and substituted for with nonteleological ones run afoul, not of philosophical arguments, but of the sheer complexity of the empirical facts with which biology must deal. Teleology is unavoidable in biology for contingent and nonconceptual reasons. The world is just much more complicated than provincialists have allowed.

3.7. Functional Explanation Will Always Be with Us

The provincialists' charge that teleological attributions and generalizations are impoverished by the standards of nonteleological descriptions does turn out to be to some extent well founded. If all teleological systems are directively organized in the way suggested above, then the assumption of observable teleology is a highly theoretical undertaking, committing us to the existence of a complex of unobservable, interconnected, directively organized subsystems. This commitment to the existence of *some* package of subsystems *or other* is neutral between which, among the vast number of potential sets of such internal components, is realized in the discharge of a given function. But without any knowledge of exactly how the function is discharged, how the cat is skinned, it is impossible to make more than the most generic and qualitative predictions about the precise topography of the functional or teleological behavior explained. Nor can we make precise predictions about its effects on measuring devices or other detectors, or its effects on the teleological system's environment. Without the restrictions on the way in which a goal is attained that are provided by successively more detailed accounts of the operative internal mechanisms, *no improvement* in the predictive powers of biological theory is forthcoming. This fact has led some autonomists to reject increasing predictive power as a progressive requirement on the scientific adequacy of biological theories. It has led provincialists to decry the continued employment of functional theories in the discipline.

The real choice that biologists must face is between, on the one hand, improving explanatory coherence and manageability by employing teleological approaches, and on the other hand, improving the predictive precision of theories about subsystems, theories too restricted to plug into general explanations of processes above the level of biochemistry. This choice is forced upon biologists by the nature of the phenomenon. They need not of course make the same choice, sacrifice one aim for the attainment of the other, in all contexts, but they are constantly faced with it. The fact that it is sometimes sensible to choose explanatory coherence and forsake, for the immediate or foreseeable future, predictive improvement, accounts for the permanence of teleological ways of thought in biology. One reason it is crucial to choose explanatory intelligibility is that identifying functions for structures we already know a good deal about is the route by which biological knowledge, both functional and nonteleological, is expanded.

To see the importance of pursuing teleological enquiries to answer nonteleological questions, consider another example from molecular biology: the intron. An intron is a portion of the genetic material whose content is not translated into polypeptide products. After a length of DNA is transcribed into a homologous length of RNA, certain portions of the RNA are excised and remaining polynucleotide segments ligated to constitute the messenger RNA. This splicing entirely excludes the segments of DNA that code for the removed regions of the RNA from having any effect on the products of RNA translation. Such segments are called introns. In certain cases, we know as much about the chemical mechanism and structure of particular introns, splicing endonucleases and ligases, as we do about the physical character of any biological process. Nevertheless, biologists persist in posing functional questions, even where the interacting internal subsystems have been identified: Thus Crick writes:

> It is impossible to think about splicing for long without asking what it is all for. In particular, what would happen to the functioning of a gene if a particular intron were removed completely? This leads us to ask how splicing arose in evolution. . . . It might be thought rash to inquire too closely about the origins of a mechanism when we do not yet know exactly how it works at the present day. This gap in our knowledge does not deter speculation, and for good reason, for such speculation may suggest interesting ideas and perhaps give us some general insight into the whole process. (Crick, 1979:268)

It is evident from this passage, and from the discussion of the origin and persistence of introns in terms of adaptive value for genomes that contain them, that the functional approach is crucial to biological investigation. (Notice, however, how Crick links adaptation to questions about the present role of the intron in the directively organized system of the genetic material.) The intron was itself discovered by hybridization – template matching – between messenger RNA for β-globin and the β-globin gene. In this procedure, base pairing takes place between the two polynucleotides in a way that leaves a kink or loop of DNA in the middle unpaired with the messenger RNA. It therefore is not expressed in the RNA. But the entire experiment is shot through with functional thinking – on which the identification of macromolecules, as genes, as messengers, and as enzymes with functional capacities, is based. Without this way of describing the phenomena, and without tools for isolating the interacting elements that attribute to them regularly realized functions, the very existence of the intron would in fact have gone unnoticed. Of course, its existence must be a consequence of the physical regularities that govern nucleic acids and their interactions. But it seems an obvious fact about the current limits of our own knowledge of these laws and of our present experimental and computational techniques that we would not yet have uncovered the intron had we been limited to nonfunctional, purely physical conceptualizations of molecular phenomena. Similarly, the conviction that introns must have a function led to the structural discovery that introns for β-globin across a range of mammalian species have much the same location, size, and sequence and that they mutate at rates much slower than DNA that has no protein producing role at all, so-called junk DNA. Not only was the structural discovery contingent on the conviction that introns have an evolutionary function, but it begins to suggest that they may also have local, immediate functions in the regulation of gene expression during the lifetimes of the individual organisms endowed with them (for further discussion of the evolutionary function of introns, see Section 8.1).

Functional and teleological attributions and explanations will persist in biology wherever the terms of the trade-off between consistent improvement in predictive accuracy and growth in explanatory systematization favor the latter. In many parts of biology, this should be the case for the foreseeable future.

Introduction to the Literature

The volume of philosophical writing on this subject is staggering. Contributions to it made over the last decade are surveyed in Alexander Rosenberg, "Causation and Teleology in Contemporary Philosophy of Science," in *Contemporary Philosophy* (The Hague, Nijhoff, 1982), pp. 51–86. Students should begin with Carl Hempel, "The Logic of Functional Analysis," in Hempel, *Aspects of Scientific Explanation* (New York, Macmillan, 1965), and Ernest Nagel, *The Structure of Science* (Indianapolis, Hackett, 1979; first pub. 1961), chap. 12.

Influential biologists have appealed to the role of teleological thinking in biology as a bulwark against its assimilation into physical science. Among the most prominent are F. Ayala, "Teleological Explanation in Evolutionary Biology," *Philosophy of Science,* 37(1970):1–15, and Mayr, *Growth of Biological Thought.*

The notion that functions are adaptations has been advocated by several philosophers and criticized by an equal number. Hull, *Philosophy of Biological Science,* summarizes much of this literature, and Ruse, *Philosophy of Biology,* expounds this view and defends it. A summary of criticisms is found in Rosenberg, "Causation and Teleology." Ruse contrasts his view that functions are adaptations with Nagel's analysis in *Structure of Science.* This analysis is one of the best treatments of teleology in terms of directively organized systems. Others influential in advancing this approach include R. B. Braithwaite, *Scientific Explanation* (Cambridge, Cambridge University Press, 1953), and G. Sommerhoff, *Analytical Biology* (Oxford, Oxford University Press, 1950).

The most influential account of the character and alleged autonomy of teleological laws is Charles Taylor, *The Explanation of Behaviour* (London, Routledge and Kegan Paul, 1964). Taylor's analysis of teleological statements is elaborated in Larry Wright, *Teleological Explanation* (Berkeley, University of California Press, 1976). Wright, however, argues that such statements are not autonomous, but at least in biology are to be understood as causal claims, making an implicit etiological appeal to evolutionary origins. Taylor and Wright are concerned with biological teleology and the intentionality of human action as well. This is also the concern of another recent book-length treatment of the subject, Andrew Woodfield, *Teleology* (Cambridge, Cambridge University Press, 1976). Perhaps the best treatment of functional and teleological theorizing about animal behavior is Jonathan Bennett, *Linguistic Behavior* (Cambridge, Cambridge University Press, 1976), especially chap. 2.

The account of functional explanation sketched in this chapter is due to R. Cummins, "Functional Analysis," *Journal of Philosophy*, 72(1975):741–65, which also contains a critical discussion of the views of Hempel, Nagel, Ruse, and others. Nagel has defended his original conception in "Teleology Revisited," *Journal of Philosophy*, 74(1977):261–301, in which many alternative views are cogently criticized.

CHAPTER 4

Reductionism and the Temptation of Provincialism

The revolution in molecular biology may be dated from the April 25, 1953, issue of *Nature* in which Watson and Crick's "Molecular Structure of Nucleic Acids" appeared. In what reads like an afterthought at the very end of the paper they wrote:

"It has not escaped our notice that the specific base pairing we have postulated immediately suggests a possible copying mechanism for the genetic material." (Watson and Crick, 1953:737).

Afterthought or not, Watson and Crick's postulation soon led to much more than just a suggestion for how the hereditary material could be transmitted. In fact, the base pairings of the DNA double helix provide a mechanism for much that had previously been known about heredity and genetics. Because of its success in stimulating the explosive growth of our knowledge of the chemistry of life, Watson and Crick's breakthrough spawned anew the oft-rejected conviction that eventually, or at least ultimately, all biological phenomena could be best and most fully explained by theories about their chemical constituents. Doctrines of this form are labeled both by their proponents and opponents as "reductionism," to suggest that biological systems are nothing other than chemical ones and that biological findings, laws, and theories are most clearly expressed, most fully justified, or best systematized through their connection to physical science.

4.1. Motives for Reductionism

To be sure, reductionism is a conviction that long antedates the discovery of the chemical mechanism underlying genetic phenomena. Its appeal is based on an assessment of the history of science as one that reflects progress in our understanding of nature. Since Galileo the natural sciences have encompassed more and more phenomena; their theories have become deeper and more accurate in description and prediction; these theories have been bound together more and more closely; and their technological applications have enabled us to control more and more of our environment. The sequence of theories in physics, in chemistry, and now in biology, has in this view been a succession of increasingly close approximations to the truth about the universe. Our current theories do not constitute *the* truth about the world, but they are closer to it than their predecessors, and these predecessors were themselves improvements on their predecessors. The method whereby this progress has been achieved is said to be that of the reduction of less accurate theories to more accurate ones, the reduction of narrower theories to broader ones, the reduction of diverse and apparently different theories to unified ones. The classical example of this course of scientific progress through reduction is the history of mechanics.

By the seventeenth century, Galileo had produced a theory of terrestrial motion that accounted for such diverse phenomena as the pendulum, the inclined plane, and the motion of projectiles. At about the same time, Kepler had formulated laws of planetary motion. The Newtonian revolution consisted in showing that the regularities Galileo's and Kepler's laws describe are both special cases of the operation of a single set of laws governing the motion of all objects in the universe. Newtonian mechanics is said to have effected the *reduction* of terrestrial and celestial motion by showing how the special cases of Kepler's and Galileo's laws are deducible from those of Newton's theory. This mathematical demonstration was considered revolutionary because it overthrew a world picture that had dominated science since the time of Aristotle. In the picture, the imperfect and ever-changing earthly domain was contrasted with a perfect, incorruptible, changeless heavenly one. This change had two great consequences for biology. First, it suggested that other apparently irregular earthly phenomena might be explained by the same principles governing the regular appearance of the heavens. Second, unlike Aristotle's fundamental explanatory scheme, these principles accord no role to final causes, to teleology of any kind.

Throughout the course of the succeeding centuries, other advances in physical science were shown to be successively reducible to Newton's laws: the theories of the tides, fluid dynamics, harmonic motion, and the mechanics of flight; and the classical kinetic theory of gases and parts of thermodynamics. Indeed, it was by dint of exploiting Newton's laws that advances were made in all these areas, so demonstrating their theories to be diverse applications of a single broad theory came as no surprise. There remained of course both physical phenomena and theories not brought within the ambit of Newtonian mechanics: optics and electromagnetism, for instance. But it was nevertheless widely held that these were only temporarily separated from the unified world picture that physical science seemed to be elaborating.

The twentieth-century revolutions in relativity and quantum mechanics have not undermined the notion that science proceeds by the reduction of narrower, less accurate, diverse theories to broader, deeper, more accurate, and more unified ones. The strengths of Newtonian mechanics are now credited to the fact that it is reducible to quantum mechanics and the theory of relativity. We can, it is often said, logically derive Newton's laws from these more accurate theories by assuming that certain constants of these theories are negligible or infinite: If the speed of light were infinite, the laws of Newton would hold; similarly if energy came in continuous and not in discrete amounts. In addition, of course, quantum mechanics is held to be the ultimate source of laws discovered in physical chemistry, in organic chemistry, and in the examination of biological macromolecules. One reflection of the grip in which reductionism holds contemporary science is the fact that at present great efforts are being made to reduce quantum mechanics and the general theory of relativity to some third unified theory, which will stand in relation to them as Newton's theory stood to Galileo's and Kepler's. Theoretical physicists will remain dissatisfied with the structure of their subject until this reduction is effected.

The reasons are not hard to identify. It is obvious that the physical theories scientists adopt must at least be logically compatible with one another. Two theories cannot be embraced as true or well-confirmed if they jointly commit the scientist to a contradiction. But scientists demand more than the mere logical compatibility of their theories. They demand active interanimation and articulated coherence. The theories we accept should sustain one another: The fundamental assumptions of one should be the derivable consequences of the other. Alternatively, both together should so well describe the same domain that they make joint predictions of greater detail and accuracy than either could provide alone. Scientists refuse to compartmentalize their theories. They insist on finding connections between the concepts different theories employ and the regularities these theories express. The reduction of one theory to another is the best assurance of the existence and significance of these connections.

The history of science has suggested a definite hierarchical structure for the scientific disciplines and their theories. And it has suggested a mechanism through which this hierarchy was constructed and improved. At its foundation are the basic theories of microphysics and space-time, which themselves await unification. Above them are the physical theories of mechanics, thermodynamics, and electromagnetism. Their character as good approximations to the truth is assured by their coherence with, and their reduction to, the fundamental physical theories. Immediately above these theories, and often indistinguishable from them, are the theories of physical chemistry, which in turn underwrite organic chemistry. Beyond this level, matters become cloudy. It will be controversial to say that theories in molecular biology, cellular physiology, embryology, development, behavioral biology and ethology, population and transmission genetics, evolutionary theory, and the other divisions of biological science arrange themselves in a neat hierarchical structure. But one reason for this controversy is the degree to which these subjects are theoretically underdeveloped by comparison to theories in physics and chemistry. Furthermore, proponents of reductionism as a scientific strategy will argue that the only way these subjects can advance is by exploiting the strengths of the more advanced theories below them in the hierarchy. Because scientific theories must be coherent with one another, and because the progress of science has consisted in the reduction of theories that reflects this coherence, the way to proceed in biology seems clear: Advances will consist in the reduction of biological phenomena to nonbiological phenomena. Part of the significance of molecular biology consists in its apparent reflection of this reductionistic scientific strategy.

Reductionists do not hold that all biology should cease until its chemical foundations have been established; such a policy would be as silly as the demand that no physical theory should have been investigated, formulated, or accepted until quantum mechanics was established. Indeed the policy would be worse than silly, it would be incoherent, for quantum theory itself may rest on more fundamental considerations, so that it could not be accepted as a basis for further work until its foundations were discovered, and so on ad infinitum. The reductionist counsels that the scientist "attack wherever he can, whenever he can see that the subject matter may yield and produce those wonderful patterns of organized knowledge which we call 'science' " (Crick, 1966:14). But the reductionist also warns that these patterns may not exist above the level of molecular genetics. If they do, we may not be able to discover them unless we use concepts and techniques that are known to be coherent with those of the rest of science. It is only through such links, forged by reduction to the "hard" sciences, that biology will lose its "weak sister" appearance of amorphousness. Only through reduction can it acquire a structure or a hierarchy of theories of its own, one that will divide its subject matter into natural subdisciplines.

Because of what is perceived as the amorphousness of contemporary biology, the reductionist is inclined to be a provincialist as well. Provincialists are eager to see as much nonmolecular biology as possible absorbed by the theoretical structure erected in molecular biology. But those divisions of the life sciences whose results cannot in principle be reduced to claims about the structure and interactions of macromolecules are to be jettisoned as scientifically fruitless. Disciplines that do not eventuate in laws or approximations to them, theories whose descriptive concepts are incommensurable with the language of chemistry, simply do not have the prospects of providing increasingly useful knowledge. The fact that they are not reducible to molecular biology explains why they have such bleak prospects, for their resistance to reduction is a reflection of their incoherence with the rest of science. It shows that these divisions of biology are dead ends in the history of scientific progress.

Underneath the reductionist's reading of the history of science as the cumulation of successive approximations to the truth, there often stands a strong metaphysical commitment as well. This metaphysical commitment is an important component in the scientist's epistemological and methodological optimism about the continued success of reduction of scientific theories to deeper and/or more accurate successors. Reductionism is a commitment to the methodological unity of science: to the notion that methods and concepts that have succeeded in one scientific enterprise's search for knowledge will succeed in others. This methodological success requires explanation, and the most direct explanation is that the subject matter of each of the sciences differs not in substance, but only in organization. The deductive hierarchy of science betrays the organization of matter, because patterns of organization of matter at each level are but the arithmetical aggregation of patterns of organization at lower levels. Accordingly, the science of the lower level should underwrite the science of the higher level in the way the behavior of the stuff of nature at the lower level gives rise to the regularities manifested at the higher level. Thus, the methodology of reductionism is founded on a metaphysical materialism.

This assurance of the metaphysical as well as methodological unity of science provides the reductionist with a sort of moral support when one comes face to face with important biological findings, and explanations, that are not amenable to immediate reduction to the level of biochemistry. For the metaphysical assurance that such findings are reports about nothing but complex arrangements of molecules, the stuff of molecular biology and ultimately chemistry, fosters the inference that all that is necessary to effect such a reduction is patience and industry, that there is no logical obstacle to eventual reduction. All failures of reduction are to be treated as temporary, reflecting a strategy that can only be criticized as premature, never as wrongheaded. When reductionism fails to succeed, the metaphysical unity of the subject matter of science puts the blame for failure not on the tactic of reduction but upon the nonmolecular theory or finding whose reduction is sought. Perhaps the biological facts are not as the nonmolecular finding alleged, or the irreducible theory is not after all on the right track. To the extent that metaphysical and epistemological commitments represent simply the most general theoretical commitments of scientists, this attitude that unity-of-science theses foster is a perfectly respectable, sober, responsible one.

4.2. A Triumph of Reductionism

Those who have followed the headlines in molecular biology over the period since the Second World War are likely to locate its scientific test in genetics, in the localization of the gene to chromosomal nucleic acids, and in the discovery of the mechanisms of RNA transcription and translation into proteins. Certainly the entire philosophical literature on reductionism in biology has focused almost exclusively on the immediate upshot of the work of Watson and Crick for the biochemical explanation of Mendelian assortment and segregation. In the philosopher's literature, whether reductionism succeeds as a scientific strategy seems very much to turn on whether this explanation can be systematic and complete.

Molecular biologists, however, do not see the matter entirely in this light. Their confidence in the powers and prospects of reductionism as a metaphysical truth and practical strategy hinges in large part on events elsewhere and much later in the postwar history of their subject: It is based on the study of how chemical structure *exhaustively* explains nongenetic, biological function, the behavior of proteins, and especially the operation of enzymes, and in the payoff of these studies for the discovery of complications in the behavior and control of nucleic acids – genes – beyond the wildest expectations of 1953. In absolute terms, what is now known about these matters is still very little: Only a few hundred proteins have been entirely broken down to their constituents. The action of only a few dozen enzymes have been entirely elucidated from a chemical perspective. And yet, relatively speaking, the progress has been immense; younger molecular biologists can remember when the structure of only one protein was known: insulin, a particularly simple one. More important, molecular biologists believe that the methods that have so far shown that chemistry alone suffices to explain biological function in these two-dozen cases can explain the thousands of remaining biological functions through nothing more than great industry and application. Let us trace the basis of this confidence.

The watchword of reductionism in biology has been the slogan that "function is a consequence of conformation, and conformation is specified by sequence." That is, the "sequence" or linear one-dimensional order of the atomic components of a biologically significant molecule causally determines its three-dimensional structure, its shape or "conformation"; its three-dimensional structure causally determines its effects, and in particular all its biological functions. What is more, because all biological phenomena are the consequences of the interaction of large assemblies of these biologically significant molecules (cell, tissues, organs, whole organisms), the linear chemical sequences of the atoms that constitute them are the determinants of, and ultimately the explanatory foundation of, all biological phenomena. This last claim about all biological phenomena consisting in the assembly of the macromolecules with which biochemists work, is, in the reductionists' view, the less interesting and less controversial claim. Less controversial, for they consider that the only alternative to it is some version of a vitalism or holism that no responsible biologist could seriously credit. The assertion is less interesting because the crucial question for reductionists is not how vast assemblies like organs and tissues interact to produce the behavior of animals, or even how large sub-assemblies of cells conspire to effect the functions of these organs. Such questions are important in physiology, and much headway has been made in answering them. But they are less interesting from the reductionists' point of view because these questions already presume the attribution of biologically significant (e.g., functional) properties to organs or cells, without actually explaining in physical detail how this is possible. It is easy to say that the blood or the isles of Langerhans have a role in circulation or digestion. What is controversial is whether they do so just in virtue of the linear sequence of the amino acids out of which their cells are composed; whether what the body and its parts do can be described and explained in terms that appeal to nothing more than the chemical bonds between atoms and molecules that compose the proteins and other macromolecules out of which the body is made.

The answer to some of these reductionist questions has been yes. And the confidence that these answers have given molecular biologists is strong enough to encourage them to generalize their claims far beyond their current data. This is not a criticism, but reflection of how scientific results underwrite a metaphysical conclusion, which in turn motivates subsequent work.

Research on the hemoglobin molecule provides the best illustration of the powers and attractions of reductionism. Molecular biologists view reductionism as having met its scientific test because they have learned so much about hemoglobin's chemistry and because its chemistry has been able to give a *complete,* physical explanation of how and why the blood comes to discharge its most important biological functions.

What molecular biology has discovered about the hemoglobin molecule demystifies the underlying mechanism for a biological function known since the time of Harvey. Since his time, it has been known that the blood's most important function is to carry oxygen (or, as Harvey thought, air) from the lungs to the tissues. How this is accomplished can now be explained on the basis of the linear sequence of amino acids that constitutes the hemoglobin molecule. What is more, on the basis of the same molecular data the molecular biologist can systematize a wide range of hitherto unexplained and unrelated regularities: how unborn fetuses provide themselves with oxygen even though their lungs are not exposed to the air; how and why hereditary disorders lead to several different blood diseases; and why the blood's inability to perform its assigned functions can have selectively *adaptive* consequences.

As with many successful reductions, the research program that led to these molecular explanations of biological phenomena did not begin with these explanations as their explicit aim. It started with the limited aim of uncovering the chemical structure of an important macromolecule. The program then turned to explaining how its parts interact with one another and with other molecules. Only ultimately, to the surprise of many of its own participants, did the research program eventuate in a direct and beautiful explanation of why the blood does what it does. The story is sketched below, beginning with the chemistry and concluding with the biological functions it explains.

Research on the hemoglobin molecule was motivated by the fact that it is an important constituent of the blood and is therefore presumably important in carrying out the blood's function. More important, however, has been its availability

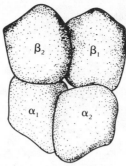

Figure 4.1. Normal adult hemoglobin molecule is a tetramer composed of 4 molecules of 2 slightly different kinds, called α and β. The α and β sub-units are held together by intermolecular forces.

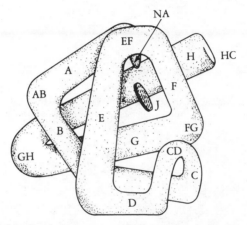

Figure 4.2. Hemoglobin β sub-unit surrounding its heme molecule (shown as a disk labeled J). Single letters mark regions of helical structure; non-helical regions are marked by double letters. At these regions, amino-acid molecules are conserved, the same across several species.

since the nineteenth century in reasonably large quantities, from several different species, and the fact that its crystalline state is very similar to its soluble state. Hemoglobin was one of the first proteins whose molecular weight and amino-acid composition became available. This enabled Max Perutz to construct an x-ray crystallographic map of the molecule in 1959, after twenty-odd years of work. The map shows that every hemoglobin molecule is a tetramer − a molecule composed of four subunits, themselves molecules composed of amino acids in a peptide linkage, surrounding four heme groups − porphyrin molecules containing iron atoms. The four subunits are of two slightly different types, called α and β respectively. Each subunit folds around the heme group in a specific nonsymmetrical way (see Figures 4.1 and 4.2). It was only in the mid 1970s that two striking things were established about these subunits and their shape, their "conformation":

1. The shape of the whole molecule of α-hemoglobin or β-hemoglobin is determined exclusively by the order of the amino acids that compose it and by the chemical properties of these amino acids. That is, given a specification of the order of

Primary structure. Linear order of amino acids.

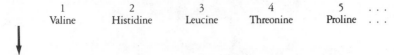

1	2	3	4	5 . . .
Valine	Histidine	Leucine	Threonine	Proline . . .

Secondary structure. Shape of molecule determined by chemically permissible covalent bonding angles between amino acids. Numbers denote amino acid sequence above. Letters indicate regions illustrated in Figure 4.2.

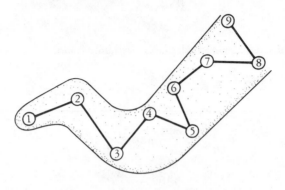

Tertiary structure. Shape imposed on secondary structure by nonatomic intermolecular forces between amino acids widely separated in primary sequence but brought together by secondary structure.

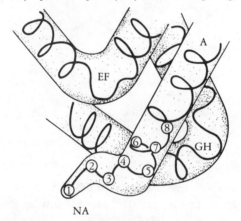

Figure 4.3. Sequence specifies conformation. Primary, secondary, and tertiary sequence of first several amino acids of human α hemoglobin. Quaternary structure of the hemoglobin tetramer is illustrated in Figure 4.1.

the 140 or so amino acids of an α- or β-hemoglobin, we can deduce the shape of the whole molecule from the fact that some amino acids are hydrophobic – not water soluble, some are hydrophilic – water soluble, some are charged negatively and some positively, and that some contain larger and some smaller atoms in their side-chains. All we need to do this is information about the chemical milieu and about the chemical bond already available in physical chemistry. This is what is meant by the claim that sequence determines conformation. The sequence of amino acids is called the *primary structure,* and the folding it dictates is called the *secondary structure* (see Figure 4.3).

2. Equally striking is the fact that although the α- and β-subunits of human hemoglobin, and the hemoglobin of a score of other species, differ in the order of *most* of their constituent amino acids, that is, have different primary structures, they have remarkably similar secondary structures. But so far from disconfirming the generalization that primary structure determines secondary structure, it establishes it more firmly. For the various hemoglobins are "homologous," identical in structure, at 9 of the approximately 140 amino acids, and at just the 9 whose chemical properties determine its secondary structure. Substitutions can be found at all the other positions in the linear sequence that makes up the primary structure, though not just any substitutions, of course: One polar amino acid may replace another; one hydrophilic amino acid may replace another without affecting the secondary structure. But at the points of secondary structure where the linear sequence begins and ends, bends, turns, overlaps on itself, changes its shape, or comes into contact with the iron group it contains, the amino acids of all hemoglobins, within and across species, are "conserved," are the same or closely similar. It is principally to the chemical properties of the conserved amino acids that we can credit the shape of the hemoglobin subunit secondary structure.

The hemoglobin molecule has not only primary and secondary structure, it has tertiary and quaternary structure. Tertiary structure is a consequence of the chemical interaction of the molecules distant in linear sequence that are brought close to each other by secondary structure. Quaternary structure is the structure produced when individual polypeptide chains of amino acids are packed together. It too "follows sequences"; the four subunits of a single hemoglobin molecule pack together in a specific way that is determined by the linear order of their amino acids and by ionic bonding between the amino acids of one subunit and the amino acids of the others (given a molecular milieu). It is the quaternary structure that produces hemoglobin's remarkable functions. And it is the detail in which biologists have come to understand the molecular mechanism underlying these functions, only since about 1975, that has produced the optimism of the reductionists among them.

The discoveries of Watson and Crick initially excited molecular biologists about the prospects for their field and provided much of the needed theory, the tools of model building, and the impetus to develop laboratory techniques that were required to advance the discipline. But, as we shall now see, its achievements in the understanding of the details of a highy significant *nonhereditary* physiological phenomenon have provided a much profounder conviction that biology is ultimately reducible to chemistry. H. F. Judson expresses its depth in his history of the revolution in molecular biology: "The unity of structure and function in hemoglobin is dynamic – whereas the unity in deoxyribonucleic acid is static, passive. The unity of hemoglobin is not, perhaps, so instantaneously obvious as that of DNA. It is similarly ineluctable" (Judson, 1979:604).

This unity of structure and function arises most conspicuously at the quaternary level – in the chemical interaction between the four subunits of the hemoglobin molecule. For the quanternary structure, generated by the linear sequence, suffices to explain the most characteristic functions and dysfunctions of the blood in mammalian organisms. The function of the blood earliest identified and most well known is its role in the exchange of CO_2 and hydrogen ions (in the form of water vapor) for oxygen in the lungs and in the reverse exchange in the muscle tissue of the body. That this exchange occurs in the hemoglobin was well known and yet mysterious. It

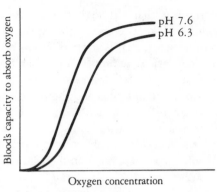

Figure 4.4. Oxygen-binding curve for hemoglobin. Lower oxygen capacity at lower pH is the Bohr effect. Adding DPG shifts oxygen-binding curve to right, reducing oxygen-carrying capacity of hemoglobin, as does adding CO_2.

was quantitatively identified as early as 1904 by Christian Bohr (father of Niels Bohr) and it is called the Bohr effect. What Bohr showed was that the amount of oxygen the blood can hold is a function of the availability, and therefore the pressure, of the oxygen: In the lungs it is high; in the capillaries it is low. So it was inferred that the large quantity absorbed at the lungs must be released at the capillaries. Some time later, J. S. and J. B. S. Haldane showed that the blood's capacity to absorb oxygen is inversely related to pH – that is, to the concentrations of hydrogen ions and of CO_2. Thus, at the capillaries in muscle tissue, where CO_2 and hydrogen ions build up as waste, oxygenated blood must surrender even more of its oxygen than required merely by the lower oxygen pressure; similarly, at the lungs, when the CO_2 and hydrogen escape, its oxygen-carrying capacity must increase. The striking thing about the curve relating the oxygen capacity of the blood and the oxygen pressure is its sigmoidal shape. Although the relation appears to be linear at intermediate ranges of oxygen capacity and pressure, at extreme values it is flattened. What this means is that up to a certain point of oxygen concentration in the blood, the more it carries, the more it can carry, and beyond a certain point the more it gives up, the faster the remaining oxygen is given off. The affinity of the blood for oxygen is a function of the amount of oxygen it is carrying (see Figure 4.4).

It was long known that the blood carries oxygen because oxygen molecules bind to iron molecules in hemoglobin. Each of the four subunits of hemoglobin contains one iron atom bound in a heme group surrounded by the globin molecule. Given the linear sequence of the latter and the chemical structure of the former, we can completely explain the Bohr effect. First of all, it was inferred from the sigmoidal curve for oxygen binding that the hemoglobin molecule exhibits "positive cooperativity," that the binding of oxygen to one of the four heme units' iron atoms facilitates the binding of another oxygen molecule to the iron atom of the second, and so on. But this is just a name for the phenomena to be explained. What is wanted is a molecular mechanism for positive cooperativity, one that shows why each of the four binding sites for oxygen are not independent of each other.

In 1967, Benesh and Benesh isolated a molecule that rests on the symmetry axis between the four subunits of hemoglobin and whose chemical interaction with them

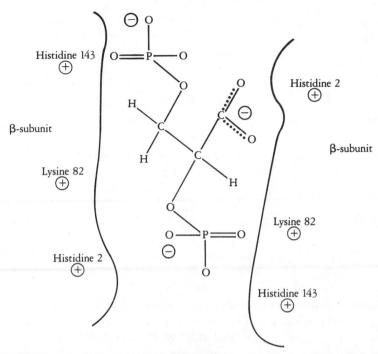

Figure 4.5. DPG's negatively charged phosphate groups bond to positively charged amino-acid molecules in β-chains, bringing them closer together and reducing oxygen-binding capacity of heme groups. Numbers give primary sequence positions of amino acids.

explains their positive cooperativity: 2, 3-diphosphoglycerate, DPG (see Figure 4.5). Given the linear sequence of DPG, we can identify three positively charged amino acids on each β-subunit to which its negative molecules bind. This binding of DPG to the β-subunits changes their shape and that of the α-subunits to which they are attached. The result is a new structure that shifts the iron atom in each α-heme unit into a position where it cannot easily bind to an oxygen molecule. But the effect is symmetrical. Once an iron atom captures an oxygen molecule, its heme group is shifted to a new shape by the effects of the presence of the oxygen. These shifts in the heme group cause changes in the shape of the α- and β-subunits, pushing the positively charged amino acids at the interface of the β-subunits away from the DPG molecule. The change in shape of each subunit narrows the cavity between them and extrudes the DPG (see Figure 4.6). The departure of the DPG allows each β-subunit to take on a configuration that exposes its iron group to oxygen more readily. Thus, the oxygenation of one subunit causes molecular changes in it, which are structurally transmitted to the other subunits; these changes weaken the binding of DPG within the whole molecule; and its departure causes further oxygen binding by other subunits. The result is "positive cooperativity." The change in shape of the whole molecule, which determines its rate of oxygenation, *its function*, is an "allosteric effect" or transition. (These are similar to those involved in maintaining blood-glucose levels, as illustrated in Section 3.4.) The general phenomenon of allostery, of the feedback and feedforward of changes in molecular structure and biological

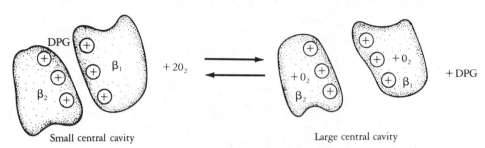

Figure 4.6. Interaction of oxygen and DPG. Locations of positively charged amino acids shown by "+." The α sub-units of hemoglobin tetramer not shown for clarity.

function, was first elucidated by Monod, Jacob, and Changeux in 1963. It is crucial to the molecular understanding of the ezymatic regulation that gives directively organized systems at the molecular level their biological character. The feedforward in this case, in which DPG is driven from the hemoglobin molecules by oxygenation at the lungs, is matched by a feedback in the muscle tissue. There the release of oxygen changes the shape of the hemoglobin so that DPG can reenter and prevent the uptake of oxygen by the blood in the tissue.

But exactly how is oxygen prised from the hemoglobin in the capillaries? Again, the primary structure, the linear sequence of the subunits, explains this. The presence of hydrogen ions and CO_2 molecules at the capillaries reduces the blood's affinity for oxygen (the effect discovered by the Haldanes). They do this just because in the acidic environment of the capillaries they bind to three particular amino acids in each subunit, one at the end and two others in the middle. Their attachment forces the subunit's configuration into a state that allows DPG binding and so reduces the oxygen-binding strength of the iron atoms in the heme groups. In the lungs, the CO_2 and hydrogen bind less tightly to their three amino acids because the pH is higher, and DPG is again squeezed out of the hemoglobin molecule.

Like almost every other functional system, however, the circulatory system can accomplish its activities in more than one way. For instance, in an unborn fetus there is no difference between oxygen pressure in the lungs and capillaries, for it does not breathe air, yet the blood still performs the same functions. How does the fetus's blood exchange oxygen for CO_2 and hydrogen ions under these conditions? Here again, the mechanisms have been elucidated down to the molecular level, which shows how the diverse phenomena brought together under a functional ascription have distinct but closely similar underlying mechanisms. Fetal blood was long known to have a higher affinity for oxygen than adult blood. The functional explanation for this difference is its role in oxygen transfer to the fetus. Thus it can take up oxygen from the mother's blood at the placenta: Oxygen released by adult hemoglobin at the placenta is absorbed faster and held more tightly by the fetal hemoglobin than by the mother's blood. This functional explanation has now been cashed in: Hemoglobin molecules continually release and bind oxygen molecules as they travel through the circulatory system. At areas of high levels of oxygen concentration, the hemoglobin molecules reabsorb the oxygen they release. In areas of low oxygen concentration, oxygen release leads to even more release and no reabsorption (see Figure 4.4). This is a consequence of their "allosterism," the positive cooperativity of their subunits, and their interaction with DPG. But the primary sequence of fetal hemoglobin is just different enough that it does not bind DPG as tightly as adult hemoglobin. Thus, among the allosteric effects that determine the binding and release of oxygen by fetal hemoglobin, none is as great as the allosteric effect in adult hemoglobin. Consequently, the fetal molecule has a higher molecular affinity for oxygen at the placenta, where it comes into contact with adult hemoglobin, than the latter does. So it absorbs any free oxygen at the placenta. And the positive cooperativity of oxygen released by the subunits of an adult hemoglobin molecule provides more and more oxygen at that point because of the allosteric interactions of the adult subunits.

The discovery of two distinct kinds of hemoglobin, fetal and adult, led to the realization that the genes that produce one are turned off shortly after birth and the genes that produce the other are turned on. It thus led not only to a reduction of the physiological phenomena of breathing but also to important discoveries about gene regulation: how the genes that produce proteins are regulated by molecular interactions with their protein products. These are feedbacks that foster further the reduction of apparently goal-directed functions. Turning off hemoglobin suited only for the womb, and turning on hemoglobin needed for the "real world," is one of those complex functions of several interconnected parts of the mammalian organism that we are coming to understand in molecular detail.

Thus, molecular biology enables us to trace the phenomenon of breathing back well past its functional explanation in terms of oxygen and waste exchange by the blood, through the localization of the mechanism in the red corpuscles, past the hemoglobin protein – the macromolecule at which the oxygen and the wastes are chemically bound, all the way down to the purely chemical changes whose interplay results in the teleology, the goal-directedness, of the circulatory system. The strategy involved is the persistent attempt to show how sequence specifies conformation, and conformation determines function, so that the primary sequence and the laws of chemistry together exhaust the causal determinants of the biological phenomenon. The chemistry involved is of course neither trivial nor did it appear independently of the attempt to answer biological questions. The two key theoretical achievements were Linus Pauling's theory of the chemical bond, which enables biochemists to specify a three-dimensional secondary structure given a one-dimensional primary structure; and Monod's appreciation that biological macromolecules, especially enzymes, and proteins engaged in transport, can and do reversibly change their molecular shape as a consequence of interactions with other molecules and that these changes in shape determine their effects – regulate their functions. The importance of this allosterism cannot be overemphasized: The existence of this phenomenon provides molecular biologists with a tool of the greatest power in their attempts to explain the biological in terms of the chemical.

In this case of the reduction of the biological phenomenon of breathing to the chemical phenomenon of hemoglobin allosterism, the tactic is one of opportunism and of tracing a particular mechanism. Opportunism in the sense that no one decided that the first priority of a chemical approach to biology was the explanation of how the blood performs its function. Rather this program of research was the result of the availability of material – in this case hemoglobin; the rise of techniques, like x-ray crystallography; the data of Bohr and the Haldanes; the theories of Pauling and Monod; and years of industry lavished on the structure of molecules like DPG without a definite view as to their payoff for the whole story. Reductionism here is not a matter of deducing one set of laws from another, as reduction so often appears to be in the philosopher's reconstructions. Because it is just one mechanism, it by no means guarantees either that there are chemical mechanisms for every organic phenomenon, nor that this is the only mechanism of oxygen transport. (We have seen the significance of both these points in connection with the analysis of teleology, especially in Section 3.6 above.)

Meanwhile, the success story of the molecular understanding of the function of the blood is by no means over. Sickle-cell anemia long exercised a morbid fascination for biologists and philosophers. For it is such a clear example of balanced polymorphism, of the persistence of a trait that combines maladaptive and adaptive properties in a combination that is on balance only just adaptive and therefore persists in some environments. Without knowing much about its molecular basis or molecular genetics, it can be identified as a recessive trait lethal in the homozygote and adaptive in the heterozygote. Accordingly, its persistence was explained by the synthetic theory of natural selection. But their knowledge of the primary sequence of hemoglobin now enables biologists to identify *one* and only *one* amino-acid difference between normal hemoglobin and sickle-cell hemoglobin. They can explain how this one difference generates malarial immunity in heterozygotes and lethal anemia in homozygotes. They can identify the exact genetic locus of the recessive trait, and cash the whole evolutionary story of the persistence of sickle-cell anemia in for a molecular one. This kind of elaboration of a fundamental theory to provide new insight into hitherto identified regularities is characteristic of reduction in physical science.

Electrophoretic studies showed that the hemoglobin molecules of sickle-cell carriers have a different electrical charge from normal hemoplobin. Further chromatographic studies revealed that sickle-cell hemoglobin contains a valine amino acid instead of a glutamate at the sixth position on the β-chain. The location of the sixth position is on the outside of the molecule. Valine is nonpolar – sticky, unlike glutamate; therefore it will bond with an amino acid on the outside of another sickle-cell hemoglobin molecule, whereas normal hemoglobin molecules do not so bond to each other or to sickle-cell molecules. However, this bonding only occurs when the quaternary structure of the hemoglobin molecule is in its deoxygenated configuration and contains no oxygen. The bonding among these molecules produces long chains of oxygenless hemoglobin molecules in the small blood vessels where oxygen concentration is lowest and therefore deoxygenated hemoglobins are found. These chains block the vessels and deform red-blood corpuscles, giving them a sickle-shaped cross-section. These sickle cells block the capillaries, further decreasing local oxygen concentration, and they foster further bonding, until the blood loses almost its entire oxygen-transport capacity.

The culprit in sickle-cell anemia is a single amino acid, whose chemical properties produce a biological effect, indeed two effects: anemia and malarial resistance. This latter effect is now understood at the level of individual molecules. The malarial parasite spends part of its life in red-blood cells, where it lowers the cellular pH. As Figure 4.4 indicates, reduced pH lowers oxygen affinity, and this fosters sickling. However, sickle-cell membranes are more permeable to potassium molecules than normal ones, so that the sickle-cells' potassium concentration is lower. The parasite requires high potassium concentrations and so cannot survive the sickling it encourages. Thus anemia and malarial resistance go hand in hand.

Because the difference between normal and sickle-cell hemoglobin is just one amino acid, and because the primary sequence of amino acids in a protein is determined by the primary sequence of nucleotides in the genome, we can trace a case of balanced polymorphism back from evolutionary theory through population genetics all the way to its biochemical basis. Two sickle-cell hemoglobin genes produce lethal anemia. Two normal ones provide no malarial resistance. One of each provides resistance and nonlethal anemia. And the differences between them have been narrowed down to a difference in the DNA primary sequence. In fact, there are over a hundred hereditary blood disorders that can now be traced back to abnormalities in the primary sequence of hemoglobin; the genetically determined substitution of one or more amino acids either blocks the normal function of hemoglobin by altering the exterior shape, or hindering oxygenation at the iron atom, or destroying tertiary structure – the folding of the molecule, or altering the quaternary structure. In each case the result is destruction of the hemoglobin's allosteric potential.

In a sense, there is nothing more to explain about how the blood carries oxygen, about what the function of the red-blood cells in oxygen transport is, or about the function of the hemoglobin molecule itself for that matter. The biological phenomenon has been reduced to the chemical mechanism; it has been shown to be nothing more or less than the chemical interactions of the components of macromolecules. And the demonstration seems perfectly generalizable to every other functional phenomenon of organisms. Naturally, there will be differences in the structure and regulation of enzymes and other proteins from organism to organism. But these differences between the biochemistry of a mammal and a bacterium are held to be differences of degree only, and not of kind.

The requisite differences, however, are huge in number and daunting in difficulty. When we add them to the problem of analyzing the function of proteins down to their primary amino-acid sequences, the molecular biologists' claims about reduction begin to sound like the motto of a philosophical program instead of the watchword of a biological methodology with substantial immediate prospects. After all, it took Fredrick Sanger and a team six years of arduous experimentation to provide the primary sequence of the first protein so characterized: insulin. At that rate of progress, the molecular biologists' claims about the reduction of biological functions to chemical structure would be at best hopes for the next millenium. For insulin is a small molecule, comparatively easy to extract from a wide variety of mammals, with some readily detectable functions. None of these things is true about most proteins, especially the enzymes that regulate biological function.

Sanger's technique for identifying the primary sequence of insulin required the degradation of the whole molecule, amino acid by amino acid. The procedure requires considerable care and is subject to error. For very large proteins, the probability of error multiplies astronomically. Learning the primary sequence of the protein is the crucial step in an effective reduction. But if the sequence will be available only in the long run, when, as J. M. Keynes said, we are all dead, reductionism is likely to be stigmatized as mere philosophy and not biology at all.

4.3. Reductionism and Recombinant DNA

At this point in the story, the industrial, technological revolution of contemporary genetic engineering comes on stage and converts the reductionists' millenial hopes into practical attainments of the present and the immediate future. Biotechnology has provided a new, reliable, high-speed method of determining primary structure and its effect on biological function. Its impact on the thesis of reductionism in biology should prove as decisive as the computer for the reduction of atomic theory to quantum mechanics.

Genetic engineering is based on the ability to produce large amounts of the genetic material, the DNA, that controls the production of any protein or enzyme of interest. Isolating the genetic material that codes for any given protein product and determining its primary sequence – the nucleic acids that convey its code – was at one time even more daunting a task than analyzing the protein products themselves. After all, the quantities of DNA that code for a gene product are even smaller than the product. They require the presence of a large number of other gene products to actually generate the product, and they contain three nucleotides for every one amino acid in the product it codes for. The genetic code is based on a triplet of nucleotides specifying a single amino acid. Moreover, the nucleotide sequence contains regions that do not code for gene products at all. At first glance, it is even less amenable to analysis than proteins themselves.

The new technology attacks the problems of analyzing the genetic material and its protein products by starting in the middle of the causal chain from genetic DNA to proteins: at the RNA. Proteins are produced at cellular organelles called ribosomes. Messenger RNA from the genes directs this synthesis of proteins at the ribosomes. In any cell there are large numbers of ribosomes, and any given gene produces large numbers of the same messenger RNA molecules, which then diffuse to the cytosol and link up with the ribosomes. Other RNA molecules, so-called transfer RNA, bring amino-acid molecules to the ribosomes. There, the presence of the messenger's nucleotide sequence determines the permissible molecular reactions among the amino acids that result in a protein (see Figure 4.7). It is not difficult to isolate large numbers of ribosomes all producing the proteins and to separate the messenger RNA that controls the order of synthesis of its sequence of amino acids into a protein. The amount of RNA thus produced is still too small to analyze, but it can be used to produce large amounts of *synthetic* DNA, and the synthetic DNA can be rapidly analyzed for its primary sequence.

After the messenger RNA is isolated, an enzyme can construct a synthetic DNA molecule using the RNA as a template. This enzyme, reverse transcriptase, is employed in nature by certain viruses, whose genetic material is RNA, to transcribe (in reverse) a DNA molecule, which then combines with a host cell's genome to enable the virus to parasitize it. The discovery and isolation of reverse transcriptase has been crucial to almost every breakthrough in recent genetic engineering. It can be employed to produce synthetic, so-called complementary DNA, cDNA. The cDNA is then annealed to a plasmid, a small independent circular DNA molecule normally found in bacterial cytosol.

The DNA can be spliced onto the plasmid at a precise location through the use of still another crucial enzymatic discovery of the 1970s: restriction endonucleases.

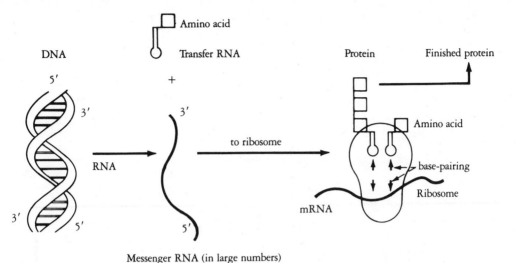

Figure 4.7. Messenger RNA and protein synthesis. Large quantities of messenger RNA can be isolated at ribosome – site of protein synthesis. RNA isolated from ribosome carries protein primary sequence in genetic code.

These enzymes are produced by the cells to protect themselves from foreign genetic material, like that of a virus, that attempts to insinuate its own genome into that of a host. Each different endonuclease operates by cutting up foreign DNA at a particular sequence, say, at the first thymine after three successive cytosine bases. Many dozens of these endonucleases have been isolated, and their specific sites of cleavage identified. They will operate on the DNA from any organism, provided that it bears the correct base-pair sequences. Using a particular endonuclease, the plasmid can be sliced at a precise location; the DNA is then spliced onto the plasmid by another enzyme, DNA-polymerase, which fuses DNA strands. The plasmid, with its new DNA component, is then introduced into an *Escherichia coli* (*E. coli*) bacterium, which reproduces, multiplying the number of plasmids with the DNA's sequence. Eventually a large number of such *E. coli* can be isolated and the now very large number of genetically identical plasmids removed. The plasmids have been *cloned*, in effect (see Figure 4.8). Now the splicing process is reversed, and large quantities of the DNA can be isolated.

If the primary sequence of the DNA can be discovered, the primary structure of the protein with which the procedure began can be read off, because we know the code that translates triplets of nucleotide bases into particular amino acids of the protein. Until the discovery of endonucleases, the determination of even twenty bases was the most that could be hoped for in a year of industrious laboratory work. There was in effect no hope for securing the primary sequence of already localized genes of only moderate size, let alone sequencing the entire genomes of even the simplest organisms. But now we can expect the primary sequence for humans to be provided within the foreseeable future. For with endonucleases as a tool, a thousand or more bases can be sequenced in a fortnight. Because there are only four bases in a DNA molecule, the number of different combinations of two, three, or four bases is small, smaller than the number of endonucleases already discovered. There are

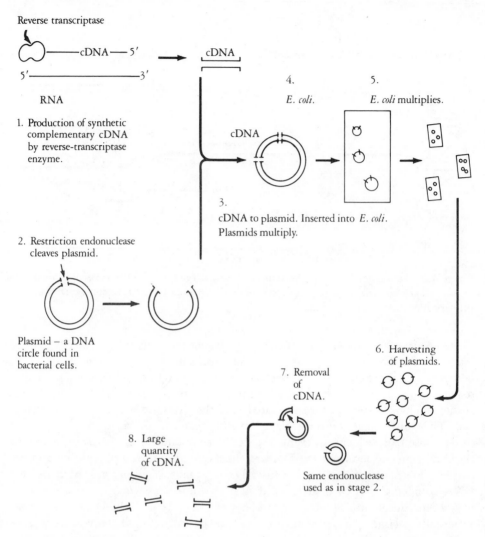

Figure 4.8. Recombinant technique for producing large quantities of cDNA. There are 8 principal steps.

enough of these enzymes already known to sever a given DNA molecule at almost any particular place in the sequence. By applying these enzymes to a DNA molecule and then separating the small pieces produced by the endonuclease, the primary sequence is readily and accurately determined. The method involves electrophoresis, a technique long in use for the separation of different compounds by their molecular weights. The total number of separated products gives the number of bases, and the order of separation reveals the primary sequence position of the last base in each fragment in the order of bases (see Figure 4.9).

But sequencing a thousand base pairs of DNA for a known protein product yields information about the sequence of more than three hundred amino acids in the protein, with accuracy and speed that are orders of magnitude greater than the direct

DNA

5'_____3'   Unknown sequence: 5'—CTACGTA—3'

1. Label 5' end with radioactive marker*:

   5'*_____3'

2. Divide into 4 samples:

   (a) 5'*_____3'   (b) 5'*_____3'   (c) 5'*_____3'   (d) 5'*_____3'

3. Treat with endonuclease enzyme that cuts DNA (a) before alanine bases, (b) before guanine bases, (c) before cytosine bases, (d) before thymine bases

4. Results in radioactively marked fragments:

   (a) two fragments   (b) one fragment   (c) two fragments   (d) two fragments

   5'*— CT 5'*— CTACGT   5'*— CTAC   5'*—   5'*— CTA   5'*— C 5'*— CTACG

5. Separate samples on electrophoretic gel. Electric current attracts segments, which move toward electrode depending on their size. Smallest and lightest fragment moves farthest. Because it is smallest, it must contain shortest fragment, that is, earliest base in sequence. A radioactivity sensitive photograph will reveal all fragments that include the 5' end. The smallest labeled fragment is the first base. The second smallest is second base, etc. Thus the whole sequence can be read off the photograph.

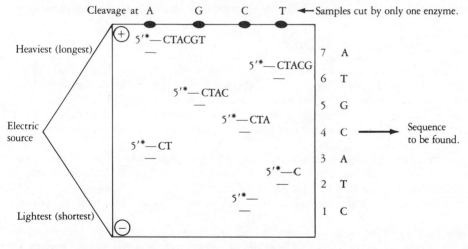

Figure 4.9. Schematic diagram of rapid DNA sequencing, showing role of electrophoresis and endonucleases in this technology.

approach through degradation of the protein. It is worth noting that the methods of rapid sequencing were developed by the same molecular biologist, Fredrick Sanger, who first sequenced insulin. Moreover, cloning and splicing produce these results without requiring any direct knowledge of the location or character of the gene actually responsible for the protein under analysis. It operates by focusing on the RNA that is isolated away from the genome, in the ribosome.

   In fact, it is through the construction of cDNAs that the most accurate knowledge about the location and character of the genes has been acquired. For in addition to providing the primary sequence of the messenger RNA and of the protein, the cDNA can be made radioactive and introduced into the genome. There it will bind to genomic sequences homologous with it, thus radioactively marking the region of the chromosome in which the real protein-producing gene resides.

Cloning and sequencing are the heart of the technological revolution that has made gene splicing, genetic engineering, and industrial biosynthesis of otherwise unobtainable compounds matters of excitement and concern in the popular press. But they are also the core techniques in the theoretical revolution that has converted reductionism from the rhetorical to the literal as a description of biological method. Ultimately it is the rapidly expanding application of these techniques that stands behind the reductionist's confidence that we can specify the chemical mechanism underlying any biological function. For this attitude is no longer a millennial hope; it is a medium-term expectation.

4.4. Antireductionism and Molecular Genetics

Technological breakthroughs have made some biochemists more optimistic than ever about the prospects for reduction. By contrast, Ernst Mayr has written:

When a well known Nobel laureate in biochemistry said, "there is only one biology, and it is molecular biology," he simply revealed his ignorance and lack of understanding of biology. (Mayr, 1982:65)

Of course Mayr does not deny that

the functioning of the genes was not understood until Watson and Crick had figured out the structure of DNA. In physiology, likewise, the functioning of an organ is usually not fully understood until the molecular processes at the cellular level are clarified.

There are, however, a number of severe limitations to such explanatory reduction. One is that the processes at the higher level are often largely independent of those at the lower levels. . . .

Extreme analytical reductionism is a failure because it cannot give proper weight to the interaction of the components of a complex system. And an isolated component almost invariably has characteristics that are different from those of the same component when it is part of its ensemble, and does not reveal, when isolated, its contribution to the interaction. . . .

The most important conclusion one can draw from a critical study of explanatory reductionism is that the lower levels in hierarchies or systems supply only a limited amount of information on the characteristics and processes of the higher levels. (1982:60–1)

How can the place of molecular biology be the subject of such almost total disagreement? Mayr describes his claims as conclusions, but they are no more the result of an exhaustive survey of the powers and limits of the molecular approach than are the reductionist's conclusion. Both are more in the nature of manifestos. Behind them stands philosophy and not laboratory success or failure. In the case of reductionists, the commitments are metaphysical; in the case of their opponents, the presuppositions seem to be epistemological: Mayr and other antireductionists do not deny that life is but matter in motion. But they assert that our knowledge of molecular matters, no matter how exhaustive, cannot generate, justify, explain, or replace our knowledge of the biological phenomena that these macromolecular processes constitute. Mayr writes: "Reduction is at best vacuous, but more often a thoroughly misleading and futile approach" (1982:63).

Arguments for this conclusion can be constructed on what has been reported about hemoglobin above. Recall that although the primary sequence of the hemoglobin subunits determines the function of the tetramer they compose, only 9 of the 140 amino acids in each subunit seem to be "conserved," or remain the same, across mammalian species. These 9 are conserved because they are the only ones that preserve the secondary and tertiary structure. But this means that no single, particular total linear sequence is required to perform the function of hemoglobin. Any of a huge number of alternative sequences will do, just so long as each preserves the 9 crucial amino acids in the right places to generate the secondary folding (provided the substitutions among the other sequences preserve their chemical properties: hydrophobic versus hydrophilic, polar versus nonpolar, bulky versus compact, etc.). Structure may determine function, but any of a large number of structures at the level of primary sequence will generate the same secondary structure and the same biological function. Thus the function of a hemoglobin molecule is in a sense *independent* of the particular chemical composition of the molecule. If this independence of function from structure can be found at the level of primary sequence, it is sure to arise at higher levels of functional organization. This indeed is one of the chief arguments of the anitreductionist: that function is independent of particular structure, that it can be identified independent of our knowledge of structure, and that when this latter knowledge becomes available it not only does not increase our knowledge of function but cannot even underwrite it completely, still less replace it.

Moreover, the anitreductionist will ask, why are the nine conserved amino acids of the hemoglobin subunit conserved? The reductionist cannot answer that it is because of their contribution to the secondary structure, and thus to the function of oxygen transport. For this is to provide a functional, teleological explanation of just the kind that ensures the autonomy of the biological even at the level of molecules in motion. This explanation of the character of a primary sequence, a structure, in terms of its contribution to a complex system is just what the antireductionist claims reductionism cannot provide. The explanation cites, as Mayr says, properties of a component that it has "because of the interactions of the components of a complex system." Of course, the explanation of why these nine amino acids are conserved is "natural selection." Evolutionary forces operate on a protein primary sequence through a feedback from its effects, via its genetical program, in the globin genes. Because natural selection for these effects cannot, it is said, be itself given a reductive explanation, it follows that the breakthroughs of molecular biology are shot through not only with a functional but in particular a Darwinist perspective. And this is a perspective that cannot be cashed in for the laws of chemistry. Accordingly, the great successes of molecular biology, so far from showing the derivative character of biology, in fact show its irreducible autonomy in stark relief.

Outside of genetics, biological knowledge seems too piecemeal to decide a dispute between reductionists and their opponents. Although reductionists point to the molecular detail they can provide for a biological phenomenon, their opponents will harp on the relatively small number of isolated successes so far attained. They will identify elements in the molecular mechanisms that seem to violate reductionists' self-denying ordinances against using biological notions in the explanation of biological phenomena. The reductionist replies that this strategy has not yet had time or resources enough to do more than fill in the molecular details of a few processes, but that there is no obstacle to a complete account of all of them. He claims that the remaining biological residue in reductive explanations is but a convenience that will be eventually eliminated or translated into chemistry by deeper theories at the molecular level. The reductionist will admit that some parts of contemporary biology may never give way to molecular mechanisms, but one is inclined to be a provincialist about them, to write off the forever irreducible as scientifically discreditable. Of course what the reductionist considers unscientific because irreducible the autonomist antireductionist considers irreducible because biological.

The debate between these two camps requires a terrain that both agree will provide a suitable test of what is in dispute between them: an area of biology sufficiently advanced on the nonmolecular level so that doubts about its scientific respectability are implausible. It must be an area sufficiently general and precise that all parties may agree on what will constitute a reduction of it, and an area central enough to the rest of biology to justify relatively sweeping reductionist or antireductionist conclusions. The only division of biology that currently meets these conditions is genetics. So, in spite of the reductionist's great successes in uncovering the molecular mechanisms of physiological function, the test of reductionism has widely been held to be found in the reducibility or not of nonmolecular, Mendelian genetics to molecular biology. Certainly for the philosopher, genetics provides the most manageable context for this debate. For genetics is the area of biological science that seems most immediately amenable to the philosopher's conception of scientific reduction. The main reason for this is that in genetics there are two clearly identifiable theories, bodies of law, expressions of regularity – one molecular and one nonmolecular, Mendelian. Both are agreed by biologists to reflect well-confirmed causal regularities that can be employed to explain and predict biological phenomena. One of these theories is agreed to explain at least part of the mechanism that gives rise to the regularities expressed by the other. What is in dispute is how much of those regularities is explained or explainable at the level of biochemistry.

The philosopher treats reduction as a relation between theories and not between things. This is partly because the philosophy of science has traditionally been concerned with language, leaving the world to science per se, and partly because a precise sense can be given to intertheoretical reduction more easily than to claims that one sort of thing is "nothing but" some more basic kind of thing. Another reason is that no contemporary philosopher or biologist seriously doubts that some version of reductionism is correct when it comes to things and their constituents. Reductionists insist on it, and antireductionists affirm with equal vigor their rejection of vitalism. For them, the denial of reductionism, as noted above, is an epistemological point, not a metaphysical one. The biological reductionists' conviction that theories are reducible ultimately boils down to their metaphysical belief that things are reducible. As we have seen in the case of functionalism, in general, and will see again here, the admission of metaphysical reduction generates for one side the mystery of an epistemological double standard. For the other side it provides a largely theoretical but potentially hollow victory over autonomism.

The reduction of one theory to another is a form of explanation, in which the reduced theory is explained by the reducing theory. Explanation has long been viewed in the philosophy of science as a species of *deduction,* mainly because whether two statements meet the logical relation of deduction is an objective, decidable matter of fact. Scientific explanation is supposed to be a matter of objective relations among statements, rather like mathematical proof. Deduction guarantees the relevance of the explanation to the phenomenon explained, because whether the explanation constitutes a deductively valid inference is a matter of purely mechanical decidability. As a form of explanation, reduction must be deductive as well. And indeed, the philosopher's reconstructions of the great reductive advances of post-Galilean science have treated them all as deductive. If we assume the force of small bodies near the earth to be zero, close approximations to Galileo's principles follow deductively from Newton's. If we assume the speed of light to be infinite, the mathematical formulae of Newton's laws may be inferred from Einstein's special theory of relativity. If we assume gas molecules to be point-masses without interactive forces, the ideal gas law follows deductively from van der Waal's equation of state. So the first and most critical requirement for a successful reduction is that the general statements of the reduced theory follow by deductive logic from the laws of the reducing theory. This requirement of deduction gives expression to the idea that the wider reducing theory "contains" the narrower reduced one as a special case. Moreover, the special premises that are needed to effect the deduction reveal the crucial differences between the theories, especially the improvements and corrections that distinguish the broader theory from the narrower one. For it is the insight that all bodies exert forces on each other that leads from Galilean to Newtonian mechanics; it is reflection on the finitude of the speed of light that enables physicists to formulate a wider theory from which Newtonian theory is deducible.

The second requirement for reduction is not independent of the first; it is a logical consequence of it, but it is usually mentioned separately, because most of the controversies surrounding reduction arise from attempts to meet it. This second requirement is the demand that the terms of the reduced theory be defined or otherwise systematically connected to the terms of the reducing theory. This requirement follows from the requirement of deducibility, because one set of statements cannot be logically derived from another unless the terms in which they are expressed have the same meaning. In logic, the failure to meet this requirement is known as the fallacy of equivocation. It is aptly illustrated in the following "argument":

> The end of a thing is its goal or purpose.
> Death is the end of life.
> _____
> Therefore, death is the goal or purpose of life.

Although this set of sentences has the formal properties of a syllogism, it certainly does not prove its conclusion, even if we treat the premises as true by definition. For the term 'end' is used in different *senses* in the two premises: In the first, 'end' means 'goal' or 'aim'; in the second, it means 'termination.' For this reason, the "argument" is not a valid deduction: Similarly, Newtonian mechanics could not be validly deduced from relativity theory unless the term 'mass,' which both employ, has the same meaning in both theories. If it does not, then although statements of the same logical form as Newton's laws may follow from those of Einstein's theory, these statements will not be any part of Newtonian mechanics, because they are not about the same subject, Newtonian mass, that Newton's laws are about. In the philosophy of science this has been a serious problem, for Newtonian mass is an absolute property of things, which does not vary as their observed velocities change, whereas in Einstein's theory masses do vary as observed velocity changes. Similarly, for Newton velocity is rate of change of absolute position in absolute intervals of time. These differences have led some philosophers of science to suggest that, in this and other cases, what appears to be reduction is actually the *replacement* of one theory by another. At a minimum, the problem reflects the fact that although the reductive requirement that terms of the two theories be equivalent is derivative from the requirement of logical deducibility, it is the locus of most of the controversies about the actuality and possibility of successful reduction.

The trouble with the requirement that the terms of the two theories be the same, or at least that the terms of the reduced theory be connected to those of the reducing theory, is that it can be met too easily — just by artificial and trivial changes in the meaning of the terms of the theories. However, if such artificial changes destroy the character of the theories in which these terms figure, then nothing is proved by successfully deducing one theory from another. For example, if we simply stipulate that 'mass' has the same meaning in Einstein's theory and in Newton's theory, then we can meet the conditions of formal connectability and logical derivability between the formulae of Newton's theory and Einstein's. But this will not show that Einstein's theory contains Newton's as a special case. For, as noted above, Newton's theory makes appeal to absolute space and time, and absolute, not relative, velocities that are independent of matter, which itself cannot be created, destroyed, or transformed into anything else (such as energy, for instance). Einstein's theory denies all these claims. We cannot introduce the proposed stipulation without making one of these two theories unrecognizable at best and incoherent at worst; the result would be an argument trading on a pun, like the one illustrated above. The connections established between the terms of reduced and reducing theories must be natural, reasonable, scientifically motivated extensions of the theories they bring together. They may be definitions, or they may themselves reflect contingent factual relations, but they must pass tests of coherence and compatibility with the remainder of science that are established independently of any claim that one particular law follows from another.

A good example of such connections is the one established between temperature and mean kinetic energy. According to the ideal gas law, $PV = rT,$ where P is pressure, V is volume, T is temperature, and r is a constant. According to the kinetic theory of gases, gases are composed of point-masses, molecules, moving in accordance with Newton's laws. We may reduce $PV = rT$ to the kinetic theory of gases by connecting temperature and the other terms of the law with the Newtonian properties of the molecules. For instance, we may establish, as a definition, or perhaps as an empirical regularity: Temperature = mean kinetic energy of the molecules; that is:

$$Temperature = \frac{1}{2} M \bar{V}^2$$

where M is the mass of a molecule and $\bar{V}$ is the mean velocity of the molecules of the gas. This statement, whether treated as a definition or empirical generalization, helps connect the ideal gas law and the kinetic theory, so that the latter explains the former and thus reduces it to the operation of Newtonian molecules. It also opens up the whole of thermodynamics to reduction through deduction from statistical mechanics. Thus it has a warrant independent from its role in connecting the terms of the kinetic theory and the ideal gas law.

In the discussion of whether nonmolecular genetics is reducible to molecular genetics, the crucial question is whether the connections that must be established are artificial or natural, whether they reflect the character of the phenomenon or are just "hoked up" to establish a point of mere philosophy. Antireductionists argue that the independent, autonomous laws of Mendelian genetics and its successors can only be connected to those of molecular biology by hoking up definitions of terms like 'gene,' 'dominant,' 'phenotype,' etc. in terms of molecular concepts. According to autonomists, the impossibility of reduction is shown by the impossibility of justifying these definitions independent of an artificial attempt to deduce Mendelian-like laws from molecular ones.

4.5. Mendel's Genes and Benzer's Cistrons

The debate about genetic reductionism in philosophical contexts has surrounded the prospects of reducing to molecular biology the Mendelian "laws" of genetic segregation and independent assortment:

Segregation: In sexual species, each parent contributes one gene at each locus, and the probability of an offspring's having genes from one parent at a locus is equal to the probability of having a gene from the other parent at that locus.

Independent Assortment: In sexual species, the segregation of genes at any one locus of a parent is independent of the segregation of any other gene from any other locus of that parent.

Logically deriving these laws from principles of molecular genetics is considered important, for two reasons: As generalizations, they fit the philosopher's conception of what scientific investigation should produce in biology. Moreover, they are obviously crucial to the explanation of the distribution and transmission of phenotypic traits, identified and observed in breeding experiments, in artificial selection, and, more rarely, in field observations. If a molecular explanation for these regularities can be given, then we shall have shown the observed phenomena of the breeding experiment, and much else of peculiarly biological interest, to be reducible to biochemistry. Now in one respect all biologists agree that this reduction obtains. Mendelian genes are composed of long chains of DNA, which constitute the chromosomes whose behavior in meiosis and mitosis generates the segregation of genes Mendel discovered. Similarly, the physical independence of chromosomes from one another within the nucleus guarantees the independence of assortment he reported (except in the case of genes located in the same chromosome). But this sketchy story is not enough to settle the hard questions about reduction. For, though all agree it can be told, it does not constitute the sort of unification and coherence-generating connection between different theories that reductionists cite to justify their demand. It permits antireductionists to insist that, though a connection can be sketched, the theories, laws, concepts, and methods of Mendelian genetics and its successors remain independent of those of molecular genetics – because the story cannot be improved. It cannot be converted into a full and complete deduction of all the generalizations of consequence in Mendelian genetics, as theorems, special cases, or even reasonable approximations to laws about the genetic material couched in the language of molecular biology. In this view, the aims and the methods differ utterly between these theories, so that the deduction of one theory from the other is either impossible or pointless. Reductionists must accept the challenge of giving more detail than this sketch. For if they cannot then the failure to systematically connect these theories at differing levels of biological organization will undermine their general account of the unity of science and of the explanatory hierarchy of scientific theories beginning with quantum mechanics and eventuating in biology.

On the surface, the job does not seem very difficult. The two laws are couched in terms that have no structural presuppositions or consequences; they describe only what happens to genes, not what they are composed of. Mendel's laws leave entirely open the structure of genes; they specify only their causal role, their function. Because Mendelian laws are about only the function of the gene, they leave us a free hand in specifying the structure, composition, and location and mechanism of operation of genes in the organism. All we need then to reduce the functional laws to statements of molecular biology is a description of the biological macromolecules that entail the behavior, the effects, that Mendel's laws ascribe to genes.

It is easy to view the early history of genetics as a start along this road. Through the interaction of results of breeding experiments that revealed the effects of genetic differences, and studies of cellular nuclei that revealed concomitant differences in the chromosomes, the gene was localized to the chromosome. It soon became apparent, in breeding experiments, that the law of independent assortment was false. But this would be a natural consequence if some genes were located on the same chromosome and so segregated together. The localization of genes to chromosomes could be inferred from the frequency with which they assort together in breeding experiments. But further morphological study of the chromosome enabled researchers to map the particular locations, within single chromosomes, of genes revealed in breeding experiments, because they were associated with detectable chromosomal abnormalities. The result was that within forty years of the rediscovery of Mendel's laws a great deal of structural and compositional knowledge about the Mendelian gene had been acquired. What was now needed was the molecular specification of the genes so located. And isn't this just what Watson and Crick provided?

Unfortunately, matters cannot be described in such neat terms, and the upshot of molecular genetics is nothing like as comforting to reductionists as they might have hoped. To see this, we need actually to consider the construction of the required connection between the Mendelian gene and the molecular gene. Recall that the reduction of Mendel's laws requires such a connection. What is needed is something like the connection effected between a gas and its constituent molecules that allowed physicists to deduce the ideal-gas law from the kinetic theory of gas molecules. To acquire our connection, we need of course to begin with the Mendelian definition of the gene. This is often said to include three features: The Mendelian gene is the smallest unit of mutation, of phenotype determination, and of recombination. The first two criteria reflect the conceptual dependence of the Mendelian notion of the gene on the phenotype, as identified in breeding experiments. For wild-type genes and mutations are identified by appeal to observable anomalies. The last criterion reflects both consideration of chromosomal location and observed phenotypic differences. Genes recombine through the exchange of material by the chromosomes during the first division of meiosis. The closer together two genes are on a chromosome, the greater their rate of recombination. Distance along a chromosome is, however, measured not in spatial units, but by the frequency of co-occurrence of phenotypes for these genes in breeding experiments. So, the Mendelian gene is said to have three functions: mutation, expression, and recombination. But the course of reduction from Mendel's gene to the polynucleotide revealed that there is no single molecular item that performs all three of these tasks. That is, although 3 nucleotides are involved in mutation, only 2 are required in recombination and 900 to 1,500 in gene expression. Because different amounts of DNA do the three jobs of the Mendelian gene, any definition or description of the latter in molecular terms will have to be *disjunctive,* will have to consist in a set of alternatives. Thus, consider the gene for red eye pigment in the fruit fly, *Drosophila melanogaster.* This is a Mendelian gene, a unit of inheritance that can mutate, say, to white eye color; it is a gene that controls the expression of an observable trait and that can recombine with other Mendelian genes. A molecular description of such a gene will be at least as complicated as follows:

R is a Mendelian gene = Either R is the minimum number of nucleotides that can be recombined between DNA strands – that is, 1 or 2 nucleotides,

or R is the minimum number of nucleotides within which a base substitution will produce a mutation and change the genetic message – the codon – that is, 3 or more nucleotides,

or R is the minimum number of nucleotides that codes for a phenotype – that is, 900 or more nucleotides.

Notice that we cannot simply identify the Mendelian gene with the last disjunct, even though it describes an amount of DNA sufficient for mutation and recombination, because the Mendelian gene is by definition the *smallest* unit of each of these functions. Nine hundred nucleotides will suffice for hundreds of mutations and recombinations. Accordingly, the relation between a Mendelian gene and DNA molecules is not one to one, but at least one to three. A statement about Mendelian genes will follow from one about DNA only if we treat the same type of Mendelian gene as being sometimes instanced by a handful of polynucleotides and sometimes by a thousand of them. The disjunctive character of a molecular description of the Mendelian gene makes any formal deduction of Mendelian regularities from molecular ones extremely complicated, though still possible in principle: We will have to show that the Mendelian statement follows from one or more of the distinct molecular descriptions, while being compatible with those it does not follow from.

But there seems no good biological motivation for going to such trouble. This definition is clearly the sort of "hoked-up" construction that does no biological work, but simply serves to prove a philosophical point. In fact, reductionists did not embrace this hoked-up construction. Instead, they treated developments in genetics as the abandonment of the Mendelian gene notion in favor of three notions: one for each of the functions originally attributed to the Mendelian gene. Appealing to terms coined in the fifties by Seymour Benzer, reductionists substituted for the Mendelian gene the 'muton,' the smallest unit of mutation; the 'recon,' the smallest unit of recombination; and the 'cistron,' the smallest unit of phenotypic expression. With this substitution, we can erect a one-to-one correspondence between various amounts of DNA and items with a Mendelian function. Though now we need to reinterpret the Mendelian laws as asserting the segregation and the independent assortment of (unlinked) cistrons.

Striking out the term 'gene' from these laws and substituting 'cistron' seems to enable us at least in principle to deduce them from molecular theory. Given polynucleotide duplication by semiconservative replication through the action of DNA nucleases, and polymerases, acting on segments of DNA large enough to constitute cistrons, we will be able to deduce the phenomenon of recombination from what we know (and will come to know) about chemical regularities governing crossover of intact double helices of DNA. By combining these portions of molecular genetics with the theory of mutations as nucleotide replacements that change the genetic code, we should in principle be able to explain Mendel's laws; we should be able to show why they are good approximations confirmed in breeding experiments and why they need to be corrected when the effects of phenotypic expression, recombination, and mutation diverge. This correction of Mendelian laws is only to be expected in a successful reduction. After all, it was already known that the law of independent assortment had to be corrected for linkage and that, even so corrected, the frequency of recombination required further qualification. Now we know in detail why these two corrections are required.

Most reductions involve similar corrections. For instance, Einsteinian limitations on the speed of light and quantum-mechanical limitations on the minimum unit of energy constitute corrections of the Newtonian theory that reduces to these theories. Improvements in the kinetic theory of gases, which treat gas molecules as more than point-masses, enable us to correct the ideal gas law that the kinetic theory explains. The power to correct Mendelian theory is one important reason to accept the substitution of cistrons, mutons, and recons for the Mendelian gene as a scientifically well-motivated step in its reduction to molecular genetics.

But there is a crucial objection to the reduction of Mendelian genetics through the substitution of the notions of cistron, muton, and recon for gene. These terms are themselves still not molecular, so that they do not constitute a reduction, but simply a refinement, of Mendel's notion. True enough, the definitions of muton, recon, and cistron make mention of varying amounts of DNA. But these clauses are empty gestures in the direction of molecular genetics. The amounts mentioned do not provide any means of actually dividing up the genetic material into cistrons, mutons, and recons. These means are still provided by the same breeding experiments that inform the Mendelian gene. That is why the notions of recombination, mutation, and phenotypic expression appear in their definitions. As such, the muton, recon, and cistron are no more biochemical in content than the Mendelian gene when defined as the amount of DNA that recombines, mutates, and controls phenotypical traits. Although Benzer's notions constitute a real advance, something more precise and theoretically consequential must be provided to effect a scientifically significant reduction. What is needed is a characterization of the gene or cistron and its companions that is *entirely* biochemical. This would enable us to identify them structurally and compositionally. Then they can be linked to the breeding experiments by generalizations that thereby reduce Mendelian genetics to molecular biology.

4.6. Reduction Obstructed

Can we add enough detail to actually effect the reduction of Mendel's genes or Benzer's cistrons to DNA? Can we show in principle how it can be attained? Unfortunately, no. For the actual relation between Mendelian and molecular theories is far too intricate and indirect to construct the required biochemical definitions of the Mendelian gene or the cistron. It is especially the indirectness of the relationship that makes for the obstacles to anything like the reduction hoped for. In brief, the problem is this: Molecular considerations explain the composition, expression, and transmission of the Mendelian gene. The behavior of the Mendelian gene explains the distribution and transmission of Mendelian phenotypic traits revealed in breeding experiments. But the identity of Mendelian genes is ultimately a function of these same breeding experiments. Mendelian genes are *identified* by identifying their effects: the phenotypes observationally identified in the breeding experiments. So, if molecular theory is to give the biochemical identity of Mendelian genes, and explain the assortment and segregation, it will have to be linked *first* to the Mendelian phenotypic traits; otherwise it will not link up with and explain the character of Mendelian genes. But there is no *manageable* correlation or connection between molecular genes and Mendelian phenotypes. So there is no prospect of manageable correlation between Mendelian genes and molecular ones. The explanatory order is from molecules to genes to phenotypes. But the order of identification is from molecules to phenotypes and back to genes. So to meet the reductive requirement of connectability between terms, the concepts of the three levels must be linked in the order of molecules to phenotypes to genes, not in the explanatory order of molecules to genes to phenotypes. And this, as we shall see, is too complex an order of concepts to be of systematic, theoretical use.

Return to the definition of the Mendelian gene. The philosophically common assumption that the Mendelian gene is simultaneously a unit of mutation, recombination, and phenotypic expression is certainly not historically accurate. As early as the work of T. H. Morgan before the First World War, geneticists were already employing recombination and mutation as guides to what counts as a unit of phenotypic expression. Few geneticists believed that the whole gene was the smallest unit of mutation: Multiple alleles for a single gene-locus like eye color were considered the result of mutations at different places on the same gene. Furthermore, by the time geneticists in Morgan's group came to focus on *Drosophila*, and especially on traits like eye color as phenotypic, they had already recognized such single Mendelian traits to depend on genetic features located in a number of chromosomes. To hold, therefore, that there was a single gene for a single phenotypic eye-color property would have committed them to viewing the gene as a highly scattered complex instead of as a unified object located at only one place in the chromosome. In short, neither mutation nor phenotypic control seemed to define gene in classical Mendelian genetics. Mutation, recombination, and phenotypic expression were all involved in the meaning of gene, but not in the simple way that makes them three criteria that cut up the chromosome at exactly the same points. Their interconnection can be constructed in the following complex and idealized way (due to Kitcher, 1982), which shows the degree to which the Mendelian notion of a gene is inextricably based on breeding experiments with phenotypic phenomena that the gene is supposed to explain:

Following Kitcher, let us call a *gene complex* a quantity of chromosomal material that determines some phenotypic trait, like eye color in *Drosophila*. This complex may be divided into any of a number of chromosomal segments. Now consider a set of data on the distribution of, say eye color, bristles, lung size, etc., from a breeding experiment. The data will ideally reflect a distinct number of observed phenotypes. Among the alternative ways of dividing up the gene complex associated with this set of phenotypes, one results in a number of chromosomal segments equal to or greater than the number of phenotypes observed. As Morgan and his followers employed the term, a *gene* is a member of this division of the chromosomal material: It is one among that number of segments equal to or greater than the number of separate phenotypic traits detected in the experiment. Thus, a *wild-type* gene complex is a set of chromosomal segments that together produce a set of normal phenotypes. A mutation site is a segment within a complex that, together with other segments in it, produces a departure from the normal phenotype detectable in breeding experiments. If a set of breeding data from recombination experiments cannot distinguish two mutation sites, then it does not assign them to two different genes. The notions of genes and gene-complex make recombination crucial. It is clear that how finely we divide the genetic material, following these definitions, will depend entirely on the sort of breeding experiments done and the sort of organisms employed. Distinguishing phenotypes through the unaided optical examination of *Drosophila* will result in a much coarser segmentation of gene complexes into genes than will the biochemical identification of proteins in their bodies, for example.

Whereas the means of discriminating phenotypes improved greatly in the half-century following the work of Morgan, the identification and enumeration of genes on the basis of the recombination of phenotypes in breeding experiments persisted and became entrenched. Revisions in the identification and enumeration of genes were the result of breeding experiments that more and more closely approached the ideal in discrimination among phenotypes and therefore multiplied the number of genes that figure in a gene complex.

But experimental discrimination among phenotypes never attained the ideal. It improved in fineness over the decades, through experiments that revealed recombinations where previously it was thought there were indivisible phenotypic units. But it nevertheless remained the fact that some phenotypes and therefore some distinct genes that could recombine, were escaping the design of experiments and therefore escaping the notice of experimentalists. Inevitably, geneticists were assigning distinct mutation sites with the same phenotypic effect to the same gene, even though undetected recombination between these sites remained possible and indeed actual. Every time an improvement in experimental design revealed such apparently *intra*genic recombinations, a choice was faced: Either redraw the lines in the gene complex to allow for another gene, or permit recombination within a single gene as a permissible phenomenon. How was one to decide which course to take?

What was needed was a criterion that would distinguish a single gene with two mutations from two neighboring genes that together produce a combinatory phenotypic trait, each of which can bear a distinct mutation. A criterion for this task was provided by Benzer, employing the so-called cis–trans test. Benzer coined the

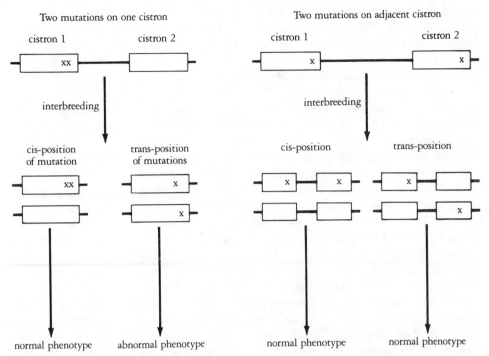

Figure 4.10. The cis–trans test for intragenic mutations.

terms muton, recon, and cistron, and his test enables us to discriminate multiple cistrons with single mutations from single cistrons with multiple mutations. A cistron is of course large enough to accommodate many mutations. But in most cases recessive mutations in one cistron will not reveal themselves in phenotypical differences if the paired homologous cistron is free of mutations. Now, consider two mutations that show up in breeding experiments. The question is whether they are mutations within a single cistron or in two different adjacent cistrons. To decide, we crossbreed these genetic lines. Recombination should separate the intracistronic mutations so that only one of the two appears in some copies of the cistron. This separation will make no difference to phenotypic expression if the mutations are in different cistrons. If they are in the same cistron, however, there will be a difference in phenotypic expression, for separating the mutations to two copies of the same cistron that then pair will produce the mutant phenotype because both cistrons bear mutations (see Figure 4.10). And, once we have isolated several mutations to a small portion of the chromosome, we can determine whether they are located in a single cistron or in several by the cis–trans test. Thus, we work backward from the phenotype, first by recombination of compound phenotypic traits, to a segmentation of the chromosomal material into a gene complex, and then through further recombination of distinguished phenotypes to a gene or cistron. Then, should matters prove experimentally discriminable, we can work back to the enumeration of mutation sites within or between distinct cistrons by recombination of mutant phenotypes.

It is crucial to note that phenotypic differences that cannot show up in breeding experiments, no matter how ideal, will never be detected, and the cistrons that control them cannot be identified. This is the sense in which the Mendelian gene — and its successor, the cistron — is *tied* to experimental discrimination of its phenotype. Genes are counted and distinguished by teasing out phenotypes from one another through recombination in breeding experiments; to this extent, such experiments determine the identities of Mendelian genes.

But if many phenotypic traits are compound products of gene complexes, there must be a molecular pathway from the individual genes and their immediate products up to the construction of the compound traits. The biochemical approach to the individuation and enumeration of genes hinges on this fact. Instead of starting with the final complex phenotypical product and working back, it begins somewhat earlier in the causal chain from the gene and works back to the gene. Such working backward from less complex phenotypic traits is obviously simpler in organisms whose final construction reflects far fewer specialized tissues and behaviors than, for example, *Drosophila*. The causal chain between the gene and the final phenotype in the bacterial mold *Neospora*, for instance, is much shorter. And if these short chains are also to be found as parts of long chains in complex organisms, nothing is lost by focusing on them. This notion was first reflected in the work of George Beadle, who in 1941 advanced the thesis that each gene controls the synthesis of one enzyme (Beadle and Tatum, 1941). The thesis was advanced on the basis of breeding experiments with *Neospora* that employed biochemical tests to distinguish between molecular phenotypes. But identifying and counting genes on the basis of recombination of enzymatic phenotypes reveals a serious shortcoming in the criteria previously employed, including Benzer's refinement of the cis–trans test. For in these cases a mutant gene can still give rise to a "normal" enzymatic phenotype and therefore be hidden in breeding experiments. In consequence, our enumeration of genes or cistrons will remain incorrect.

In the primary sequence of a polynucleotide long enough to be a cistron, to produce one enzyme, a point-mutation will sometimes result in an enzyme with an abnormal primary sequence of amino acids. This sequence may prevent the formation of the secondary, tertiary, and quaternary structures required for full enzymatic functioning, but it may permit a close enough approximation for sufficient catalytic activity. As a result of this approximation to normal function, the mutation may not show up in a biochemical assay or test for the phenotypic enzyme. Benzer's cis–trans test will not enable us to decide whether two mutations of this attenuated kind are on different cistrons or on the same cistron. Because, no matter how arranged, their mutant enzymatic product passes the biochemical test for presence of the normal phenotype. Of course, we could still use the cis–trans test with experimental techniques that discriminate not just the presence or absence of a phenotype enzyme but also its amount, or perhaps its primary structure.

But in doing this we are working all the way down from one end of the causal chain to the other. Having started with the notion of the phenotype as detectable by unaided observation of whole animals, our present stop is the primary sequence of the polypeptide for which the genes code. We may, indeed, treat any of the links in the causal chain back to the gene as phenotypes at different levels of organization, provided only that they are independent enough of environmental forces to assort in roughly Mendelian ratios.

Before going any further, let us reconsider the problem of reduction. Recall that the relatively superficial equation of the Mendelian gene with the cistron, muton, and recon generated a complex disjunctive characterization of the gene that seemed unmotivated, even though it might enable us to deduce principles of segregation and assortment for cistrons from findings of molecular biology. Our historical tour shows first that the real definition of the Mendelian gene is not simply a conjunction of three notions reflecting recombination, mutation, and phenotypic expression, but that each of these processes plays a role in the definition of the other two. So the definition, if it is to reflect the veritable notion of the Mendelian gene, will be far more complex than a mere three-part disjunction. Moreover, if we substitute for the Mendelian gene the notion of a cistron, then, depending on how finely we discriminate phenotypes, we shall be faced with yet further disjunctive definitions of varying complexity.

The coarser our discrimination of phenotypes, the more complex and disjunctive will be the descriptions of their genes. The fewer disjuncts we allow to count as a single type of gene, the further will the phenotypes we can explain be removed from what we set out in Mendelian genetics to explain. For instance, suppose we have set out to explain the inheritance of normal red eye color in *Drosophila* over several generations. The pathway to red eye pigment production begins at many distinct molecular genes and proceeds through several alternative branched pathways. Some of the genes from which it begins are redundant, in that even if they are prevented from functioning the pigment will be produced. Others are interdependent, so that if one is blocked the other will not produce any product. Still others are "ambiguous" – belonging to several distinct pathways to different phenotypes. The pathway from the genes also contains redundant, ambiguous, and interdependent paths. If we give a biochemical characterization of the gene for red eye color either by appeal to the parts of its pathway of synthesis, or by appeal to the segments of DNA that it begins with, our molecular description of this gene will be too intricate to be of any practical explanatory upshot. We could of course describe a gene just in terms of one branch along the pathway, or of its immediate product, so that our definition would be quite simple, on the order of the one-gene–one-enzyme strategy, but then we will be unable to explain the inheritance of red eye color, we will only be able to explain the inheritance of an obscure and perhaps biologically insignificant immediate gene product.

But the situation is actually much worse than this, for what counts as a unit of genetic control in molecular genetics is nothing like as simple as the amount of DNA that generates an enzyme, or even a much smaller protein. Even the review of complications to be given here underestimates its degree and will be soon superseded by further discoveries. To begin with, the genes that compose the lion's share of the DNA do not produce even indirectly any enzyme or protein that can reveal itself in even the biochemical analysis of a Mendelian phenotype like eye-color pigment, wing size, etc. Indeed, not all genes are composed of DNA; some viral genomes are RNAs instead. But all DNA genes function as templates for the production of RNAs. Of these, there are three types: ribosomal RNA, which constitutes along with proteins the site of protein synthesis; transfer RNA, which carries the amino acids to the ribosomes for protein synthesis; and messenger RNA, which carries the primary sequence of the protein's amino acids in a code of ribonucleotides synthesized on the DNA template. Ribosomal and transfer RNA are respectively the ends of two lines of biosynthesis; they do not code for proteins at all, but they provide the necessary tools for this synthesis. Accordingly, the molecular genes that code for them cannot be identified with any protein or enzyme, even though they are among the most well understood and plentiful of genes in the chromosome. The genes for transfer and ribosomal RNA are not structurally distinguishable from the genes for protein-coding messenger RNA (mRNA). In fact, all three RNAs are often transcribed from the DNA together and separated by nucleases, enzymes whose sole function is to divide these RNAs from one another.

In addition to genes that produce no proteins, but only the equipment for making proteins, there are the so-called regulatory genes. Their activities consist in turning on and off the genes that do make messenger RNA, mRNA, for proteins, the so-called structure genes. These regulatory genes include: repressors, which produce proteins that bind to the DNA, blocking mRNA synthesis; control or operator genes, whose binding by still other enzymes permits the transcription of mRNAs; and promoter sites, distinct from structural genes, which serve as initiator sites for the enzymes that initiate the structural gene's activity. Added to these on–off control genes, there are genes that regulate mRNA production by attenuation. All these repressors, operators, promoters, and attenuators are distinct genes, subject to mutation and recombination, which have specific effects on the biosynthetic pathways that ultimately result in Mendelian phenotypes. But none of them can be identified as *the* source of an enzyme, protein, or any component of a phenotype open to classical breeding experiments of the most ideal refinement.

Moreover, there are a whole host of genes that do code for protein-producing mRNAs. However, their proteins immediately return to the DNA, where they regulate RNA synthesis; DNA replication and repair; and the activation and deactivation of repressors, operators, attenuators, promoters, etc. These enzymes are endpoints in the biosynthetic pathways that lead to them. They are not way stations to the ultimate construction of Mendelian phenotypes. So, the genes producing them are no more accessible to breeding experiments than the RNA genes or the regulatory ones. (See Figure 4.11 for a summary tabulation of the types of genes required to produce a single protein.)

Even when we turn to structural genes that produce mRNA for proteins that ultimately do figure in the biosynthesis of phenotypic traits, there are still serious complications. It is well known that the genetic code is redundant. Each amino acid in the primary sequence of a protein is expressed in the mRNA by a triplet of nucleotides, which are produced from a triplet in the DNA template. Transfer RNAs collect specific amino acids, depending on their primary sequence, and bring them to ribosomes. There they are annealed together in accordance with the sequence of triplets of nucleotides in the mRNA. But for many amino acids more than one triplet specifies the same amino acid. For some, in fact, any one of four different combinations of purine and pyrimidine bases will do. This means that two different primary sequences of DNA can give rise to exactly the same protein; that some mutations in the primary sequence will not be expressed under any circumstances, because they make no difference whatever to the protein product. Thus, if we specify a structural gene as "the determinant of a protein," then the complete molecular characterization of that gene will be a vast disjunction of alternative sequences of nucleotides, any one of which will generate an mRNA that produces it.

Types of genes required to make a single protein:

1. Genes for equipment to make each protein

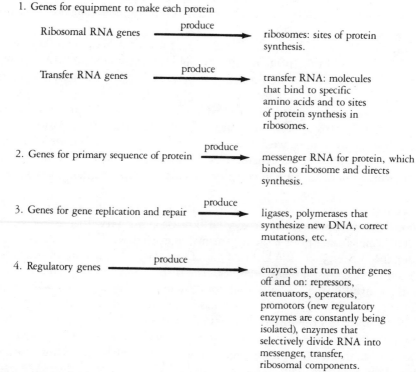

Ribosomal RNA genes —— produce ——▶ ribosomes: sites of protein synthesis.

Transfer RNA genes —— produce ——▶ transfer RNA: molecules that bind to specific amino acids and to sites of protein synthesis in ribosomes.

2. Genes for primary sequence of protein —— produce ——▶ messenger RNA for protein, which binds to ribosome and directs synthesis.

3. Genes for gene replication and repair —— produce ——▶ ligases, polymerases that synthesize new DNA, correct mutations, etc.

4. Regulatory genes —— produce ——▶ enzymes that turn other genes off and on: repressors, attenuators, operators, promotors (new regulatory enzymes are constantly being isolated), enzymes that selectively divide RNA into messenger, transfer, ribosomal components.

Figure 4.11. *Partial* list of types of molecular genes, all of which are required to produce a single protein phenotype.

But of course, as we have seen, for any one protein product to be produced, the operation of many regulatory genes, RNA-producing genes, as well as the operation of genes for RNA- and DNA-enzymes must also be set in motion. Many of these genes can come in multiple disjunctions of primary sequences as well. Further, many of them come in multiple copies, so that if one were not to function, little or no effect on protein synthesis might be felt. These facts multiply the complexity of a complete, disjunctive, molecular description of a gene for a given biochemical phenotype beyond our intellectual powers. We cannot fully express or employ this complete description of a molecular gene in the explanation of the expression, transmission, and distribution of its phenotypes. But this is not the end of the complications.

It has been discovered that structural genes, indeed most eucaryotic mRNA-producing genes, contain regions that are not expressed in the final mRNA. That is, in addition to redundancy in primacy sequence, long portions of these genes are first transcribed onto the mRNA and then snipped out of it, after which the remaining segments of RNA are ligated to constitute the finished mRNA product. These uncoded regions of the DNA are called 'introns.' Why they exist, and what their function is, remains a mystery (see Sections 3.7 and 8.1). According to some researchers, one function of the intervening sequences is to control or direct the elimination of their own transcripts from the mRNAs. If the primary sequence of introns can vary greatly between structural genes that code for the same mRNAs, we have still another source of disjunctions in any complete molecular description of a Mendelian gene.

Leaving the DNA complexities aside, consider the biosynthetic pathways from the DNA via the RNA, to the proteins they produce, thence to the cellular organelles, the cells, all the way up to the tissues, organs, and biological functions that are their ultimate products. These pathways are also disjunctive. From any molecular gene there are a dozen overlapping and alternative pathways to its ultimate phenotypic consequence. Most biosynthetic pathways look more like fishnets than causal chains. For they link many inputs with many outputs through many different channels, some of which double back on themselves in positive and negative feedback loops, whereas others control, supplement, or preempt complementary or competing pathways.

To see just a small portion of this complexity, we can trace back one thread of this network. We cannot trace it back very far because as yet very little is known about the detailed biochemistry of large organisms, and what we do know is spotty. But consider the property of skin color. One of its phenotypical determinants is the dispersion of melanin-pigment granules. The dispersion of these granules is partly under the control of the crucially important molecule adenosine, 3', 5'-monophosphate, or cyclic AMP. But cyclic AMP has a huge number of other equally significant roles: It mediates the intracellular effects of at least thirteen distinct hormones; degrades glycogen in the liver; increases the production of hydrochloric acid by the gastric mucosa; diminishes the aggregation of blood platelets; etc. Thus it is a node on each of the pathways that eventuate in these distinct functions. Stepping back to its immediate origins, we discover two enzymes: adenalate cyclase, a plasma-membrane protein that synthesizes cyclic AMP, and a phosphodiesterase that degrades it. Both these enzymes are the products of biosynthetic pathways themselves. Consider the adenalate cyclase. Its activity is inhibited by another important class of molecules, the prostaglandins. But these molecules themselves have several other effects: They stimulate blood flow to particular organs; control ion transport across membranes; modulate synaptic transmission; and probably have several other even more important effects as yet undetected. Moreover, if we move still further back in the biosynthetic pathway, we discover that whereas prostaglandins inhibit the formation of cyclic AMP, their formation is itself controlled by the precursors of the cyclic AMP it affects. For cyclic AMP is derived from ATP, adenosine triphosphate, and ATP is produced, among other places, in the mitochondria through the oxidation of fatty acids. But prostaglandins are built up out of fatty acids, so the quantity of fatty acids oxidized to produce ATP, and therefore eventually cyclic AMP, affects the amount available for the production of prostaglandins. But of course ATP is produced in several different ways, in addition to oxidation of fatty acids. So this portion of the pathway to cyclic AMP, to prostaglandins, and to the distribution of melanin skin-pigment granules, is just one of several alternative branches in the pathway to this phenotype. Moreover, the molecular genes that control ATP production from fatty-acid oxidation are not even to be found in the chromosomes of eucaryotic organisms; they are located in the mitochondria, separate self-reproducing cellular organelles not contained in the nucleus, but in the cell cytosol. (Mitochon-

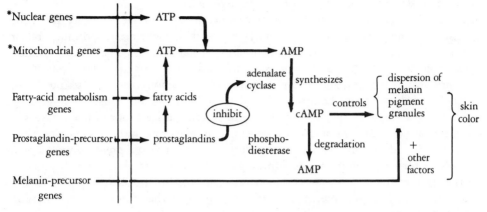

Figure 4.12. A portion of bio-synthetic pathway to skin color. Genetic control for phenotype of skin color includes all genes in Figure 4.12 and several more; starred genes are mutually redundant to this pathway. Many of reactions in pathway (beyond gene transcription and RNA translation) are reversible, thus making pathway a nested set of feedback loops, further enhancing redundancy and interdependence of every component in pathway.

drial DNA constitutes another complication in the conception of the molecular gene.) Thus, to trace skin color back to its biochemical causes, and eventually to its molecular genes, will involve moving along several overlapping, competing, alternative networks, crossing many pathways. No one of them will be either necessary or sufficient for the production of any of the large-scale effects at their end-points, or even for many of the small-scale molecular stages along these pathways. Again, the causal connection between the set of molecular genes that constitutes the gene complex for some phenotypic end-product will be as heterogeneous and disjunctive as the set of genes is itself (see Figure 4.12).

The relation between the molecular gene and the Mendelian phenotype is not a one-to-one relation, or a one-to-many relation, or even a many-to-one relation. It is a many-to-many relation. Anything identified as a gene by the molecular biologist is a starting point on a pathway to many end-points, each identified in Mendelian genetics as a phenotype. Almost any feature identified, even on biochemical grounds, as a phenotype is the end-product of pathways that begin with many different molecular genes, and indeed different sets of different genes. This means that although there are many individual causal chains we can identify between a particular segment of DNA and a particular phenotype, these stories will be both unavoidably incomplete and relatively isolated and ungeneral. For they will cite conditions in the genes that are not causally sufficient or even necessary for their effects, but only sufficient ceteris paribus; and the ceteris paribus − other things equal condition − will be too large to bother filling out. There will be natural groupings of these pathways that bring together phenomena that are closely similar. But, like teleological laws, these groupings will trade off generality against precision and completeness. The more detail they include, the more exceptions they will have to permit, thus reducing their generality and undercutting the naturalness of the grouping of pathways into homogeneous kinds; the greater their generality, the more they must gloss over differences and exceptions, which destroys their precision, predictive accuracy, and explanatory completeness.

Thus, there cannot be actual derivation or deduction of some regularity about the transmission and distribution of phenotypes, say their segregation and independent assortment, from any complete statement of molecular genetics. The sense of 'cannot' here must be specified. The 'cannot' is not a logical one; it has to do with limitations on our powers to express and manipulate symbols: The molecular premises needed to effect the deduction will be huge in number and will contain clauses of staggering complexity and length. For they will have to describe at least four different layers of disjunctions: disjunctions of all the biochemical interactions among the members of a disjunction of all the different types of molecular genes. To these premises will have to be added further, enormously complex general statements about the disjunctive pathways that the gene interactions eventuate in; finally, we shall require that the phenotypic general statements be expressed in terms that figure in the pathway descriptions, and these too will require that disjunctive biochemical descriptions be substituted for the original Mendelian phenotype terms.

The result, for the simplest deduction that actually meets serious conditions of accuracy and completeness, will be no shorter, in small type, than a medium-sized telephone directory. The information on which such a reduction rests might be storable and accessible on a reasonably powerful computer. It might even be programmed to generate particular pathways for any of a large number of molecular genes or Mendelian phenotypes, provided we added enough constraints to the program to prevent it from printing out every node in the network of pathways to which the first step was linked. For such a printout would probably be the entire *Encyclopaedia Brittanica*–sized network.

4.7. Qualifying Reductionism

Even if we traced out the complete pathway through all the causally relevant macromolecules from DNA to phenotype, we would not have yet succeeded in reducing Mendelian to molecular genetics. For we would only have linked molecular genes to Mendelian phenotypes. But the aim of reductionism is not to reduce phenotypic regularities to ones about the molecular gene. It is rather to reduce regularities about the Mendelian gene to the molecular one. It is regularities about Mendelian genes that are to directly explain the phenotypical phenomena. As we have seen, however, Mendelian genes are identified, distinguished, counted, and described by reference to phenotypes. The chromosomal material is not divided by Mendelian genetics on the basis of its structure. For all the variation in chromosomal shape, size, banding patterns, etc., chromosomes are segmented into Mendelian genes by the mutations and recombinations that show up in breeding experiments. Mendelian genes are identified by their effects in breeding experiments.

Over the course of more than a century, these breeding experiments have become increasingly sophisticated in the fineness of phenotypic differences they reveal. But the data they provide are always about the *effects* of the Mendelian genes, and these genes are cited to explain the data. To identify genes in some other way involves either specifying their structure and composition or specifying them by appeal to other effects that do not show up in traditional breeding experiments. Molecular biology pursues both these routes and has ended by multiplying the genetic determinants of phenotypic distribution and transmission beyond the wildest flights of imagination. Following this route leads away from any manageable regularities that could *in practice* explain what Mendelian genetics in fact does explain. In the end, for anything to do the job of the Mendelian gene with respect to Mendelian phenotypes, it must be individuated as the determinant of the phenotype.

Of course, phenotypic determination is only part of the meaning of the Mendelian gene. Its localization to the chromosome, its composition out of DNA, helps specify the gene's character independently of the phenotype that identifies it. The individual molecular descriptions that we can give to stand behind and specify the exact mechanism for any particular Mendelian gene are among the most important things we may expect to learn about it. On the other hand, we can expect no general theory of the molecular gene to provide a systematic explanation of the Mendelian gene's behavior. For it could do this only if it bore such systematic, manageable relations to the phenotypes that Mendelian genes themselves do. The incongruity between general molecular characterizations of the genetic material and useful, manageable regularities about phenotypes infects the relation between Mendelian and molecular genes as well. These relations are just as heterogeneous because the Mendelian gene is nothing if not the unit of phenotypic control. It marches in lockstep with the phenotype it determines and is no more manageably connectable to the molecular gene than is the phenotype.

In the end, the thesis that we can in fact deductively reduce Mendelian genetics to molecular genetics founders on the impossibility of meeting the criterion of connection between the terms of the two theories. Such vast, unwieldly, general statements as we might construct – in which a Mendelian gene is equated with the molecular one – will be full of disjunctions, conjunctions, negations, exceptions, qualifications. It will make so many appeals to stages in the pathway between the DNA and the phenotypic end-point of the pathway that it will be without any independent scientific standing. It will do no other work than substantiate a purely formal possibility without payoff for the actual advancement of either molecular or Mendelian genetics.

There are two potential responses to this conclusion that may tempt the reductionist: One is to say that reduction is not after all a matter of deduction or logical derivation, so that the admitted complications in the relation between these two theories are no reason to deny the thesis of reduction. The other temptation is the provincialist's gambit of saying that the impossibility of logical derivation casts a pall over the theoretical respectability of Mendelian theory. It should be read out of the scientific corpus now that we see it is incapable of a manageable unification with molecular biology and therefore physical science. Let us consider this second option before the first.

Recall the original motive for reductionism: It reflects a conviction about the nature of the theoretical organization of science and about its progressive, cumulative character. In this view, a theory that cannot be made actively coherent with more fundamental accounts of its underlying mechanism is a scientific dead end. The notion that the body of scientific theories displays a unity of more basic common phenomena underlying disparate and apparently dissimilar domains demands more than mere logical compatibility between theories about closely related phenomena.

But between Mendelian and molecular genetics we seem to have at most mere compatibility. Accordingly, improvements in molecular genetics cannot be expected to have much effect on improvements in Mendelian *theory;* progress in one area cannot be expected to carry with it progress in the other. Increasingly, Mendelian genetics can be expected to be left behind in the effort to improve the explanatory depth and predictive strength of our accounts of heredity. The regularities we have already in hand about Mendelian genes and phenotypes are riddled with exceptions, qualifications, and conditions. It is no surprise they cannot be subsumed under molecular regularities that are free of such restrictions on their generality and truth.

Provincialists may thus advocate the abandonment of Mendelian genetics and offer the impossibility of reduction as one good argument in favor of its abandonment. Because Mendelian genetics cannot be unified with molecular genetics, it simply is not respectable science after all. This does not mean it is pseudoscience. It is not that Mendelian genetics is no better than alchemy; rather, it is a dead end on the route to a completely general theory of biological phenomena that the natural science of biology aims at. The only way to provide general theory is to proceed in accordance with the tools and concepts of physical science, as molecular biology attempts to do. Because the concepts of Mendelian biology cannot be manageably linked with those of molecular biology, there is something amiss with these concepts. Provincialists may not deny that the concepts had an important heuristic function. Thinking about matters in terms of Mendelian genes and phenotypes was an important stepping-stone to the ultimately correct approach to these matters, the molecular one. But, they may reckon that the Mendelian notions and the theory in which they figure have outlined their instrumental usefulness, and their failure to be manageably linked to the rest of science, via molecular biology, shows their deficiencies. The fact that the laws of segregation and of independent assortment are exception ridden, and cannot be entrenched in a wider theoretical network that actually derives them from more fundamental principles, shows that at best they describe local and restricted regularities. They are gappy uniformities thrown up by the operation of universal and exceptionless regularities at far lower levels, operating on the accidental conditions of life on this planet. Mendel's laws obtain here on earth to the extent they do only as a result of the operation of selective forces on the biochemical origins of life. If the basic constituents of the primal soup had been different, or the conditions under which they interreacted had been different, then natural selection would have thrown up a different mechanism than meiosis and mitosis or none at all. Because Mendel's laws represent the consequences of this biological accident for the mechanism of heredity, it is no surprise, in the provincialists' view, that they should not be unifiable with the real, universal regularities of chemistry on which molecular biology rests. (What is correct in this view is developed further in Sections 5.3 and 7.9.)

However consistent, this attitude really sounds like sour grapes or special pleading. Reductionists have been proven wrong in their claims about what we can actually accomplish by way of reduction. So the provincialists among them respond that the outcome discredits the theory that cannot be reduced, instead of the philosophical thesis of reduction. Even though reductionists are correct to treat Mendel's laws as restricted local regularities that cannot be improved or manageably entrenched, the conclusion that they should be written off is tantamount to throwing the baby out with the bathwater. For these laws have indeed proved valuable heuristic instruments, essential to many practical improvements in plant and animal breeding, in the control of disease and dysfunction, and in the maintenance and improvement of habitats and environments. Their predictive power may not have the fine-grainedness of chemistry's best general laws, nor the improvability of successive versions of the kinetic theory of gases. But the practical applicability that they do have is more than enough to ensure the permanent entrenchment of Mendelian genetics, both in the textbook presentation of biology and in the theoretical edifice of life science. Moreover, though their connections to molecular biology are not clear and simple enough to underwrite claims of actual derivability, these connections are nevertheless strong enough to enable us to give piecemeal explanations both of instances in which the Mendelian laws apply and of instances that are exceptions to them. We can give at least schematic molecular explanations of their approximate truth and their particular failures.

Not only can we give general schematic explanations of the molecular basis of Mendelian phenomena, we can provide actual explanations of particular Mendelian-like regularities from molecular ones in restricted cases. In these cases we are dealing with very small genomes of organisms that breed in very large numbers, very rapidly, under conditions we can subject to laboratory control. In experiments with certain viruses and plasmids the primary sequence of the genome has been entirely sequenced and many of the products of its genes' expressions are known in equal detail. Enough of this information can be coded and stored in current computers so that regularities governing the transmission and distribution of the phenotypes these genes express are in fact deduced through the implementation of the computer's program. Naturally, these cases are very much simpler than those of sexually reproducing eucaryotic systems to which Mendel's laws apply. But the structural similarity between these simple genomes and the most complex we know of provides some reason to believe that a molecular account for these more complex ones is at least possible in principle. In this respect, the explanation of genetic transmission and distribution in simple laboratory model systems plays a role much like that of simple systems in physical science. In physics, for example, one reason we believe that Kepler's laws of planetary motion are reducible to Newton's is that we can deduce a general statement about the motion of one planet around the sun from Newton's laws. This general statement is close enough in form and empirical content to Kepler's principles to explain them. On the other hand, from Newton's theory we cannot deduce the motion of the planets around the sun in the same kind of detail. There is a notorious three-body computational problem in Newtonian mechanics that has never been solved. It reflects limitations on our information-storage and computational powers. But no one suggests that it either discredits Newtonian mechanics or threatens the claim that laws of planetary motion can be reduced to it. A similar point can be made about the Schrödinger wave equation of quantum mechanics. Although this equation is held to govern the behavior of all atoms, it can only actually be solved for the simplest of them, hydrogen. For helium and beyond, physicists must employ approximation methods to determine the states of the atomic systems. Again, no one suggests that this restriction on our actual computational powers restricts the generality and applicability of the Schrödinger wave equation. Seen in this light, the fact that we can effect the derivation of Mendelian-like regularities from molecular genetics for the simplest model systems seems as much as biological reductionists need. For they may combine it with the possibility of general schematic explanation of Mendel's laws and their exceptions, and the possibility of piecemeal tracing out of causal chains within the complex cases, to substantiate their reductionist convictions. These three possibilities make provincialist suggestions that we rid ourselves of Mendel's laws unacceptable.

It is the fact that our best molecular explanations of Mendel's laws, and of their exceptions, are inevitably piecemeal, partial, and restricted when complete, and incomplete when at all general, that provides the real autonomy of Mendelian genetics. This autonomy is, however, instrumental and heuristic, not ultimately epistemological or metaphysical. In this respect, the reductionist is right after all. To see this, let us consider the first of the two responses that reductionists make to the failure to actually effect the derivation: the response that actual logical derivation is not required for reduction, that something else will satisfy the demands of reductionism.

Admittedly, it also sounds suspicious to change the standards of reduction from the actual provision of a derivation that satisfies conditions of logical deduction and connection of terms. But the original demand, as propounded by Logical Positivists for reduction in physics, did not envision the sort of complexity that has shown itself in the last decades' work in molecular genetics. After all, it was only this explosion that showed no deduction could in fact be accomplished between Mendelian laws and molecular ones. This complexity has not shown that such a deduction is logically ruled out as impossible in principle. The fact that the premises of the deduction will be too numerous and too long to permit any actual reasoner to do the derivation is no reason to conclude that a more powerful reasoner, possessed of all the relevant generalizations, could not do it. It is true that the required connections between Mendelian terms, like gene, phenotype, dominant, recessive, mutation, etc., and their molecular realizations, may be far too complicated to do any actual scientific work for limited creatures like us. But for a really powerful thinker they may well prove as unifying and synthesizing a set of connections as the statement "temperature equals mean kinetic energy" proves for us.

What we have already discovered is not enough to show that some omniscient creature in possession of all the relevant facts *could* effect this deduction; but, by the same token, no information currently in hand precludes this abstract possibility. What is more, the two commitments, to the finitude of nature and to determinism, sketched out in Chapter 3, provide reason to expect this derivation is not just logically but physically possible. And they do so in just the same way that they provide reason to believe that behind every functional relation there lies a disjunction of causal relations that constitutes its mechanism. For if these metaphysical commitments are right, then despite the instrumental autonomy of nonmolecular descriptions of phenomena, these descriptions must be backed by a finite, if unmanageably large, disjunction of molecular mechanisms. This means that the reduction of Mendelian genetics to molecular genetics is at least in principle possible; that the impossibility with which biological complexity confronts the reductionist is a practical, instrumental one. It is one we cannot overcome only because of limitations on our own powers to collect and manipulate the findings and theories of molecular biology.

There are then no metaphysical or epistemological consequences for reductionists to resign themselves to, in the actual facts of the matter. For their metaphysical commitment to reduction is left untouched by any practical limitations on our power to prove Mendelian conclusions, given molecular premises; nor is the abstract epistemological possibility of justifying our Mendelian beliefs on the basis of molecular findings called into question. The concrete possibility exists in many admittedly isolated cases, and their completion is restricted by the fact that our brains must be small enough to be portable and therefore are not big enough to do the whole epistemological job.

The theoretical possibility of reduction may not be enough for meeting Logical Positivist standards, but it is rather too much for the dyed-in-the-wool epistemological autonomist. For it allows the practical, heuristic, instrumental autonomy of Mendelian genetics. But it provides a reason to believe that the piecemeal molecular accounts of it will increase in number and perhaps also in generality, that in places Mendelian accounts will be replaced by molecular-genetic descriptions and explanations. Moreover, where it remains unreplaced this will reflect no unbridgeable differences between physical science and life science, but rather practical, instrumental obstacles to scientific unification.

4.8. The Supervenience of Mendelian Genetics

The reduction of Mendelian to molecular genetics is a logical and physical possibility. But whereas reductionist biologists may be satisfied by this account of the matter, it will not suffice for philosophers. They shall want a far more precise statement of how matters lie than the assurance of a reducibility-in-principle underwritten by a commitment to vague and vast theses about the finitude of nature and determinism. For this reason, reductionist philosophers have over the last decade and more revised their accounts of reduction to accommodate the successive complications revealed by radical improvements in biological knowledge. The cottage industry of formulations, counterexamples, reformulations, and new counterexamples that for long characterized the philosophical treatment of teleology came to be paralleled in treatments of reduction. Starting with the Positivist requirement that laws be derivable and concepts connected, philosophers have added more and more qualifications to the thesis to reflect the actual relations between Mendelian and molecular genetics.

The first qualification added to the account of reduction was the need to correct the reduced theory before reduction. This addition reflected the recognition that since crossover, recombination, and linkage falsify Mendel's laws, they have to be corrected before reduction. After all, we did not want to require the reducing theory to entail propositions already known to be false. But correcting these laws might altogether change their identities, so that the deduction of corrected laws would show nothing about the reduction of the original theory. So it was suggested that the correction meet a requirement of strong analogy with the uncorrected original target of reduction. For example, cistron theory must be very much like Mendel's theory.

Then the reduction of scientific theories came to be viewed as a relatively long process, in which incremental steps are taken that involve changes in both theories. Often the result is either a blending of two research programs or a third intermediate theory. Thus the biochemical theory of the pathways between Mendelian genes and

Mendelian phenotypes could be viewed as a third theory that elucidates the connection of Mendelian to molecular genetics. Certainly the actual century-long history of the localization of the gene — the interaction of breeding experiments, cytological studies, and ultimately biochemical analysis — reflects a gradual process in which both theories have influenced one another. The gaps between them have been bridged by a relatively distinct theory like Benzer's account of cistrons, recons, and mutons, which combines both Mendelian and molecular insights. Finally, philosophers have begun to add historical and sociological conditions to the purely logical conditions on reduction; these attempt to capture the process of scientific change instead of its artificially frozen products. Reduction in this new view is a relation between ongoing research programs, instead of one between sets of sentences that constitute the theories in question.

These revisions are likely to improve the philosopher's understanding of theoretical change and of reduction as it is actually accomplished in science. But two crucial things need to be noted about the studies that eventuate in this improvement. They all begin with the assumption that there has been and is a reduction of Mendelian genetics to molecular genetics. They all result in modifications of the philosophical view of reduction to accommodate it. As such, the philosophers are pursuing a largely philosophical question without any direct payoff or methodological prescription for biology, where the argument about whether a reduction has taken place is still moot. Secondly, the complications added to the philosopher's account in order to reconcile it with the actual character of reduction in biology are of just the sort of sociological and historical character that scientists certainly cannot and probably should not be expected to allow to influence their science. Moreover, the kind of reduction that philosophers ultimately identify in biology does not touch on the issue in dispute between biologists like Monod and Crick in the reductionist camp and their opponents like Mayr. For it is no longer a claim about the nature of things and the theoretical relations between alternative accounts of them.

By way of taking part in this dispute, the most reductionist philosophers can provide is a relatively precise statement of the metaphysical unity of the objects of Mendelian and molecular genetics; a statement in which both they and their biological allies can found their epistemological conviction that real, full-blooded intertheoretical derivations are a least in principle possible. Such a statement will be quite complex, as we shall see. But the task of expounding it is worth undertaking. For it not only underwrites this possibility but reflects the relations between theories at many levels of biological thought. The relation to be expounded is also reflected between functional or teleological attributions and the causal ones that underlie them (elucidated in Chapter 3), and between evolutionary claims and their evidential and explanatory bases (to be developed in Chapter 5).

The relationship between Mendelian phenomena and molecular phenomena is one of *supervenience,* not in the ordinary sense of this term but in a special and formal sense. Mendelian objects and properties supervene molecular ones: The latter fix the former ones completely, even though they cannot be connected to them in any manageable way. The relation of supervenience has been expounded in contemporary philosophy as a relation between properties and sets of properties that have the two following characteristics:

1. If a set of properties A supervenes on another more basic set of properties B, then no two distinct objects that share identical properties from the more basic set B can differ in the properties they share with set A.
2. If the set of properties A supervenes on the set B, then it may be that no property in set A can be defined or manageably connected to any set of properties in set B.

For example, consider the set of all clocks and their time-telling properties. This set will include circadian processes like tree rings; hearts that beat; optical systems like sundials; mechanical systems like water, sand, and pendulum clocks; wristwatches; thermal clocks like marked candles; electronic clocks; electrical clocks; quantum-mechanical oscillators; atomic clocks; radioactive materials with various half-lives; etc. Among the time-telling properties are: counting seconds, counting nano-seconds, counting hours, counting seasons, counting years, counting geological epochs, etc. Now of course all clocks are nothing but physical objects. The trouble with establishing this obviously true claim is that there is no finitely manageable list of physical properties that all clocks have in common, which could be cited to substantiate our obviously true claim, all clocks are physical objects. We neverthe-less believe that all clocks are physical objects. The reason is that every clock does have a certain set of physical properties and that if anything else had just these physical properties it too would be a clock. No two clocks may be alike in all their physical properties; indeed, if we discriminate finely enough we are bound to find physical differences between any two clocks no matter how similar they appear to direct observation. Despite their lack of identical structure, what makes them all nothing but physical objects, nothing but complex aggregations of physical micro-particles, is the supervenience of the property of being a clock on microphysical properties. Every clock has a set of physical properties. These physical properties are what its being a clock consists in, because anything else with these properties would also be a clock. On the other hand, another clock will almost certainly not have exactly the same physical properties as the first; this is what the differences between clocks – their sizes, colors, mechanisms, accuracy, energy consumption, etc. – consist in. Because clocks differ from each other in physical properties, there will be no physical laws about all clocks. For the vast number of different sets of properties that being a clock supervenes on will not figure in any single general law or even a small number of them. Nevertheless, the physical behavior of a particular clock is explainable on the basis of physical law, because the set of properties on which its being a clock supervenes does figure in a small enough body of physical laws to explain its behavior. This same set of laws may not suffice to explain the behavior of another clock, even though it too is nothing but a physical object, because it is a clock by virtue of having a different set of physical properties governed by at least some different physical laws.

We need a somewhat more formal account of supervenience. Let us continue to use the example of clocks to construct it. Consider all the time-telling properties of clocks: Some clocks, like trees and their rings, can discriminate years, decades, and even centuries, but not millennia. Other biological clocks can discriminate the seasons, lunar months, and days. Then there are clocks that can do no better than discriminate hours, others that tell seconds apart accurately, and still more advanced systems that count milliseconds or even nanoseconds, etc. Call the set of time-telling properties that actual clocks have M. M will include the properties of telling the years, of telling the days, of telling the hours, and so on down to nanoseconds and up beyond millennia. Now, from M we may construct another set M' that will include all logically possible conjunctions and disjunctions of the M properties and their complements (their absence). Thus, for example, there will be the property in M' of "telling seasons, and not telling years, or decades, or months, or days or hours, etc." This property will be true of leaves that live only one year, growing and turning colors through the seasons. But many of the properties in M' will be true of no clock. Thus no clock will have the property of "telling decades, nanoseconds, and nothing in between." So now we have M and M', all the properties constructible from M. Among the members of M' is a special set, the set of properties that provides the most complete and exhaustive descriptions that any time-telling system could satisfy. Each of the members of this set of "maximal" properties will contain every property in M or its complement, and each will be incompatible with any other maximal property (that is, every time teller must have one and only one of these maximal properties). Call this subset of M' the set M^*.

Now consider another set of properties, N. This will be the set of all basic physical properties of objects — their size, composition, shape, chemical properties, electrical and thermal characteristics that clocks actually have. N is meant to include any physical property that effects the operation of any clocks. Out of this set N, we may construct a set N' similar to M': It is the set of all properties logically constructible out of the basic members of N. We may specify a further set N^* of N-maximal properties, each a mutually incompatible, complete description. Now, every time-telling physical object has one and only one of these N-maximal properties.

We are ready to express the supervenience of "being a clock" on "being a physical object." The set of time-telling properties, M, supervenes on the set of physical properties, N, if and only if any two objects that are physically indistinguishable, and thus share all the same properties in N', share all the same properties in M'. But, of course, if two clocks are physically indistinguishable then they must have exactly the same time-telling properties; if a clock has all the same physical properties as some other object, then the second object is a clock as well. Accordingly, time telling, being a clock, supervenes on being a physical object: Clocks are nothing but physical objects. It is being a physical object, with certain physical properties, that makes something a clock. This will be so even though it may be the case that no two clocks have any physical properties in common, even though we cannot reduce 'being a clock' to any manageably short list of physical properties necessary and sufficient for being a clock. Moreover, if clocks are nothing but physical objects, then for every time-telling property in M, there is exactly one N-maximal property, one complete physical description in N^* that gives the physical properties that constitute the time-telling property in question. So if two objects have the same N^* property, they have the same M properties, although two objects may have some time-telling properties in common, say, telling seconds, whereas they share only a few N properties and have utterly different physical mechanisms.

Of course, because the set of all physical properties, N, is extremely large, we cannot express the true universal generalizations that link every M property to one and only one N^* property; we cannot even write down the description of such a physically maximal property. This is why we cannot give a manageably long physical definition of 'clock,' even though all clocks are nothing but physical objects.

Let us apply this account to Mendelian and molecular genetics: In brief, reductionists hold that Mendelian phenomena are supervenient on molecular phenomena, that the set of Mendelian properties and the set of molecular properties satisfy the relations of the sets M and N in the illustration above. Because they hold that the number of properties to which molecular biology need appeal in its descriptions is finite, there is an (extensional) equivalence of any Mendelian property and some molecular maximal property, just by virtue of the supervenience of Mendelian genetics on molecular biology. We may demonstrate this equivalence as follows: Consider any Mendelian property, such as "being the gene for red eye pigment in *Drosophila melanogaster*," symbolized as R. Any organism that has R does so by virtue of having some vast and complex disjunction of conjunctions of molecular properties and their complements. This disjunction will be a maximal property, constructible from the set of molecular properties. Call this molecular maximal property P_1. Because Mendelian properties are supervenient on molecular ones, any other organism with molecular maximal property P_1 will also have Mendelian property R though many, perhaps all, other organisms with R will not have P_1, but rather some other molecular maximal property, P_2 or P_3, etc., in virtue of which they have R. If the number of basic molecular properties is finite, then the number of molecular maximal properties is as well. Therefore having Mendelian property R is the result of having a finitely long disjunction of molecular maximal properties, and there is a general statement to the effect that anything with properties P_1, or P_2, or P_3, etc., has property R:

$$P_1 \text{ or } P_2 \text{ or } P_3 \text{ or } \ldots \text{ or } P_n \rightarrow R$$

Any R must be the result of one or another complex of molecular properties. Now suppose for any enumeration of molecular maximal properties that give rise to a Mendelian property, R, we discover an apparent counterexample: an organism, o, with R but without any of P_1 through P_n. Organism o must have some molecular maximal property by virtue of which it has R. Call this property Q. If any other object, s, has Q, then of course s will have R as well, because o has R and R supervenes on molecular properties. But this means that Q is in the set P_1 through P_n of molecular maximal properties after all. Because this conclusion contradicts our assumption that o lacks every property in the set P_1, P_2, $\ldots P_n$, the assumption must be false. In other words, for any Mendelian property R we can establish that

$$R \rightarrow P_1 \text{ or } P_2 \text{ or } P_3 \text{ or} \ldots P_n$$

But (1) and (2) jointly imply the equivalence of Mendelian properties with molecular maximal ones:

$$R \leftrightarrow P_1 \text{ or } P_2 \text{ or } P_3 \text{ or } \ldots P_n$$

Although it appears to be a bit of purely formal, symbolic reasoning with no special upshot for biology, this is an important result. For consider, all biologists agree that there is nothing more to complex systems than their components, that constitutive reductionism, as Mayr labels it, obtains. One way to express this general agreement is to say that if an organism has a given molecular microstructure, then any other organism with this microstructure must have the same complex non-molecular biological properties. Antireductionists will agree with this claim. To deny it is to deny the uniformity of nature at so fundamental a level that general scientific systematization at any level will be threatened. But having agreed to this weak claim, antireductionists go on to insist that it is compatible with the irreducibility of higher-level properties to microlevel ones. Now we can show that the apparently harmless admission of this metaphysical part-to-whole compositional determinism entails that there is a systematic reductive relationship between theories about different levels of organization.

Theories supervene, in the sense that the properties they mention do. If the number of basic properties mentioned in a more basic theory is finite, then there are equivalences between properties mentioned in the derivative theory and properties *constructible* in the basic one. These equivalences are enough to underwrite the actual reduction of the supervened theory to the supervening one, when the number of the latter's properties is small. What "small" comes to in this case will vary as our calculational and expressive powers change and as our theories become more and more complete and accurate. Indeed, we may describe the historical process of reduction in molecular biology as that of increasing the size of the set of molecular properties much more rapidly than our computational and expressive powers have increased. That is why the practical difficulties surrounding actual reductions are at present insuperable. Perhaps hereafter, as these powers increase relative to the size of the set of molecular properties, the difficulty of actually effecting the reduction will decline.

The supervenience of Mendelian on molecular genetics makes precise both the role of metaphysical assumptions in reductionists' convictions and also their epistemological consequences. These assumptions make equally plain the instrumental, heuristic sense in which reductionists will admit the lack of a reduction. The assumption of the finitude of nature is nothing more or less in this case than the belief that the number of molecular properties of organisms is finite – not a very dangerous speculation, considering the alternative. The assumption of determinism is here nothing stronger than the belief in supervenience – that no two things can be exactly alike in microproperties while differing in macroproperties. Anyone granting these two claims is committed to a precise conclusion about the reduction of Mendelian to molecular biology: There is no epistemological obstacle to it; the only obstacle to it is the relationship between the actual complexity of things (the number of properties they manifest) and our powers to express and manipulate symbols representing their properties. At present these powers are relatively weak. Accordingly, we cannot construct or manipulate the equivalences that supervenience guarantees, and therefore we cannot effect the derivations we might hope to.

Even if and when we can effect the derivations, the derived, reduced theories may continue to have lives of their own. This of course is just what has come to be expected in physics. Thus, Newtonian mechanics is viewed as a theory that is no longer fundamental and whose applicability as well as exceptions are explained through its correction and derivation from relativistic and quantum theories. Nevertheless, for most purposes in the immediate vicinity of the solar system, at speeds well below that of light and energy levels above that of the quantum, Newtonian mechanics is not just central but irreplaceable as an instrument of prediction, explanation, and pedagogy. It may turn out to be exactly the same for Mendelian genetics and for any other equally well-articulated body of generalizations in biology. In this respect, the unnamed Nobel laureate whom Mayr quotes as saying "there is one biology, and it is molecular biology" was wrong. Mayr was partly right: There are severe limitations to explanatory reduction. But Mayr's reasons are not the right ones, and the strength of his conclusion must be greatly reduced. According to Mayr, the limitations on reduction include the allegation that "the processes at the higher level are often largely independent of those at the lower level . . ." This the reductionist can deny: The problem is not one of independence, but rather of total dependence on too large a range of lower-level variables to make for reduction. Contrary to Mayr's general indictment, reductionism is far from vacuous: Whether, as he goes on to say, it is misleading and futile are separate questions.

At present, an attempt to actually derive Mendelian from molecular genetics would be futile. Should geneticists actively attempt to remove the obstacles to such a reduction? Should they attempt to construct the molecular maximal predicates needed for actual connectability? Only if doing so can be expected to improve the predictive precision and practical application of Mendelian theory, only if the course of corrections required for reduction uncovers or incorporates such practical improvements – or if evidence begins to suggest that there is more to Mendelian phenomena than just molecular interactions, that the former are not supervenient on the latter but require the operation of other forces as well. In this case, the way to identify such further forces left out of molecular genetics would be to attempt the reduction and to identify exactly where and why it goes wrong.

Short of these two eventualities, reduction should certainly not be a direct aim of any biologists. If even accomplished, it will be a by-product of attempts to solve other less formal, more material problems in genetics. Thus, reductionism as a methodological prescription is, as Mayr says, misleading: It is not an end to be sought in and for itself. But then it has rarely been urged as one. Reduction is supposed to be a reflection of the unity of science as a whole. Its attractions are those of any accomplishment that helps us understand nature. The reductionists' aim, like that of any scientists, should be to deepen the explanation and strengthen the scope and precision of their theories. Reduction is a means to this end, *sometimes,* but not always. For it is only when a given theory has reached a particular stage of its development, characterized by the attainment of some explanatory power and by the exposure of serious impediments to predictive improvements, that reductionism becomes a useful strategy for predictive improvements. Until the theory to be reduced reaches such a stage, reductionism is indeed, as Mayr has it, misleading and futile. In many areas, Mendelian genetics is already prepared for at least piecemeal reductions; in others, it remains autonomous.

4.9. Levels of Organization

The fact that Mendelian phenomena are supervenient on molecular interactions would lead us to expect the kind of disconnection between function and structure to which antireductionists point: Recall that the function of the hemoglobin molecule is not affected by limited variation in its primary sequence, except at 9 out of the 140 or so amino acids. This makes the function of the whole molecule independent of the most particular details of its structure. Now we see why, however. The molecule's function is the result of its having a finite but large number of disjunctions of different sequences, all of which have certain chemical features in common. We know that there are only 20 amino acids, and we know their relevant chemical properties (hydrophilic versus hydrophobic, charged versus uncharged, bulky versus small, acidic versus basic). Therefore, we can come close to constructing full equivalences between the property of being a hemoglobin molecule and the property of having one or another particular primary sequence. And we can therefore reduce the oxygen-CO_2-hydrogen-ion-transport function of the blood to biochemistry alone. In this case at any rate, reductionism seems vindicated in practice as well as theory.

We can generalize on this vindication to explain the autonomy of levels of organization in many areas of biology. Levels of organization strike one as obvious in biology. Organisms are composed of parts: relatively independent interconnected organs, tissues, and cells; cells are composed of parts: nuclei, mitochondria, ribosomes, etc. These organelles are composed of parts themselves, and so on down to the level of primary sequence. Some generalizations about these parts seem easy to come by; others are the result of the most painstaking research. Often generalizations about organisms, or about their parts, have no obvious explanation in terms of generalizations about their constituents. Even when we know everything we need to know about the constituents, we may be unable to construct a manageable explanation of the general behavior of the large part. One reason is of course the sheer complexity of the way the constituents are put together to make up the part in question; the other is the apparent imperviousness of the larger part to changes and disturbances among its constituents. But imperviousness and complexity are just what we should expect if the relation between these apparent levels is that of supervenience. Imperviousness is to be expected because many disturbances or changes to components may keep the whole part within the class of maximal underlying properties to which its supervening properties are equivalent. Complexity is to be expected whenever the number of properties that characterize the constituents and the part they compose is so large that the equivalences are beyond our powers of our needs to construct.

It should come as no surprise that the problems of reduction and of teleology turn out to be one and the same. Both reflect the supervenience of phenomena at one level on those of another. Thus, functional attributions in biology are supervenient on the disjunction of directively organized systems that constitute them. Therefore, they cannot be expected to be eliminated in favor of nonfunctional descriptions of the directively organized systems. Physiology can be expected to remain in places permanently autonomous from the molecular biological phenomena that it supervenes on. The Mendelian study of populations will not be *replaced* by any molecular surrogate. This will be true even though in each of these cases the first level or organization is nothing but the aggregation of the second.

There remains one serious unanswered objection to this view of the matter. Recall the independence of the function from much of the structure of the hemoglobin molecule. The antireductionist points out that the explanation of that part of the structure that is indispensable for function – the nine amino acids that are conserved across species and determine its tertiary structure – appeals to the forces of evolution, making reference to natural selection, fitness, and ultimately Darwinian theory. What molecular biology leaves unexplained is why this particular means of oxygen-CO_2-hydrogen-ion-transport and exchange originated and came to predominate. After all, there may be a large class of other molecular mechanisms that could have done the job. The explanation of this fact is Darwinian, and indeed Darwinian theory figures at every level of biological thought and description. Unless it too can be shown to supervene nonselectionist theories, the reductionist's view, even weakly interpreted, must ultimately be rejected. The objection recalls the autonomist's claim that what makes teleology both permissible and ineliminable is its covert and overt reliance on evolutionary forces. Therefore, unless an account of the theory of natural selection can be provided that unifies this theory with the rest of physical science, the autonomist will have been proved right, not just temporarily and heuristically, instrumentally and practically right, but right in principle. For without evolutionary theory there really is no biology at all. It is to this task of demonstrating the unity of evolutionary theory with the rest of science that the next three chapters are devoted.

Introduction to the Literature

The best introduction to the notion of intertheoretical reduction is to be found in Hempel, *Philosophy of Natural Science,* chap. 8. A more advanced discussion, with detailed application to antireductionist biological theses, is to be found in Nagel, *Structure of Science,* chap. 11. This chapter is the largest target of opponents of postpositivist theories of reduction as the deduction of narrower theories from broader ones. The most important of these opponents have been Kuhn, *Structure of Scientific Revolutions,* and Feyerabend, "Explanation, Reduction and Empiricism." Among the most important responses to these critics has been P. S. Kitcher, "Theories, Theorists and Theoretical Change," *Philosophical Review,* 87 (1978):389–406. Kitcher has applied this response to the discussion of reduction in genetics in an important paper noted below.

Motivations for reductionism are advanced in Smart, *Philosophy and Scientific Realism,* Crick, *Molecules and Men,* and Monod, *Chance and Necessity.* The last two urge the importance of successful reduction as a strategy for advance in biology. An excellent and readable history of the reductionistic successes of molecular biology is Judson, *Eighth Day of Creation.*

A beautifully illustrated and particularly complete account of the nature and function of the hemoglobin molecule is R. E. Dickerson and I. Geis, *Hemoglobin: Structure, Function, Evolution, and Pathology* (Menlo Park, Calif., Benjamin/ Cummings, 1983). Their earlier *The Structure and Action of Proteins* (Menlo Park, Calif., Benjamin/Cummings, 1969) is informative but somewhat dated. The most accessible and informative accounts of the cloning of nucleic acids and their sequencing are to be found in any post-1980 advanced textbook of biochemistry. There were in addition several papers on this subject in *Scientific American* during the late seventies. Many of these papers have been brought together in *Recombinant DNA,* ed. D. Freifelder (San Francisco, Freeman, 1978). In addition, *Science,* 209(1980)4463: 1317–1438, contains several important introductory papers on this subject.

K. Schaffner, "The Watson-Crick Model and Reductionism," *British Journal for the Philosophy of Science,* 20(1969):325–48, is one of the earliest unambiguous advocates of the possibility of a straightforward reduction of Mendelian to molecular genetics. Ruse, *Philosophy of Biology,* chap. 7, notes the problem of distributing the functions of the Mendelian gene to different amounts of DNA, but advocates the reduction of cistrons, mutons, and recons as unproblematical. This view is rightly challenged, for some of the reasons mentioned in this chapter, by Hull, *Philosophy of Biological Science,* chap. 2. Hull itemizes many of the complications that were known to obstruct reduction when his book appeared in the early seventies. Its litany of difficulties has been overwhelmingly increased by subsequent discoveries. The best recent discussion of the subject, which like Hull's involves an appreciation of the insights of Kuhn and Feyerabend, is P. S. Kitcher, "Genes," *British Journal for the Philosophy of Science,* 33(1982):337–59. The discussion in Section 4.6 follows Kitcher's presentation rather closely. A now somewhat dated history of the gene concept is E. A. Carlson, *The Gene: A Critical History* (Philadelphia, Saunders, 1966).

Among those urging changes and qualifications in the notion of reduction to accommodate the history of twentieth-century genetics are L. Darden and N. Maull, "Interfield Theories," *Philosophy of Science,* 44(1977):43–64, and K. Schaffner, "Approaches to Reduction," *Philosophy of Science,* 34(1967):137–47. A summary of much recent work in this field is W. Wimsatt, "Reduction and Reductionism," in *Current Issues in the Philosophy of Science,* ed. P. D. Asquith and H. Kyberg (East Lansing, Mich., Philosophy of Science Association, 1978). The notion of supervenience as adapted here is due to J. Kim, "Supervenience and Nomological Incommensurables," *American Philosophical Quarterly,* 15(1978):149–56.

CHAPTER 5

The Structure of Evolutionary Theory

A theory is a relatively small body of general laws that work together to explain a large number of empirical generalizations, often by describing an underlying mechanism common to all of them. For all its brevity, this statement provides at least an important first approximation to the definition of a scientific theory. Of course several of its terms are vague, and others need considerable further analysis of their own before the statement can be accepted as fully informative. How small does the small body of laws have to be, and how large the number of empirical regularities to be explained? How do we distinguish them, and how do the small number of theoretical laws actually explain the larger number of empirical regularities? When is an underlying mechanism required, and when not? These are all hard questions in the philosophy of science, questions to which many alternative answers have been given. The aim of the answers has been to deepen our understanding of theories in general. Because our aim is to deepen understanding of just one theory, the theory of natural selection, the first approximation stated above, free of amplifications and explications, will suffice at the outset. As we proceed, we shall have more to say about theories in general, and especially physical theories, with which the theory of evolution is to be compared. Although philosophical amplifications and explications of the notion of theory shall not be our concern hereafter, our aim will be conclusions that will be minimally compatible with alternative accounts of theory broached in the philosophy of science.

Understanding the character of the theory of natural selection, its concepts and laws and their applications in explanation and confirmation, is not just a necessary condition for understanding biology. It is a sufficient one as well. For overt and covert appeals to adaptation, to fitness, and to mechanisms that secure them figure at every level and in every compartment of biological thought. We have already seen the plausibility of its role in functional attributions, even when these are treated as local claims about particular arrangements, without regard to their evolutionary origins. Equally, we have seen that thoroughgoing reductionists have to reckon with the intrusion of evolutionary considerations at the most nonbiological, purely chemical, level of their science. These facts show that understanding evolutionary theory is necessary for understanding the remainder of biological science. But evolutionary theory not only intrudes into all domains of biology, it is also characteristic of the form and content, the prospects and limits of theory in all domains of biology.

Coming to grips with this theory should shed light on all the controversies about the nature of biology, its differences from and similarities to physical science.

5.1. Is There an Evolutionary Theory?

The first problem in coming to grips with the theory of natural selection is that, a hundred years and more since it became the common coin of biological thinking, there is still no agreement on a canonical version of the theory, or even on a small set of variants all can agree express its central ideas. What is more, many biologists hold this to be no deficiency or impediment to the advancement of their science. G. L. Stebbins has written, quite recently, that "evolutionary biology is so complex that attempts in the near future to build syntheses around the framework of rigid, all-inclusive generalizations or laws, will continue to be self-defeating and will lead to disputes and confrontations that generate more heat than light." (Stebbins, 1982:14). Stebbins echoes a long tradition of denials that evolutionary theory has or can be made to have the form and structure of theories familiar from physical science. Stebbins seems to hold out the hope that someday laws and generalizations, working together in the way that the components of a physical theory do, may be possible in evolutionary biology. Many others have held that it will be permanently impossible to do this. There is a wider gulf that separates biologists like Stebbins from these thinkers, however. For Stebbins believes that the way to proceed in evolutionary theory is to formulate generalizations, to improve them up to their limits of usefulness, and then to supplant them with others, treating none as a fixed and final expression of evolutionary laws.

On the other hand, arguments that such systematization will never be possible hinge on the denial that evolutionary theory can even be expressed in generalizations that might be improved to eventually constitute laws. Proponents of this view hold that the impossibility of laws of evolution reflects some fundamental facts about living systems or about our ability to understand their evolution. These facts are alleged to distinguish biological knowledge from what physical science can provide about objects in its domain. In other words, controversies about the prospects for systematizing evolutionary phenomena in a body of laws and generalizations are but reflections of the debate between reductionists and antireductionists, teleologists and their opponents, in yet another area of the philosophy of biology.

Evolutionary phenomena are doubtless far more complex and intricate than other natural phenomena. It is therefore no surprise that an acceptable general theory of evolution should have appeared much later than theories of equal import for other sciences, with simpler domains. It is no surprise that such a theory once expounded will probably be more controversial, require more improvements, experience greater revision, and be expressed in very different terms over the course of its early history. All these factors suggest that differences between the present status of Darwinian theory and theories of equal importance in physics and chemistry are to be expected and have no direct bearing on the ultimate fate, character, or canonical expression of the theory of natural selection. Accordingly, many claims that evolutionary theory is essentially different from, say, Newtonian mechanics, can be dismissed once we recognize that they rest on these relatively transient historical differences, or on the sheer greater complexity of biological phenomena as compared to mechanical phenomena. In and of themselves, these differences constitute either only practical impediments to a precise, agreed-upon, general theory of evolution, or premature pessimism about the limits on our ability to understand evolutionary phenomena.

In fact, what seems striking about the theory of natural selection is not the difficulty of identifying the generalizations of the theory, but their extreme obviousness once they are stated. They seem so obvious as to tempt one to say that Darwin did not need a five-year voyage on *H.M.S. Beagle* to discover them; he could have hit upon them by reflection on what was directly observable in any rural setting in Britain. The theory is often presented as follows:

1. The number of organisms of any one type can increase in geometrical proportions. But;
2. The actual number of organisms of any one type remains close to constant over long periods.
3. No two individual members of a type of organism are identical; variation is characteristic, and some of the variation is inherited.

Therefore we may infer that:

4. Because organisms can produce more offspring than their surroundings can support, there must be a struggle among organisms to survive.
5. In this struggle, the ones whose variations best adapt them to their surroundings, the fittest, survive, whereas the less fit organisms, with less well adapted variations, do not.

And:

6. Because the variations are heritable, there will be a change in the proportions of the variations from generation to generation: There will be evolution.

Observations 1, 2, and 3 are obvious in the extreme, and the conclusions, 4, 5, and 6, seem direct and plausible inferences from them.

There is of course a vast historical literature on exactly what Darwin himself took to be the crucial content of the theory of natural selection. In the recent past this literature has focused largely on the question of whether or not Darwin was committed to variations always being small, so that evolution had to be gradual, and whether the units subject to selective forces must be individual organisms or not. The informal presentation given here is neutral on these two interpretative questions. It says nothing about the magnitude of variations, and it is compatible with the attribution of heritable adaptive variations to collections of individuals, as well as to their members. Because it is neutral on these points, scholars may well reject the presentation as even a simplified version of Darwin's theory. Nevertheless, it does reflect the actual introduction to the theory credited to Darwin in contemporary textbooks. We shall leave aside hereafter the historical question of what exactly Darwin said and the even more vexed question of what he thought. We shall, however, return to the substantive questions of whether variation is small and evolution gradual or not, and what the units of selection may be, but only after a fuller appreciation of the theory has been provided.

Leaving aside the interpretative questions, and the complaints that transforming insights like statements 1 through 6 into a precise scientific theory are premature at best, we must face two other objections to the claim that Darwinian evolution is anything like a theory in physical science. The first is the assertion that evolutionary biology cannot consist of a body of interconnected generalizations and universal laws because it is in large part a narrative of events on our planet, joined together by local generalizations of a rough and nonuniversalizable sort. In this view, it is not like a physical theory, a set of claims alleged to be true everywhere and always. For it makes immediate reference to terrestrial phylogeny. The second objection is that the theory is indeed general, but that it is too general. Unlike theories in physical science, it has no content. It is but a conceptual scheme, a useful descriptive apparatus at best, and a vacuous, empirically empty metaphysical speculation at worst. These two objections trade on the correct view that a scientific theory must be general and cannot be a description of particular facts: It must be universal in its claims and cannot be if it implicitly or explicitly mentions particular objects and events, no matter how large the scale of such events. At the same time, scientific theory must have empirical content if it is to have any real explanatory power. If the theory of natural selection cannot be defended against these two traditional objections, then any prospect of showing the unity of biology with the rest of natural science entirely disappears. Biologists are not usually troubled by either of these objections, even though they would be hard pressed to refute sophisticated versions of them. But they should be, for if they are correct, they have profound consequences for biology's future.

It is indeed correct that evolutionary biologists are intensely, indeed principally, interested in tracing and explaining the course of evolution on the earth. And it is true that most evolutionary controversies surround questions of the actual sequence, causes, and rate of evolutionary events during the last 4 billion years. It may therefore be supposed that an account of this sequence of events should figure in any statement of the theory of natural selection. It is this sequence against which evolutionary claims are tested; it is this sequence they are called upon to explain. Surely therefore it is this sequence that must inform the content of the theory. One philosopher's version of this view has gained some biological acceptance. T. A. Goudge propounded it in the following terms:

Narrative explanations enter into evolutionary theory at points where singular events of major importance for the history of life are being discussed . . . Narrative explanations are constructed without mentioning any general laws . . . whenever a narrative explanation of an event in evolution is called for, the event is not an instance of a kind, but is a singular occurrence, something which has happened just once and which cannot recur . . . Historical explanations form an essential part of evolutionary theory. (Goudge, 1961:65–79)

Goudge, and other biologists and philosophers who concur with him, often cite Karl Popper's dictum that such general statements as we might extract or infer from evolutionary explanations are so trivial "that we need not mention them and rarely even notice them" (Goudge, 1961:76). It is true that statements like 1 through 6 above seem quite obvious and do not always figure explicitly in evolutionary explanations, even when they or statements like them are assumed to link evolutionary events. Having endorsed Goudge's view of the matter, Ernest Mayr goes on to say,

historical narratives have explanatory value because earlier events in a historical sequence usually make a causal contribution to later events. . . . Philosophers trained in the axioms of essentialistic logic seem to have great difficulties in understanding the peculiar nature of uniqueness and of historical sequences of events. Their attempt to deny the importance of historical narratives or to axiomatize them in terms of covering laws fails to convince. (Mayr, 1982:72)

The mistakes of such views as Mayr's rest mainly on the failure to see that causal claims, whether singular or general, reflect the operation of underlying laws of nature. The fact that these laws sometimes go unmentioned sometimes reflects their obviousness. At other times, it reflects our failure to have discovered and formulated them even though we believe they exist. When we embrace a narrative, causal explanation but no general law, it is because we have some evidence that there is an as-yet-undiscovered law or more probably an unmanageably large number of laws connecting the events of the narrative. As we shall see, the existence of a large number of such underlying laws inevitably makes the principles of connection implicit in the narrative, like "the survival of the fittest," appear to be trivial, if not vacuous. If evolutionary explanation is causal explanation, it must rest ultimately on laws. These laws, however, need not be distinctively evolutionary laws, and if so a claim like Mayr's may be right after all: Detailed evolutionary explanations of terrestrial phenomena can and must proceed in the absence of distinctively "evolutionary" laws. "Evolutionary theory" may be a fundamentally different thing from physical theories, which can be expressed in terms of a small number of contingent statements combining generality and precision.

The conventional biological and philosophical response to this view of evolutionary theory is the claim that it mistakes the course of evolution for the mechanism of evolution; that the theory of natural selection must provide only the mechanism; and that the particular course of evolution conditions the acceptability of a mechanism only in that the mechanism must be able to account for the actual sequence, given a specification of its starting point. The trouble with this reply is reflected in the quotation from Stebbins above. The theory does in fact never seem to be stated in ways entirely free from the actual course of terrestrial phylogeny. It is continually revised in the light of particular findings about the actual course of evolution. Propositions usually viewed as its consequences clearly make claims about particular species, groups of species, and their adaptive fates under changes in the terrestrial environment. No biologists have ever set themselves the task of constructing an evolutionary theory that could apply to evolution elsewhere in the galaxy, or indeed in the immediate vicinity of our solar system. No doubt some general restrictions on evolution might be extrapolated from Darwinian theory, but no one would call these extrapolations the core of that theory. Moreover, they may be prey to the other objection to any general formulation of the theory, that its principles are untestable tautologies.

Doubtless there is a distinction between the mechanism of evolution and its actual course on earth. But this is no answer to those who hold that evolutionary theory cannot be separated from the actual course of events on the planet. For it may be held that although reasoning reflected in statements 1 through 6 above is a useful way of linking the events in phylogenetic history, the actual underlying causal mechanisms linking events in evolutionary sequences are, as elsewhere in biology, extremely complex and heterogeneous; that the full specification of all such mechanisms is even more overwhelmingly disjunctive than is the specification of molecular regularities underlying Mendelian genetics; that in none of them do notions like fitness or adaptation figure; and that accordingly the only level of description at which these evolutionary notions have a role is one whose terms refer solely to events peculiar to the earth and its evolution.

If this were the case, it would not be surprising if we were unable to express a small number of clearly agreed-to laws of evolution that satisfied minimal conditions of generality and explanatory content. Evolutionary theory would under these circumstances be viewed as a convenient overlay for chronicling a sequence of events whose real causal connections involve complex mechanisms systematized in physical law, but not manageably explained by such laws because of their complexity and heterogeneity. So viewed, "evolutionary theory" would indeed be a misnomer in terms of the definition of theory given above. Rather, it would be a chronicle or history, with some rules of thumb, guides for linking events in the chronicle, that do not themselves reflect any particular causal regularities. Evolutionary theory would be more like a sort of speculative philosophy of history than a scientific theory. Many features of the actual character of evolutionary theory, and the debates surrounding it, can be plausibly explained by this view of the matter. Like Toynbee's or Spengler's or Marx's theory of history, it suffers from disputes about its expression, its application, its evidence. Both detractors and supporters of evolutionary theory have denied it much or any predictive power, even when they have credited it with explanatory content. This too is what philosophers of history expect of their "theories."

Some autonomists and many provincialists might be attracted by this view of the matter, though they would draw vastly different conclusions from it. Both would cite the anomalous and distinctive character of evolutionary theory as one of the most distinctive irresolvable differences between biological and physical science. The provincialist would go on to demand real physical theory in biology, and the autonomist would insist on its impropriety and impossibility.

Yet autonomists would be ill advised to embrace this view of evolutionary theory that leaves it on a par with "theories" like Hegelian historical idealism and dialectical materialism. Autonomists need to show that their science rests on a theory composed of scientifically respectable, empirically testable, explanatorily powerful, though irreducible laws. Otherwise the system of ideas whose autonomy they defend will fail to be a science at all in the eyes of their opponents.

Provincialists rest content with this view at their peril as well. For if they deny the possibility of a manageable scientific theory that explains the course of terrestrial evolution, they leave the lion's share of biological phenomena practicably unexplainable on any grounds — physical, chemical, or biological.

For both autonomists and provincialists, the only recourse is to show that a recognizably scientific theory can be extracted from what biologists want to say about the course of evolution: that is, a small body of general laws that work together to explain a large number of empirical regularities by providing an underlying mechanism common to them all.

5.2. The Charge of Tautology

Attempts to identify the body of laws that work together to constitute evolutionary theory have been taxed with a second difficulty. The charge is that, because it is devoid of empirical content, the theory of natural selection is no theory but rather a vacuous and tautological triviality, a disguised definition, an unfalsifiable piece of metaphysical speculation. This objection is almost as old as the theory itself. And it refuses to go away, even though generations of biologists have satisfied themselves that the charge is groundless and that they need not even respond to it. The charge of course rests on the assumption held by almost everyone that a real theory must be testable and cannot be merely a set of definitions, disguised or undisguised, without bearing on the way things actually are. Because, the argument goes, the theory of natural selection explains survival in terms of fitness as in statement 5 above, and defines fitness in terms of survival, the theory is circular, and no evidence can refute it.

Consider the following sample definitions of terms like adaptation, fitness, and natural selection:

E. O. Wilson, *Sociobiology: The New Synthesis* (glossary entries):

Adaptation in evolutionary biology, any structure, physiological process, or behavioral pattern that makes an organism more fit to survive and to reproduce in comparison with other members of the same species.

Fitness see genetic fitness

Genetic Fitness the contribution to the next generation of one genotype in a population relative to the contributions of other genotypes.

I. M. Lerner, *The Genetic Basis of Selection:* "Individuals who have the most off-spring are fitter in the Darwinian sense."

C. H. Waddington, *Towards a Theoretical Biology:* The fittest individuals are those that are "most effective in leaving gametes to the next generation."

G. G. Simpson, *The Meaning of Evolution:* "Natural selection is differential reproduction, plus the complex interplay in such reproduction of heredity, genetic variation, and all other factors that affect selection and determine its results."

Given these definitions, we logically could never find a case where decisions about differences in fitness, or adaptation, or natural selection diverged from measurements of differences in rates of reproduction. It follows, therefore, that the theory has no explanatory power. For to explain the survival of one group of organisms by appealing to their fitness is to explain survival by appeal to survival. But no purely contingent event is self-explanatory. Given one description of an event to be explained, merely providing another description of the same event in a terminological variant of the first description could not possibly explain the event. If someone undertook to explain why Darwin was still a bachelor in 1831 by pointing out that he was still an unmarried male in 1831, no one would accept this claim as explaining the particular state in which Darwin found himself. Although they involve more terminological variation than this simple case, evolutionary explanations have persistently been held by some to have just this vacuous character.

Few contemporary biologists take this attack on their theory seriously. Even when it is pointed out that the most distinguished leaders of the field define fitness or adaptation in terms of survival, most biologists tend to shrug their shoulders. They do not recognize that such a definition commits them to the vacuity of their central theoretical edifice, or they believe that such definitions are but incautious formulations, easily replaced by definitions that avoid this conclusion. As we shall see, providing such an alternative definition is no easy matter. Some biologists, however, accept the claim that fitness is to be defined in terms of survival and accept also the conclusion that the theory of natural selection is a grand tautology. This view in fact goes hand in hand with the perspective on the theory, canvassed in Section 5.1, that it is a congeries of particular statements about the course of evolution on the earth and of convenient rules of thumb for connecting the events of this terrestrial sequence. The rules of thumb are the very claims like statement 5 above, stigmatized as tautological in attacks on the theory. This attitude makes a virtue out of a vice, while strengthening the claim that evolutionary theory cannot be anything like a theory in physical science. It treats claims like statements 1 through 6 not as contingent generalizations about facts in the world, but rather as reflections of a convenient categorical system, a descriptive typology, with no empirical content, one that is convenient for describing particular empirical facts with which the biologist must deal. Evolutionary theory is thus a sort of "language" justified by its convenience, not by its content. For it has none. Like Euclidean geometry, it can be treated as a system of pure definitions, useful for systematizing empirical facts in the immediate vicinity of the earth. But also like Euclidean geometry, it is without applicability across interstellar distances and in worlds much different from our own. One biologist has put it this way:

> The "theory of evolution" does not make predictions, so far as ecology is concerned, but is instead a logical formula which can be used only to classify empiricisms [*sic*] and to show the relationships which such a classification implies . . . [Evolutionary] "theories" are actually tautologies and, as such, cannot make empirically testable predictions. They are not scientific theories at all. . . .
>
> If the generalities of natural selection and Darwinian evolution do not provide unambiguous or falsifiable predictions, they cannot be called a scientific theory. If however, they provide a pattern of logic based on a few axioms, they do form a tautology to which we may apply empirical correspondences. A tautology is useful when, as in mathematics, the logical relations are too complex to be recreated each time they are needed. . . .
>
> The pattern provided by some tautologies may lead to an ordering of the facts which a researcher might not otherwise see . . . (Peters, 1976:1, 4, 11)

This way of viewing the theory of natural selection is in the end unavailing, even when stated more coherently. For at best it simply postpones the questions it is meant to avoid or translates them into new and misleading terms. It is indeed true that we can treat a statement of the theory of natural selection as a body of implicit definitions of the terms that figure in it. Fitness is whatever satisfies statement 5; variations are whatever behave in accordance with statement 3; "organisms" are whatever can increase in geometrical proportions; etc. In this respect evolutionary theory is no different from Newtonian mechanics. We may define force as whatever varies with the product of mass and acceleration. Gravity is whatever obeys the inverse square law. But so treating a theory, viewing it as a body of definitions for the terms that figure in it, raises the question of whether the system of tautologies is applicable to the real world, and, if so, why the system of definitions is applicable. Thus, we may ask whether there is anything that satisfies the definition of mass or force or gravity, that is, whether by applying these definitions to things we can generate confirmed predictions about their future behavior. In the case of Newtonian mechanics, the answer is evidently yes. But this raises the question of why the

Newtonian system of definitions is applicable, whereas the Aristotelian, for example, is not. The only answer to this question is an empirical theory about the behavior of the entities to which we have applied the Newtonian "definitions," which explains why they behave in the way that makes these definitions useful. Of course, the simplest explanation is Newtonian theory, this time treated as a body of contingent claims about mass, force, and gravity, instead of definitions of them. Of course, there are other possible explanations, and indeed the history of physics has shown that there are better explanations for the usefulness of the Newtonian "definitional scheme": quantum mechanics and the special theory of relativity.

Return now to evolutionary theory, treated as a body of tautologies, albeit useful, convenient ones. If we so treat the theory, we need to explain why it is more convenient than competitors, like Lamarckian theory or nonevolutionary theories of animal and plant diversity. The explanation will have to be a body of contingent, factual regularities, either the theory of evolution itself, now treated as such a factual theory, or some other account of the matter. The first alternative is ruled out for those who claim that Darwinian theory can be true only if true by definition. Accordingly, another theory to explain evolution will have to be provided. Otherwise, it must be held that evolution is just not susceptible of theoretical explanation, of being systematized by a small set of general laws that work together to explain it by providing its underlying mechanism.

In the end, the distinction between questions of whether a body of definitions is applicable to phenomena, and whether a set of contingent claims is true of them, is an artificial one. For adopting the first attitude merely raises the question of what contingent facts make the definitions applicable and so raises the issue of the truth of contingent claims immediately. Accepting the charge that evolutionary theory is circular or tautologous but appealing to its conceptual convenience for organizing phenomena is unavailing. Biologists must face the problem of providing a noncircular account of the meaning of the key terms of evolutionary theory or forswear its explanatory power altogether.

However, any adequate account of the meaning of the key terms of the theory must also do something else at the same time. It must also explain why the theory has so persistently been misunderstood, both by biologists and nonbiologists, as a vacuous and empty triviality. For if this fact is not simultaneously explained, something essential to the meaning of these crucial terms has been left out, and the account of their meaning will be inadequate. After all, definitions of terms like fitness and adaptation, even by evolutionary theorists, engender this theory-vitiating circularity. The ease and naturalness with which biologists who should know better fall into this trap, and have done so since the theory's outset, must reflect something central to these notions. If we do not explain the natural tendency to make this mistake, we shall not have given an adequate account of their meaning and theoretical role. What is more, by failing to clearly diagnose features of the theory and the employment of these terms that make it tempting to define fitness or adaptation in terms of survival, biologists are likely to repeat the same mistakes, generating the same pointless arguments.

If we cannot provide respectable accounts of the meaning of key terms of the theory that are independent of it, then we will have to surrender all hope of showing that the Darwinian theory is in fact a scientific theory within the meaning of that term with which we began.

5.3. Population Genetics and Evolution

Many biologists and philosophers believe that identifying the small body of general laws that constitutes the theory of natural selection, or at any rate its contemporary version, is rather easy and noncontroversial. These biologists and philosophers direct our attention to the "synthetic theory of evolution." They admit that problems like those canvassed above daunt Darwin's original version of the theory. But they reckon that its connection to and interpretation by means of modern population genetics solves all the traditional questions surrounding the structure, content, and scientific status of the theory of natural selection. Their view has the additional convenience that it enables us to waive questions about what Darwin actually held and to forgo any attempt to express the ideas of statements 1 through 6 with any greater generality and precision – or indeed to defend the claim that they or statements like them capture the core of the theory of natural selection.

The "synthetic theory of evolution" is indeed a synthesis. It is constituted by the interanimation of Mendelian genetics and Darwinian-like principles. In the words of Julian Huxley, one of the founders of "the new synthesis," "Mendelism is now seen as an essential part of the theory of evolution. Mendelism does not merely explain the distributive hereditary mechanism; it also, together with selection, explains the progressive mechanism of evolution" (Huxley, 1942:26). But this is just what we want in our search for an expression of natural selection that answers to the definition of scientific theory with which we began. The theoretical laws will turn out to be Mendelian principles of segregation and assortment. They explain variation and, together with selection on the variation, generate evolution, thereby systematizing a large number of empirical regularities, and for that matter particular sequences, by providing their underlying genetic mechanism. Evolution turns out to be nothing but changes in gene frequencies. This view of evolutionary theory, which purports to solve all the problems surrounding its status as a scientific theory on a par with, say, Newtonian mechanics, has been most fully articulated by Michael Ruse. He believes it does justice to the complexity of evolutionary biology, which discourages biologists like Stebbins, while nevertheless refuting those who hold that "evolutionary theory" is just a different undertaking from physical theory altogether. Indeed, according to Ruse it can be shown to satisfy strictures on scientific theories far more rigorous than the characterization with which we have been working.

As Ruse notes,

a great many different areas fall under what are loosely called 'evolutionary studies.' There is *systematics*, the study of the distribution of organisms and the reasons why they fall into the particular kinds of groups they do. There is *morphology*, the study of the different kinds of characteristics that organisms have, together with theorizing about the reasons for such characteristics. There is *embryology*, the study of the development of organisms. There is of course *paleontology*, the study of organisms long dead and fossilized. And there are many other areas. All of these disciplines have factual claims peculiar to themselves. . . . However, what I suggest is that all the different disciplines are unified in that they presuppose a background knowledge of genetics, particularly population genetics. The knowledge that the population geneticist supplies about the way heritable variations are transmitted from one generation of a population to the next is presupposed and drawn upon by every kind of evolutionist, even (nay, particularly) those like the paleontologist who study the largest of heritable changes, from fish to reptiles, and from reptiles to mammals and birds.

. . . [It] is . . . by virtue of the fact that the theory of evolution has population genetics at its core, [that] it shares many of the features of [theories] of the physical sciences. The most vital part of the theory is axiomatized; through this part (if through nothing else) the theory contains reference to theoretical (non-observable, etc.) entities as well as to non-theoretical (observable, etc.) entities; there are bridge-principles; and so on. (Ruse, 1973:148–9)

In other words, population genetics provides what philosophers of physical science have called a hypothetico-deductive core for evolutionary theory. According to hypo-thetico-deductivism, the small number of theoretical laws that work together in a theory do so because they can in principle serve as axioms in a deductive system that logically implies the empirical regularities they explain as theorems. Indeed, the-oretical explanation on this account is deduction from laws. In this view of theories, the axiomatic system provides an underlying mechanism for the empirical regularities through its appeal to theoretical entities whose causal interactions eventuate in the observable regularities reported in the empirical generalizations. The terms in which these observationally confirmable regularities are couched provide an interpretation for the theoretical terms. This underwrites their legitimacy by contrast with the "theoretical" vocabulary of pseudoscientific theories, like astrology or alchemy, that allegedly cannot be linked to the terms of confirmed observational regularities. There is much to be said for this rational reconstruction of the nature of scientific theories. And Ruse is right to find that the laws of Mendelian genetics, its theoretical terms like gene and phenotype, and the empirical regularities about breeding experiments that it systematizes, satisfy much of this picture of theories as they figure in physical science. Ruse does go on to admit, much as Stebbins does, that

it cannot be denied that the *whole* theory does not possess the deductive completeness pos-sessed say, by Newtonian mechanics. Because many factors—the newness of the theory, the fact that many pertinent pieces of information are irretrievably lost, the incredible magnitude and complexity of the problems, and so on — many of the parts of evolutionary theory are just 'sketched in.' . . . Hence at best one can say that evolutionists have the hypothetico-deduc-tive model as an ideal in some sense — they are far from having it as a realized actuality.(1973:49)

In this respect, evolutionists are no worse off than many physical scientists. For the hypothetico-deductive model was never presented as more than an ideal explication of what is in principle possible by way of the most rigorous presentation of a scientific theory.

For all its convenience in reassuring us that the theory of natural selection is a respectable scientific one, and despite its apparent reflection of the intimate connec-tion between evolution and genetics that is enshrined in the rubric "the synthetic theory of evolution," this view of the matter misrepresents the structure of evolu-tionary theory. It blinds us to its actual relations with genetics, and either ignores or fails to solve the serious problems canvassed above that bedevil the theory of natural selection. The fact is, evolutionary theory does not presuppose population genetics. The truth of evolutionary theory does not entail the truth of Mendelian genetics. Nor is Mendelian genetics logically incompatible with Darwinian theory's explana-tory competitors. Indeed, if anything the relation is reversed: The laws of Mendelian genetics, to the extent they obtain, are themselves in need of explanation in accor-dance with the theory of evolution. Of course the two theories are intimately connected, though not as part and whole; rather, they are connected as a set of empirical generalizations about terrestrial biological phenomena explained by the operation of a general theory on initial conditions.

That the theory of natural selection requires the existence of hereditary mechanisms is obvious: Variations must be heritable if they are to result in evolution over time. This fact is apparent from the history of the subject. For Darwin himself felt impelled to provide a theory of inheritance to underlie his theory of natural selection. This was the so-called theory of pangenesis, or blending inheritance, which involved the existence of "gemmules" as the units of hereditary function. Equally, the logical independence of this theory of heredity from the theory of evolution is evident from the fact that its obviously unsuccessful character does not shake anyone's confidence in the theory of natural selection. The real independence of evolutionary theory from any particular genetic theory is even more obvious in the fact that the earliest exponents of Mendelian genetic theory took its results to be incompatible with the theory of natural selection itself. Indeed, in his exposition of "the new synthesis" Huxley begins by showing that the *incompatibility* of Mendelism and Darwinism is only apparent. We shall see why Mendelian genetics may appear to be incompatible with evolution in this section.

The principles of Mendelian or population genetics that Ruse identifies as the theoretical core of evolutionary theory are the law of segregation and the principle of independent assortment:

Segregation: For each sexual individual, each parent contributes one and only one of the genes at every locus. These genes come from the corresponding loci in the parents, and the chance of any parental gene being transmitted is the same as the chance of the other gene at the same parental locus.

Independent Assortment: The chances of an offspring receiving a particular gene from a particular parent are independent of the offspring's chances of receiving any other gene (at a different locus) from that parent. (Ruse, 1973:13, 14)

In and of themselves, these two principles can hardly represent the core of the theory of evolution. For (as Ruse in fact reports) these laws entail that within an interbreeding population gene frequencies *remain the same* (ceteris paribus). That is, they jointly entail the Hardy-Weinberg law of genetic equilibrium:

Given a large, panmictic population in which there is no net emigration or immigration and which is in mutational equilibrium, gene ratios will remain constant and after the first generation, genotype frequencies will remain constant as well.

Among philosophers, Ruse takes the most pains to demonstrate that the Hardy-Weinberg law follows from the principles of segregation and assortment alone. For to be a hypothetico-deductive system, that is, a scientific theory, Mendelian genetics must imply empirical regularities as theorems of its assumptions. The derivation is Ruse's way of establishing that Mendelian genetics satisfies this condition. The demonstration takes on increased importance when it comes to showing that evolutionary theory is a hypothetico-deductive system as well. For the way this is done is simply to announce that the core of the theory is Mendelian genetics, which has already been shown to be such a system.

But the conditions under which the Hardy-Weinberg genetic equilibrium obtains are circumstances in which there is no evolution. If gene and genotypic frequencies remain the same over generations, then the phenotypes that they control remain the same as well, over generations, and no evolution results. It is only by introducing considerations independent of the mechanism of heredity, considerations from evolutionary theory, like fitness and the causes of variation, that the Hardy-Weinberg law and the Mendelian laws from which it stems can have any bearing on evolution. Biologists have long recognized this fact:

The Hardy-Weinberg law is entirely theoretical. A set of underlying assumptions are made that can scarcely be fulfilled in any natural population. We implicitly assume the absence of recurring mutations, the absence of any degree of preferential matings, the absence of differential mortality or fertility, the absence of immigration or emigration of individuals, and the absence of fluctuations of gene frequencies due to sheer chance. But therein lies the significance of the Hardy-Weinberg law. *In revealing the conditions under which evolutionary change cannot occur, it brings to light the forces that operate to cause a change in the genetic composition of a population. The Hardy-Weinberg law thus depicts a static situation.* (Volpe, 1967:38; emphasis added)

Darwinian theory requires some law of heredity or other, and Mendel's laws (suitably revised in the light of phenomena like linkage and crossover) provide laws of the sort required. But they do not make for evolution itself. Indeed, they cannot, if they are to provide the sorts of laws that natural-selection theory requires. What it requires is an account of how traits are passed on, *unchanged,* so that they may continue to exist in a population. It is variation and selection that account for the spread or disappearance of traits throughout the population. Other hereditary laws could equally well provide evolutionary theory with its hereditary requirements. Our choice of Mendel's principles is not based on the fact that they provide the assumptions of Darwinian theory, but on the fact that they are the most well-confirmed candidates among the alternative statements that could meet the evolutionary theory's requirements. This should be clear both from the fact that the laws of independent assortment and segregation have required significant modification as a result of findings in molecular genetics, without thereby producing a change in evolutionary theory; and from the fact that we may construct new hereditary laws incompatible with Mendel's (and doubtless false ones) that logically could serve Darwinian needs just as well.

Even more strikingly, Mendel's laws can easily be shown to be compatible with the Darwinian theory's principal competitors as alternative accounts of evolution. Indeed, the laws of heredity that population genetics provides are as much required by this competitor's appeal to heredity as they are required by Darwin's theory. For example, consider the following description of a Lamarckian theory:

According to it, the course of evolution is due to the following factors: (i) a changing environment which acts on individual organisms; (ii) the consequent production of new needs in those organisms, needs which must be met if the organism is to survive; (iii) the active response of some organisms to meet these needs, i.e., the establishment of new habits (or the cessation of old habits) or use of various bodily parts; (iv) a resulting change in those parts and hence in the somatic structure of the individual organisms which possess them; and (v) the transmission of the structural changes and habits from one generation to the next, so that the organisms concerned are gradually transformed into new species . . . Thus the theory involves the assumption that the effects of use and disuse, or environmentally induced effects on the individual, are inherited in kind and become germinally fixed. This assumption has come to be known as 'the inheritance of acquired characteristics' . . . (Goudge, 1961:84–5)

It is clear that Lamarckianism is in as much need of laws to explain the nature of the hereditary phenomena it postulates as is Darwin's theory. In particular, factor (v) above claims that traits are hereditarily transmitted. Why should they not be transmitted by genes in accordance with Mendel's laws? All we need add to enable Mendel's laws to underlie the course of Lamarckian evolution is the stipulation that the changes in somatic structure described in (v) above make for changes in the germ cells and consequently are inherited in accordance with (v) above and, in particular, produce adaptive mutations in the genes of the animals that make active responses to their environmental needs. It is worth noting that such a mechanism has been mooted in recent immunology. A hypothesis much like Lamarck's has been cited to explain the hereditary transmission of apparently acquired immunological characteristics (Steele, 1979). The coherence if not the acceptability of this theory shows the consistency of neo-Lamarckianism and Mendel's laws of population genetics.

Perhaps the clearest way of marking the fact that Mendel's laws are not the foundations of evolutionary theory is to show that they are, if anything, its consequences. (Parts of the argument to follow are due to J. Beatty, 1982.) The laws of segregation and of independent assortment in fact will have to follow as empirical generalizations or particular facts about the character of terrestrial evolution, to be explained in and not assumed by evolutionary theory. To see this, consider the explanation of why Mendel's laws obtain. This explanation typically proceeds by appeal to facts of cell physiology: During meiosis, chromosomes are divided in a way that distributes complementary halves to the gametes, eggs, and sperm. The doubling and twofold division of the homologous chromosome pairs results in each gamete's receiving one chromosome from each pair and thus one gene at each locus. Because the probability of the gamete's receiving any one gene is equal to that of receiving any other gene, the Mendelian law of segregation follows a consequence of the process of meiosis. But the process of meiosis is a phenotype controlled by the genome and is subject to heritable mutations. These mutations were discovered in early breeding experiments on *Drosophila*, which showed that in certain strains genes did not segregate in accordance with Mendelian ratios. This inherited property was eventually localized to a stage in meiosis at which the chromosomes do not separate or "disjoin." This nondisjunction of chromosomes during meiosis was eventually traced to a single gene locus in the *Drosophila* chromosome.

Disjunction constitutes another exception to the operation of Mendel's laws. Together with crossing over and linkage, it constitutes a qualification on their universality. Nevertheless, Mendelian segregation is the rule and not the exception. If we now ask why, there seem to be two possible avenues of explanation. One is ultimately biochemical. It traces meiosis down through biosynthetic pathways to the structural and regulatory molecular genes that code for it. It then shows how differences in the primary sequences of these genes may result in normal meiosis or nondisjunctive meiosis. However, this explanation leaves unanswered the question of why normal meiosis predominates and nondisjunction constitutes a rare exception. The answer to this question must be given in terms of evolutionary fitness: The genes that result in normal meiosis provide a more adaptive strategy for their own transmission from generation to generation than those that produce nondisjunction. Of course, at present our knowledge of molecular genetics is insufficient to provide even the first part of this explanation, the tracing out of the biosynthetic pathways to genes with different primary sequences. The second avenue of explanation for the predominance of normal meiosis reflects this fact. It proceeds directly to the evolutionary claim that normal meiosis is more adaptive and therefore long ago came to predominate as a strategy for sexual transmission. Naturally such an explanation must be sustained by a detailing of why normal meiosis is more adaptive, or what deleterious consequences for survival follow from nondisjunction. But the crucial point is that no matter which way we proceed in attempting to explain the prevalence of normal meiosis, we come up against evolutionary considerations.

But Mendel's laws obtain because of the prevalence of normal meiosis. Any explanation of its prevalence is therefore an important part of the full explanation of the laws of segregation and independent assortment. If evolutionary theory is required to explain normal meiosis, it is also called for in an explanation of Mendel's laws. This is really no surprise. After all, the organization of the eucaryotic genome and the appearance of sexual reproduction, both of which eventuate in Mendelian hereditary regularities, are themselves profound problems of evolutionary biology. They are presumed to have arisen in the course of evolution in accordance with evolutionary regularities. The problem is to find those features of sexual reproduction and eucaryotic genomic organization considered strong enough adaptive advantages that they were able to displace alternative means of hereditary transmission − non-Mendelian means − among the complex organisms in which they appear.

The upshot is that, far from being the core of the theory of natural selection, Mendelian genetics reflects regularities that must themselves be consequences of that theory. In fact, far from being axioms of any hypothetico-deductive account of evolutionary theory, Mendelian laws cannot even be expected to figure among the theorems of such an account. For they are the result of the operation of evolutionary forces *on boundary conditions,* on the local circumstances that obtained on the earth at the time normal meiosis first appeared and came to be fixed as the predominate means of hereditary transmission among sexually reproducing organisms. From a description of these conditions, together with appropriate evolutionary laws, we should in principle be able to explain Mendel's laws. They are rough empirical regularities true for a restricted period of organic evolution on the earth. Considered as a brace of such local regularities, the exceptions, qualifications, and restrictions on Mendelian theory are just what we would expect. Indeed, we can now see why its regularities are not reducible by deduction to those of molecular genetics. For if the latter really are true, exceptionless, general laws, they could not by themselves validly imply false, exception-ridden, restricted statements about particular events on the earth during a certain geological epoch. Molecular genetics could no more by itself explain why Mendelian phenomena obtain than physical theory can by itself explain the contemporary preponderance of gasoline engines as opposed to diesel ones.

Mendelian genetics does not constitute the core of the theory of evolution in any theoretically significant sense. Its principles do not suffice to explain evolution; indeed, by themselves they imply the absence of evolution. These principles are logically compatible with theories that directly deny the distinctive claims of the theory of natural selection. If anything, they require evolutionary explanations. And such explanations tend to show that Mendel's laws are not laws at all, let alone underived axioms of a scientific theory. If we hope either to understand evolutionary theory or to show that it does meet physical sciences' conditions for being a scientific theory, we shall have to approach the theory from another direction altogether than its need for a hereditary mechanism. Its undoubted need for such a mechanism is only one aspect of the theory. Because it is a need that can be filled by any one of a number of hereditary theories, without any ramifications for the rest of the theory, the particular character of any hereditary mechanism is just not very central to the theory of natural selection.

5.4. Williams's Axiomatization of Evolutionary Theory

There is in the biological literature a formal presentation of the theory of natural selection that captures all the leading ideas of the theory and is compatible with the many disagreements about the rate and underlying mechanism of terrestrial evolution. Moreover, this formal treatment of the theory of natural selection reflects the structure, generality, and empirical content usually taken as the hallmark of a powerful scientific theory. This account has been philosophically controversial but many of the objections to it are based on misunderstandings of its claims and of the constraints on any scientific theory. In what follows I shall first set out this version of the theory in detail. Then I shall consider its ramifications and its adequacy as an expression of the theoretical foundation of evolutionary biology. It will be seen that adopting this account of the theory bids fair to solve all the problems with which natural selection is taxed.

The formalization to be presented was first advanced by Mary B. Williams. Her original paper (Williams, 1970) repays even more careful attention than I can give it here. Like many formalizations, her account is shaped to respond to considerations of a highly abstract nature. These considerations have their ultimate justification in the logical and mathematical strictures developed through the long and successful attempt to bring to light the foundations of physical theory. Although I will not trace out all the formal steps, readers will have to bear with a certain amount of apparently arbitrary circumlocution. I shall, however, attempt to motivate each step in the construction.

What is required is an axiomatic account of the ideas underlying informal accounts of evolution such as those given in statements 1 through 6 above. An axiomatic account is desired for two reasons: First, it will identify the propositions that the theory takes as reflecting the fundamental evolutionary forces and distinguish them from what it treats as the derivative phenomena explained by the operation of these forces. Second, an axiomatic presentation enables us to decide in a purely objective, ultimately mechanical manner whether evolutionary regularities describing phenomena held to be explained by the theory really are adequately accounted for by it or whether they constitute unexplained or exceptional findings that may undermine our confidence in the theory or in its completeness as an account of evolution.

It is crucial, however, to understand what is meant by an axiom in this context. The expression 'axiom' carries the suggestion of self-evidence or of a priori certainty.

This is no doubt a consequence of the fact that the most commonly familiar axiom system is provided by Euclidean geometry, and its axioms are widely held to be necessary truths. But in mathematics, and in the formalization of scientific theories, the term 'axiom' has no connotation of certainty—or even truth for that matter. To call a proposition an axiom is to make a relative claim: It is to say that, with respect to a body of statements in which it figures, the proposition is not derived from other propositions. Rather, it is used as a basis for the derivation of other propositions called 'theorems.' For any given set of sentences that bear logical relationships to one another, there is almost always more than one possible axiomatization. Thus, a proposition that stands as an axiom in one axiomatization of a given set of statements may appear as a theorem derived by logical deduction within another axiomatization from different axioms. So to identify a proposition as an axiom is merely to specify its logical role with respect to a deductive presentation of a body of statements. Consequently, no axiomatization for a given scientific theory can be logically privileged as the correct, or even the best, axiomatic account of it on formal considerations alone. Nevertheless, some axiomatizations may be shown to be preferable on nonlogical grounds. But, in any case, to offer an axiomatization must not be mistaken for an assertion that the axioms − or the theory that they systematize − are necessarily true, or a body of disguised or undisguised definitions, or insulated from modification.

But the first step in providing an axiomatization is not the enumeration of the axioms but the introduction of the subject matter of the theory. This is done by specifying the vocabulary of the theory, distinguishing the fundamental entities and processes it systematizes. Even this first step is likely to be controversial, because biologists are in sharp disagreement about the meanings of the key terms in the theory. Thus, evolutionary theory is universally agreed to be about 'species' and their evolution. But because there is no agreement surrounding the meaning of this term, any axiomatization that begins with a definition of it is likely to immediately alienate one or more of the parties to the dispute about the nature of species. What we want is a term that will be neutral as between alternative interpretations without being vacuous or, at any rate, a term defined specifically enough so parties that disagree about empirical matters can express their disagreement in a common terminology.

To distinguish the fundamental entities, and processes, the terminology must begin with terms taken as 'primitives' within the axiomatization, that is, as undefined terms from which other terms may be explicitly defined. Again, it is crucial to recognize that in this context 'primitive,' like 'axiom,' is a relative notion. A term is a primitive, with respect to the vocabulary of a theory just in case it is not defined by other terms, *in the axiomatization*. A primitive is a term with a meaning, it is just that the meaning is not given in the theory; thus 'straight line' may be a primitive with respect to certain axiomatizations of geometry, whereas 'triangle' is a defined term. This does not mean that 'straight line' is a term without meaning. The notion of primitives versus defined terms will be crucial to our account of evolutionary fitness.

It is hard to think of a less controversial term to serve as a primitive for the axiomatization of a biological theory than 'biological entity.' This is a term the theory of natural selection will share with incompatible evolutionary theories and, for that matter, with nonevolutionary biological theories. What is a biological entity? The term may be employed to refer to organisms, the traditional subjects of the theory of natural selection; or it may refer to particular genes – polynucleotide strings that control the development of organisms that contain them; or we may treat 'biological entities' as populations of organisms. If any of these interpretations is permitted, then the theory of natural selection will have to be true when the primitive term 'biological entity' is interpreted in each of these three ways. This potential for multiple interpretation is characteristic of scientific theories. Newtonian mechanics, for example, can be treated as a theory about the interactions of stars, planets, billiard balls, or molecules. It is particularly crucial that evolutionary theory be subject to multiple interpretations, under each of which it can be assessed. In this way, it can accommodate the disagreements surrounding its true subject matter and the heterogeneous uses to which it is put even by those who do not identify a unique subject matter for the theory.

The second primitive to be introduced is the relation 'is a parent of.' This relational term too figures in many alternative evolutionary theories and in non-evolutionary parts of biology. Again, it too may be interpreted in terms of the usual notion of biological reproduction, or in terms of gene replication, or for that matter in terms of colonies cut off from their original populations, in the case where whole populations are treated as biological entities.

Given these two primitive terms, we may introduce two general propositions assumed by all evolutionary theories, Darwinian and non-Darwinian. They in effect are axioms of all evolutionary theories. Therefore it would be misleading to identify them as axioms of the theory of natural selection:

No biological entity is a parent of itself.
If a is an ancestor of b, then b is not an ancestor of a.

'Is an ancestor of' is the first defined term in our account. It is defined out of the 'parent-of' relation: Organism a is an ancestor of b if a is the parent of b, or the parent of the parent of b, or the parent of the parent of the parent of b, etc. (The 'etc.' can be removed by a formal device from mathematics; it need not detain us.)

These apparently trivial statements in effect specify the domain of any evolutionary theory. They are not necessary truths, and they do set important limits on what counts as a test of any theory of natural selection. To see this, consider the following farfetched story. Suppose we discover on a planet in the vicinity of another star what appear to be biological entities, rather like organisms on the earth. They are complex, goal-directed systems with internal components organized to ensure the entities' survival in the face of environmental vicissitudes, predators, disease, etc. Suppose further that each of these "organisms" reproduces in the following way: It collects components lying around on the ground and puts them together in roughly the same arrangement as its own components. Occasionally, this reproduction results in slight improvements; when it has finished, the resulting "organism" immediately takes its "parent" to pieces, leaving the parts on the ground. If an "organism" frequently puts parts together that had in the past gone together to make up its own "parent" or some more distant ancestor, then any cumulative improvement, any evolution, will be set back. An attempt to test a theory of evolution by natural selection on this planet might well result in an apparent disconfirmation: Despite inheritance, variation, and selection, there might be no evolution. But in such a case the conclusion that the theory is disconfirmed would be unwarranted, if it were held that the frequent recombination of the same parts in the same arrangement constituted the reappearance of the same organism. For in this case the axiom – that no biological entity can be its own ancestor – would be violated.

Like any theory, the theory of natural selection must be put to the test. But the test must be a fair one. Our initial axioms, for all their apparent innocuousness, set out minimal ground rules for any such test: The entities whose evolution provides relevant tests for the theory must satisfy these two specifications.

Obviously, the theory of natural selection will treat 'biological entities' as subject to selective forces (hereafter, I will call them 'organisms' – without prejudice, however, to alternative interpretations). But organisms are not the units of evolution. So we shall have to construct these units out of 'organisms' and the 'parent-of' relation. We should not automatically expect the terms we construct to describe the units of evolution to be familiar ones, like 'species,' for example. 'Species' is a term whose meaning is not only controversial, but contains elements that do not figure in any evolutionary theory (as any creationist will attest). Like other ordinary terms, say 'air' or 'clock' or 'fish,' it may not figure in any scientific theory, unless a new meaning is legislated for it. On the other hand, any axiomatization of the theory of natural selection must enable us to express and assess claims made about evolution that employ the term 'species.' Accordingly, although we should not expect the elaboration of the axioms to follow familiar biological and nonbiological terminology, this familiar terminology may well guide us. For the result must ultimately be adequate to perform the functions of that terminology. If in the end there are straightforward translations between the terms of the axiomatization and words like 'species,' this will be all to the good.

Thus, to describe the units of evolution, we coin the biological neologism 'clan': It is clans that we shall hold to evolve through differential reproduction, and of course 'clan' must be so defined as to allow talk of clans of organisms, genes, or whole populations. The notion can be explicitly defined in terms of our primitives and certain logical as well as set-theoretical devices. For our purposes, it suffices to say that

A *clan* is a set of biological entities and all the descendants of the members of the set.

Thus, a clan has a genealogical structure describable by identifying biological entities and specifying the 'parent-of' relations among these entities. If there is only one or a small number of initial biological entities, and a progressively larger number of descendants, the genealogical structure may take on the appearance of a tree. Any branch of this tree will be a subset of members of the clan all related by the parent-of relation. Because our theory will hold that it is different branches that have different evolutionary fates, it will be convenient to have another defined term to refer to such subsets of a clan:

A *subclan* is one or more branches of a clan.

It will be a degenerate consequence of this definition that every clan is one of its subclans, but this will do no harm.

Although some further terminological work will have to be done later, we may at last turn to a statement of the axioms of the theory of natural selection proper. The first axiom states that the units of selection, whatever they may be, are composed out of or are identical to biological entities related by the parent-of relation. This axiom is so obvious that it hardly seems to bear stating. Once stated, however, complications immediately ensue:

Axiom 1. *Every Darwinian subclan is a subclan.*

'Darwinian subclan' is a primitive term of the axiomatization. The term's definition cannot be given in the theory, although the theory will tell us enough about the role of a 'Darwinian subclan' in evolution to enable us to identify one. In particular, in the theory 'Darwinian subclan' will refer to subclans whose members are all similarly affected by selective forces, evidently because they have some physiological, anatomical, or behavioral property in common and are all exposed to similar environmental contingencies, resources, predators, competitors, etc. What has just been said is *not* a definition of Darwinian subclans; it is a contingent claim about them that is ultimately to be justified by the confirmation of the theory of natural selection.

How do we know there are any Darwinian subclans? The axiom does not assure us of their existence. The only assurance of their existence will be the confirmation of the theory of natural selection itself. In this respect, the notion of a 'Darwinian subclan' is like the notion of 'positive charge' in electromagnetic theory. This theory can tell us nothing about positive charge itself, but only about the consequences of positive charge for electrical and magnetic phenomena. If one were to demand that the theory explain what positive charge itself is, or what positive charge has that negative charge lacks, one would be making a fundamental mistake about the meaning and function of these notions in electromagnetic theory. It is not that one sort of charge has something that the other lacks; rather, the terms 'positive' and 'negative' are employed simply to refer to differences that are not themselves explained or defined in the theory, but are no less real for all that. How do we know they are real in the absence of a more fundamental theory of the nature of electromagnetic charge? Only because the theory in which the notions of positive and negative charge figure is well confirmed. Because it is well confirmed, we have reason to suppose that the entities and properties it describes exist. The same must be said for the primitive terms of the theory of natural selection. It will be desirable to eventually explicate its primitive terms by defining them in another, more fundamental theory — if possible, in the way that positive charge is explicated and grounded in quantum mechanics. But we cannot await such later theories to legitimate the employment of the primitive terms of the earlier ones. The reason is that we cannot discover the later, deeper ones except by first uncovering the earlier, more accessible theories.

The second axiom of the theory is the Malthusian limitation on population reflected in the informal account of the theory given in Section 5.2.

Axiom 2. *There is an upper limit to the number of organisms in any generation of a Darwinian clan.*

Notice that the axiom asserts only the existence of a limit; it does not specify what that limit is or what in general the causes of such a limit are. The actual level of upper limit to population may be well below the thermodynamical limits specified by the finite size of available energy resources for any clan. But we know on the basis of such physical limitations that the axiom is true. Thus, the axiom of evolutionary theory may be justified independently of the theory by its derivation as a theorem of a physical theory. (It may be noted in passing that 'a generation of a subclan' is a defined term, one that can be constructed from the primitive terms by counting ancestors back to a common one. If the number of such ancestors of two organisms is equal, they are members of the same generation.)

Now we may introduce the most characteristic and most controversy-producing axiom of the theory: the axiom that attributes different levels of fitness to biological entities.

Axiom 3. *For each organism, there is a positive real number that describes its fitness in a particular environment.*

The most crucial aspect of this axiom is the fact that the notion of fitness is here introduced as a primitive term, not defined within the theory. The theory has much to tell us about the causal consequences of differences in fitness; it is after all fitness that determines the direction in which selection drives evolution. But this is a factual claim, a consequence of the theory, not a definition of fitness. Chapter 6 shall have a great deal to say about fitness, but a brief discussion of the term here will avoid needless misunderstanding at the cost of only slight duplication.

To say that fitness is a primitive term of the theory is not to deny that it can be defined, only to assert that any definition of it is not a part of the theory. Most biologists of course suppose, for reasons that will be addressed in Chapter 6, that a proponent of the theory of natural selection owes us a definition of fitness. Therefore they provide characterizations like those cited in Section 5.2. These definitions redound to the discredit of the theory because they give it a tautological, unfalsifiable character.

'Fitness' is a theoretical term, like many other terms in science: 'force,' 'magnetic,' 'charge.' Like such theoretical terms, 'fitness' is a quantitative predicate whose numerical value cannot be determined independently of the theory in which it figures. And thus, like 'force' or 'charge,' it is a term that can be characterized only through its causal role. We may offer an operational characterization of the term, but it will of course appeal to the phenomena that it is employed to account for; and such a characterization must not be confused with a definition, operational or otherwise. Thus, consider the following characterization offered to go along with Axiom 3: Amounts of fitness are measured in positive real numbers; given an organism, b, with k ancestors, Williams writes:

As a possible operational definition of [the fitness of b], $\emptyset(b)$, I might suggest the following. Let $v_1(b,k)$ be the sum over all the k-ancestors of b of the number of reproducing off-spring of each. Then let $v_2(b,k)$ be the number of k-ancestors of b. Then $v_3(b,k) = v_1(b,k)/v_2(b,k)$ is an estimate of the average fitness of the k-ancestors of b. Now let the operational definition of $\emptyset_o(b)$ be:

$$\emptyset_o(b) = \sum_{k=1} (0.5)^k v_3(b,k),$$

where k is the number of generations for which data are available. Then about 0.5 of the fitness of b is estimated by the "average fitness" of its parents; about 0.25 of the fitness of b is estimated by the "average fitness" of its grandparents; etc. . . . $(0.5)^k$ is a factor which adjusts the importance to be attached to more remote generations . . . [it] is most appropriate for sexual populations in slowly changing environments; for other situations the experimenter may wish to substitute a more appropriate factor for $(0.5)^k$. (Williams, 1970:359)

Fitness is characterized by this "operational definition" in terms of the reproductive rates of the ancestors of an organism. The reproductive rates of ancestors in each preceding generation are suitably weighted to reflect a diminishing influence on the numerical value that the characterization is employed to calculate.

If we now treat this calculating device as a definition of fitness, then explaining future differences in reproduction by appeals to differences in fitness are but covert explanations of future reproductive differences in terms of an appeal to past reproductive-rate differences. But clearly, the theory of natural selection is not committed to the conclusion that future differences in reproduction rates are determined by past differences. In fact, the theory holds that evolution results from changes in selective forces that favor reproduction among organisms whose ancestors had relatively lower rates of reproduction. Over the short run, and holding the environment constant, this operational definition is a useful way of estimating fitness for sexually reproducing organisms. But it is not generally reliable and is not a definition of fitness at all. Indeed, where new selective forces make for evolutionary change, this measure of fitness will give quite wrong measures.

We need to link fitness to heredity so that differences in fitness can generate differences in reproductive rates and thereby evolution. Our axiom must express the ideas that, although the maximum size of a Darwinian subclan must remain constant because of Axiom 2, the proportions of members of different lines of descent within the Darwinian subclan will change *if* there are differences of fitness between these branches at any given generation and if these differences in fitness persist from generation to generation. Ultimately, of course, the proportion of the members of the fittest branch must come to be fixed at 100 percent, provided the forces determining fitness remain the same. To express this axiom succinctly, we need a new defined term to refer to branches or sets of branches of a Darwinian subclan that can vary in their relative sizes from generation to generation in the lines of descent constituting the Darwinian subclan. Because any branch in the line of descent can change its proportion of members of the whole Darwinian clan at a generation, the term we need to refer to such lines of descent is simply 'subclan of a Darwinian subclan.' Instead of calling this set of branches a Darwinian subsubclan, or a sub-Darwinian subclan, Williams coins the term 'subclan'.' The only important thing to bear in mind about a subclan is that unlike a Darwinian subclan, of which it is a part, the subcland can expand and contract its size from generation to generation, while the Darwinian subclan's size remains constant. One is tempted to assimilate Darwinian subclans and subclands because both seem to be sets of organisms similarly affected by selective forces. Nevertheless, they need to be kept distinct, as a well-worn example may illustrate.

Industrialization is said to have caused a change in the proportions of dark to light members of the English moth species *Biston batularia*. We may speculate that had the species been unable to adapt to industrialization it might have become extinct. In this respect all members of the species were similarly affected by selection; additionally, the darker members of the species were also similarly affected by selection and increased substantially their proportions in the whole population of the species. In this case, the whole species constitutes the Darwinian subclan, and the network of descent of darker moths constitutes the subcland. Had this subcland not existed, the Darwinian clan might have become extinct. Employing the notion of subcland, we may now state the axiom of the survival of the fittest or, as Williams prefers, the expansion of the fitter:

Axiom 4. *If (a) any Darwinian subclan, D, has a subcland D_1, and (b) D_1 is superior in fitness to the rest of D for sufficiently many generations, then the proportion of D_1 in D will increase.*

This axiom leaves considerable work to the evolutionary biologist. First, of course we need to define 'fitness for a subcland' on the basis of the primitive term 'fitness of an individual organism.' This is largely a matter of applied mathematics, however. Second, we will have to specify what "sufficiently many" means. How many generations are sufficiently many to ensure the increase in the subcland D_1 relative to the Darwinian clan D is a function of how superior D_1 is in fitness to the rest of the subclands in D and how large D_1 is. The smaller the fitness differences, the longer it will take to produce a change in proportions; the smaller the subcland, the more likely that nonselective vicissitudes will postpone the increase in its proportions; on the other hand, the longer the superiority persists, the less likely it is that chance effects will overtake the subcland's tendency to increase its proportions.

Like many general laws in physical science, Axiom 4 asserts that a particular effect will ensue without stipulating when it will or how large it will be. In this respect, the axiom is like the second law of thermodynamics. In its deterministic or "phenomenological" version, this law asserts that entropy or thermodynamic disorder always increases toward a limit, without giving its rate of increase or the value of the limit. The phenomenological version of the second law is of course false, as is a deterministic version of Axiom 4. For just as entropy does not always increase, the fittest subcland does not always increase in proportion. Such versions of these laws are nevertheless useful approximations to their stochastic or statistical versions for very large systems over very long periods of time. The usefulness of, and the exceptions to, these phenomenological versions are explained by their statistical foundations. The statistical version of the second law of thermodynamics is a probabilistic claim about increases in entropy over a sufficiently long period of time, one that is consistent with short-run decreases in entropy. Similarly, Axiom 4 is a statistical statement about long-run increases in the proportions of subclans and their increased fitness.

Axiom 4 is particularly important for the application of evolutionary theory to population biology, for it is the foundation of the so-called fundamental theorem of natural selection, due to Fisher. According to this theorem, the fitness of a subclan population increases at a rate proportional to inheritable variability in the fitness of its subclands. This theorem follows rather obviously from Axiom 4. For if the proportion of the fittest subcland must increase, the greater the fitness differences between the fittest subclan and the rest of the clan, then the faster will be the rate of increase of the whole subclan's fitness. (The statistical character of evolutionary theory is treated at greater length in Chapter seven, Section 7.8.)

Notice that Axiom 4 does not assert that there are subclands with superior fitnesses compared to their containing Darwinian clans. Indeed, it does not even assert the existence of these latter. It is equally silent on whether there are subclands that retain their fitness superiority from generation to generation. It says only if there are such subclands, then certain consequences ensue. Still less does Axiom 4 express or presuppose any mechanism for the retention of fitness differences from generation to generation between subclands.

The last axiom makes the existence claims of the theory. It tells us that if there are Darwinian subclans, then there are subclands, that these subclands retain their fitness superiority from generation to generation, and that the superiority is always great enough to have impact on the future character of the Darwinian subclan.

Axiom 5. *In every generation of a Darwinian subcland* D *(that is not on the verge of extinction), there is a subcland,* D_1*: and* D_1 *is superior in fitness to the rest of* D *for long enough to ensure that* D_1 *will increase relative to* D*; and as long as* D_1 *is not fixed in* D*, it retains sufficient superiority to ensure further increases relative to* D.

The proviso excluding Darwinian subclans on the verge of extinction is required because in populations that are only a small number of generations away from extinction, the fitness differences may not persist for long enough to result in an increase in their subclands. More important, notice that Axiom 5 makes no claims about heredity or its mechanism. Indeed, the regularity it describes could obtain no matter whether there were a dozen different hereditary mechanisms operating at the same time, or at different times. For that matter, Axiom 5 would obtain if fitness differences persisted from generation to generation as a result of a long-term fluke without any physical foundation whatever. The axiom does not even require that fitness differences be completely heritable; partial concomitance of fitness values between successive descendants will suffice to generate evolution. This neutrality with respect to hereditary mechanisms is, we have seen, an important attribute of any account of the theory. For such mechanisms will be required by all competing evolutionary theories, must operate in the absence of evolution, and are themselves open to evolutionary explanation.

This concludes our axiomatic presentation of the theory of natural selection. For all its apparent formality, our presentation is a considerable simplification of Williams's original, one that leaves out its mathematical and set-theoretical super-structure and thus a fair part of its rigor and expressive power. On the other hand, much of the reluctance to adopt this axiomatization may be attributed to the highly abstract appearance that these formal features give the theory. Once convinced of the virtues of this axiomatization, readers are urged to consult the original.

5.5. Adequacy of the Axiomatization

We must now turn from the exposition of the axiomatization to a consideration of its adequacy: Does the axiomatization actually express the theory it is intended to axiomatize, or is it a sterile exercise in formalization without biological payoff? There are several dimensions along which these questions must be addressed. The most obvious one is whether all, most, many, or at least the distinctive claims treated as integral parts of the theory follow from these axioms as theorems by the laws of logic alone. But, in addition, we will want to consider whether the axiomatic system is easily adaptable to the heterogeneous classes of phenomena described as evolutionary; whether or not these all-inclusive generalizations can account for the terrific complexity that led Stebbins to deny their possibility. We shall want to determine whether the axiomatization enables us to solve outstanding disputes about biological problems: Can it shed light on the controversy between gradualism and punctuated evolution, as well as on the relations of each of these views to evolutionary theory; can it lay to rest controversies about the levels and units of selection? Then there are methodological and conceptual problems about evolutionary theory that provide a test of the axiomatization's adequacy. In particular, does it dispose of the charge of circularity under the constraint of simultaneously explaining why this charge persists even among the biologically sophisticated?

It would be humanly impossible to show that all the statements correctly viewed as consequences of the theory of natural selection follow from these axioms as theorems. The sheer number of such statements is too large, as is the number of distinctive theorems that can actually be proved from these axioms. In general, the axiomatization of a theory is tested for adequacy by showing that the most distinctive and important features of the theory are reflected in its theorems. One trouble with this approach is that there is disagreement about what constitutes these features, and it is this disagreement in part that motivates the search for an axiomatization. Nevertheless, Williams holds that it is, for example, relatively easy to see that such leading ideas of Darwinian theory as differential perpetuation and descent with modification can be derived from the axioms. In particular, she deduces the following theorems from the axioms:

Differential perpetuation. *In every generation of a Darwinian subclan, there are subclands D_1 and D_2 such that during the next several generations D_1 increases faster than D_2.*

She also shows, by deducing the following theorem, that differential perpetuation is determined by fitness differences and leads to the eventual disappearance of the less-fit subcland and the fixity of the fitter one:

Fixity and extinction of subclands. *In every generation of every Darwinian subcland, there is a fitter subcland that is not yet fixed but that is expanding and will become fixed in the Darwinian subcland.*

Employing this theorem, Williams proves a theorem of:

Descent with modification. *For any Darwinian subclan that does not die out, there is an infinite sequence of subclands, such that each subcland is contained within its predecessor, and each is fitter than its predecessor for long enough to ensure that it becomes fixed in the Darwinian subclan.*

Williams proves another theorem that is particularly powerful in showing the adequacy of this axiomatization to the very different kinds of domains in which biologists apply it and in showing how these domains are unified in a theoretical core constituted by this axiomatization. The theorem takes the following apparently abstract form:

Equilibrium theorem. *If a Darwinian subclan contains two subclands, D_1 and D_2, which do not contain any common ancestors after a given generation (i.e., they do not* *interbreed after the given generation), and if* D_1 *is superior in fitness to* D_2 *whenever it contains less than a certain proportion of the total members of the Darwinian subclan, while* D_2 *is superior to* D_1 *whenever* D_1 *contains more than this proportion, then there is a number* e *between zero and one, such that the proportion of* D_1 *to* D *will either stabilize at* e *or oscillate around* e.

For all its abstractness, this theorem provides the theoretical foundations for the explanation of such diverse evolutionary phenomena as the predator–prey, host–parasite, and resource–competition mechanisms of ecology; the appeal to heterozygote superiority in the explanation of balance hypothesis in genetics; and some alleged phenomena of species selection. The explanatory power of this theorem is worth illustrating in some detail.

Recall the fact that anything may properly be interpreted as the subject matter of evolutionary theory that satisfies the axioms governing 'biological entity' and 'parent-of.' Now, consider two species, like wolves and caribou. These are certainly both members of the Darwinian subclan mammal, and they long ago diverged and no longer interbreed. Accordingly, all the axioms of the theory should be true of each of them and true of the Darwinian subclan that they jointly constitute as well. Thus, the theorem should be true of them and their Darwinian subcland.. That is, as the proportion of wolves to caribou increases, the average fitness of wolves should decrease and vice versa. There should be an equilibrium level of proportions between wolves and caribou around which their respective population levels should oscillate or stabilize. But this is in effect what obtains among populations of wolves whose survival depends on predating caribou and caribou whose survival depends on avoiding wolves. The same can be shown for cases of parasitism and forms of ecological competition. What the axiomatization shows is that, first of all, predator–prey balance and other forms of equilibrium are not just intuitively explainable on evolutionary grounds. They are formal and ineluctable consequences of the theory. Secondly, the derivation of the existence of such points of stability or oscillation from the theory underwrites the search for models that will enable us to calculate their exact value. Without such theoretical assurance, we have no reason to suppose that an equilibrium value exists, and models that presume the existence of such points of stability are ungrounded. Thirdly, by asserting the existence of such points of stability or oscillation, the theorem provides a consequence of the theory that enables us to test it. By interpreting the theory in terms of particular organisms, we can make predictions open to test, which will either confirm or disconfirm the theory.

Among the most striking, and practically significant, predictive applications of the equilibrium theorem are to be found in ecological discussions of wildlife cycles. The most well-developed studies of this phenomenon have focused on animal species in Canada. In these studies, three types of cyclically fluctuating species have been found: tundra species, like the arctic fox, snowy owl, and roughlegged hawk, all of which prey on herbivorous lemmings, mice, and willow grouse; birch-forest species, the red fox, marten, roughlegged hawk, and northern shrike, which prey on the same three herbivorous species as the tundra predators and also on the snowshoe hare; and coniferous-forest species, including the lynx, horned owl, goshawk, mink, fisher, marten, and red fox, which prey on several grouse species and the snowshoe hare. The first two of these groups exhibit four-year cycles of population oscillation, whereas the species in the last oscillate around equilibria points in ten-year cycles.

The herbivorous mammals in these groups function as predators with respect to the tundra, birch, and evergreen vegetation. They have been known to so completely consume their ambient vegetation as to destroy their food supply and their concealing ground cover, thus increasing the ease of their falling prey to carnivorous members of their community. These herbivore cycles are held by ecologists to cause the carnivore cycles, but not vice versa. For instance, lemmings' well-known suicidal march to the sea seems to be caused by insufficient food in their indigenous locales. By contrast, the oscillation in grouse species does seem to be a result of oscillations in the carnivore species' population cycles. The difference in duration of the first two cycles from the third seems to be a function of the longevity of the most important species in each group. Moreover, the cycles seem to be most constant and regular in the most extremely northern environments. This is what we should expect, for under these extreme conditions the number of species is few, their selective interactions are limited by and large only to one another, and environmental vagaries of weather, geography, etc., are most narrowly limited. Some of the data gathered to uncover these cycles approach the regularity of cycles among populations competing in laboratory experiments. It must of course be recognized that among some species in competition no such cycles have been uncovered. Furthermore, the data on which many oscillations have been identified are far from unquestionable. But all this means is that testing a theory is difficult, especially a theory as general and as removed from particular phenomena as the theory of natural selection (see Section 7.7 for further discussion of the relationship between ecological models and the theory of natural selection).

One of the motivations grounding the appeal to population genetics as the core of the theory of evolution is the recognition that evolution requires a hereditary mechanism. For many organisms of interest, this hereditary mechanism is given by Mendelian genetics, and Mendelian theory enables us to make good numerical estimations of critical variables only generically specified in the theory of natural selection. Thus, for some fitness-conferring properties we know not only that they are somewhat heritable, but we can make an order-of-magnitude estimate of how heritable, given Mendelian theory; the theory also enables us to give some content to the notions of 'sufficiently long,' 'sufficiently fitter,' and 'sufficiently large' in Axiom 4 above. These considerations make it tempting to read the structure of genetics in the way Ruse does, and indeed this reading represents the orthodoxy of the "new synthesis." We have seen why this account of the matter is unsatisfactory. But once we interpret the axiomatization as a theory of the evolution of Darwinian subclans of Mendelian genes, we can see how natural such a misunderstanding is. It also becomes clear why the theory may be developed without great harm under this mistaken conception, and how evolutionary theory enters into the explanation of Mendelian genetics.

Consider two alleles of a genetic locus, A and a, where the heterozygote Aa is superior in fitness to either of the two homozygotes, AA and aa. In this case the lines of descent, of genetic replication of individual alleles, constitute noninterbreeding subclands. Because of heterozygote superiority, the average fitness of the members of the subclands A and a will be a function of their proportions in the entire Darwinian subcland composing both subclands. Therefore, their respective proportions will stabilize or oscillate around an equilibrium point. The existence of such a point of equilibrium between allele ratios shows why heterozygote superiority has the consequences Mendelian theory accords it. The derivation of the phenomena as the interpretation of a theorem of the axiomatization reveals that they are ultimately to be explained as the consequences of natural selection operating on the genome.

Once it is recognized that natural selection operates on the genome, that fitness levels can be accorded to genotypes, genes, and alleles, it is easy to suppose that the genome is the fundamental level of evolution. For the genetic material controls the hereditary character of the whole organism and provides the mechanism that substantiates the fifth axiom. Because Mendel's laws provide a description of the behavior of the genome, it is equally easy to infer that they therefore provide the mechanism not just for heredity, but for evolution as well. And in a sense they do, when and only when supplemented by the evolutionary laws couched in the axioms hitherto elaborated. But this last proviso is so crucial, the supplementation so great, and the grounds for these laws so independent of Mendelian considerations, that the assertion becomes seriously misleading, even though inevitable enough to produce the conviction that, at least for sexually reproducing organisms, population genetics is the core of evolution.

We are not yet finished with alternative interpretations of the equilibrium theorem. We may use it to explain claims about selection not at the level of the gene or the individual, but about selection at the level of populations and species. It is sometimes held that there are relations of dependency not only between the numbers of organisms of distinct species, but among the numbers of distinct species themselves. Thus it is held that if the number of distinct plant species changes in an ecosystem then the number of niches for insects or other animals may change, leading to a change in the number of animal species in the ecosystem. The reverse may also be true: The introduction of too many distinct species of competing insects, for example, may lead to competition for available plant species that eliminates some of this overabundance of species. Here the subclands are different types of species, plants and animals, which may vary in number through immigration or emigration. According to the theorem, there should evolve among them a balance in numbers of species of the two kinds, which leaves neither empty niches nor niches subject to persistent competition over time.

Subsequent chapters will add support to the conviction that the five axioms here expounded do adequately capture the content of the theory of natural selection, its logical structure and capacity for diverse application. They will expound its relation to actual and possible evidence for and against the theory and its powers to settle outstanding biological and conceptual issues surrounding evolution. But we may conclude by returning to the charge with which the chapter began: Either because of temporary obstacles or for permanent reasons, we cannot expect to find in evolutionary biology a theory of the precise and finished "framework of rigid, all-inclusive generalizations or laws" that characterizes physical theories. Stebbins is led to this pessimistic conclusion by reflecting on the fate of exceptionless generalizations offered in the course of evolutionary biology's development.

In each case, a strict and exceptionless claim has been propounded and initially confirmed; only later have exceptions to it been revealed, which cannot themselves be systematized in a way that preserves the original generalization as anything more than an unimprovable first approximation. Thus, Stebbins instances the Mendelian laws and, reflecting on their exceptions, rightly notes that "the entire history of

Mendelian genetics has consisted of discoveries that have modified these laws" (Stebbins, 1981:3). These laws are best viewed as modes of a "modal theme," a theme "designed to fit precisely the most common situations, but less common relationships can be expressed as variations of the central theme." The ultimate variations may differ so much from the modal theme that connections between them can be recognized only via intermediate, less complex variations, until simpler laws are eventually developed. According to Stebbins, the modal theme to replace Mendel's laws is given by a general statement of the molecular basis of heredity.

Among other modal themes, Stebbins constructs a *tertium quid,* an alternative between hypotheses that genetic variation is largely adaptive or mainly neutral: Roughly, sometimes neutrality obtains, but usually persistent variations are adaptive. Similarly, evolutionary speciation is usually a result of spatial isolation of small founder groups, but other sympatric, parapatric, stasipatric mechanisms occur as well. Thus, "with respect to the origin of species, the concept of a modal framework permits the evolutionist to adopt a pluralistic concept but at the same time focuses attention on the most probable course of events." (Stebbins, 1981:8). He attempts to strike a similar compromise between the allegedly punctuated character of macroevolution and the supposed gradualism of microevolution.

It is easy to criticize this ecumenical view on the grounds that the theories it attempts to strike a compromise between are incompatible and exclusive alternatives. It will therefore be no surprise that an attempt to do justice to all of them will not be expressible in general statements that do not themselves reflect these incompatibilities by being self-contradictory. Adopting "modal" themes in the description of the course of events characterizing terrestrial evolution is perfectly coherent and probably correct. The actual course of events is the result of the operation of many different forces on many different mechanisms. But adopting modal themes in the attempt to elaborate theories that will explain this course of events is a counsel of despair, that matters are too complex to extract a manageable number of useful generalizations. Or it is the mistaken idea that incompatible theories can be reconciled by splitting their differences for the course of evolution between them.

Notice, however, that the axioms hitherto stated are all neutral as between the apparently irreconcilable differences Stebbins attempts to bridge. But they are also directly relevant to the explanation of each of the phenomena these different approaches are based upon. We have already seen the bearing of evolutionary theory on any theory of heredity and on any specification of a heredity mechanism. The establishment and persistence of such a mechanism requires evolutionary explanation. The fact that no one mechanism has come to predominate, which leads Stebbins to surrender the hope of a single account of hereditary applicable to all biological phenomena, is just what we would expect, given the theory: Different selective forces advantage different modes of hereditary transmission. Interpreting the theory as a claim about natural selection among genes naturally leads to the expectation of diversity of hereditary mechanisms, just as it leads to the expectation of diversity among the organisms that bear these mechanisms.

The same point must be made about debates surrounding the question of whether most persistent genetic variations are neutral or whether they are adaptive. It is often supposed that the former alternative is incompatible with a theory of natural selection that makes adaptation the motive force of evolutionary change or persistence. But this is clearly a misunderstanding. As we know from the disconnection between any given Mendelian phenotype and any given molecular gene, or its immediate products, evolution may proceed at different rates at different levels of organization. Changes that reflect natural selection at the level of the gene may simply have no resultant cumulative effect at the level of the whole organism. Under different interpretations, the theory of natural selection can be applied to different boundary conditions, different fitness differences, and larger and smaller subclans and sub-clands. These differences will disconnect rates of change at one level from rates at another level. If a given molecular variation is neither necessary nor sufficient by itself for any change in the fitness of the organism that contains it, but is necessary or sufficient for a change in the molecule's fitness with respect to the genetic material in which it occurs, the result will be variation that appears to be selectively neutral only if the theory is interpreted *exclusively* as a claim about organisms and not as a claim about any biological entities that satisfy a parent-of relation.

Of course, in recent years the most publicized debate within the evolutionary community has concerned the "tempo and mode" of evolution. In the light of the above axiomatization, this debate can be seen to be, first of all, a dispute entirely among proponents of the theory of natural selection, not one between proponents and opponents. Secondly, it is best viewed as a dispute about the particular values of parameters and variables in the theory coupled to a question of historical scholarship. The dispute is between proponents of a theory known as punctuated equilibrium and proponents of a theory that punctuated equilibrium theorists call Darwinian evolution. The punctuated-equilibrium theorists have received considerable public attention. This is due in part to the implication of the labels they employ, which suggest that they represent a scientifically respectable movement that casts doubt on the warrant of the theory of natural selection. Public interest in disputes between creationists and evolutionists has thus brought this technical "in-house" controversy into much greater prominence than other scientific disputes. Punctuated-equilibrium theory holds that the actual course of terrestrial evolution has been characterized by geologically long periods of changeless succession of generations, without substantial evolution either within species or from one species to another; but these long periods of evolutionary equilibrium are interrupted – punctuated – by bursts of relatively rapid evolutionary change. It is these changes that result in the great species diversity characteristic of terrestrial evolution. Additionally, we may infer from the persistence of traits and the absence of evolution that their persistence cannot be explained by adaptational considerations. For they are so to speak "thrown up" in the bursts of change and then left alone by evolutionary forces through the long periods of equilibrium. This rejection of adaptationalist thinking is a consequence of, and not a presumption of, the punctuated-equilibrium theorists' view. These theorists contrast their view with the claim that evolution is persistent over time—moves at the same rate – by continually taking small steps of roughly the same "adaptive" size. Thus, evolution within a species is always gradual, and evolution from one species to another species is equally gradual. This gradualist thesis is called by the punctuated-equilibrium theorists *who reject it* the "Darwinian theory" of evolution, because they claim to find much textual evidence in Darwin's writings, and in the writings of Darwinians who followed him, that he and they held the rate of evolution to be constant over geological epochs and to be small as well. The question of whether Darwin actually held this view is of purely historical interest, and I shall not comment on it except to say that, in calling the view they reject "Darwinian," punctuated-equilibrium theorists have needlessly confused the public. For they have conveyed, intentionally or otherwise, the suggestion that natural-selection theory itself is in doubt among respected members of the scientific community.

But the first thing to note about the dispute is that it not only hinges on a particular reading of the geological record by paleontologists – a highly theory-laden piece of interpretation – but that it also seems largely to be about this record and therefore about the values of evolutionary variables and parameters. It does not seem to be a more fundamental dispute about what these variables and parameters are. Both parties hold that there is a theory of natural selection on which they are agreed. Accordingly, a statement of the theory should be neutral as between their two views. It should in addition be adaptable to the expression and testing of their incompatible views. This is exactly what the axiomatization does. It is obviously neutral as between these theories, for it makes no claim about the rate of evolution, but makes only the generic claim that evolution will occur under certain conditions. To satisfy ourselves that evolution obtains, we will have to estimate both the values of fitness differences, and the sizes of subclans and subclands. We will have to variously interpret the primitives of the theory to answer questions in dispute between punc-tuated-equilibrium theorists and gradualists to the extent that the paleontological record or other sources of evidence admit. But, seen in this light, the dispute is no obstacle to the expression of a general theory of evolution and is no motive for expressing evolutionary ideas in Stebbins's "modal themes."

Of course, those who hold that evolutionary theory is inextricably bound up with a narrative of the course of evolution on earth will not be satisfied with this conclu-sion. For it admits that the axiomatization cannot by itself decide between these competing stories of terrestrial evolution. And if either punctuated-equilibrium theorists or gradualists are wedded to this view of the theory and to the demand that it must *by itself* decide between them, then the axiomatization will be called into question.

But in the terms of the original question that this view of the theory posed, our axiomatization provides a satisfactory response to the narrationalist's challenge. In Section 5.1, it was pointed out that this challenge cannot be met merely by drawing an abstract distinction between a theory of evolution and its application to describe the course of local evolution on this planet. For the problem is not to justify the in-principle possibility of such a theory. This in-principle possibility is not seriously in doubt. The challenge is to show that the generalizations that connect the events in the sequence of terrestrial evolution on earth are manageable in length and small enough in number to work together as a single theory. We must show that evolu-tionary theory is not just a telephone-directory-sized compilation of divergent and disjunctive causal laws from which a different set of generalizations will have to be drawn on each occasion that we want to justify or explain a particular sequence. The size and the applicability of the axiomatization meets this challenge and meets it clearly.

In fact, it should be emphasized that the theory thus axiomatized has the sort of universality characteristic of theories in physical science: None of the axioms is expressed in terms that restrict it to any particular spatiotemporal region. If the theory is true, it is true everywhere and always. If there ever were, or are now, or ever will be biological entities that satisfy the parent-of relation, anywhere in the universe, then they will evolve in accordance with this theory (or else the theory is false). Not only does the form of the theory give us this assurance, but the foundations of each of the axioms is sufficiently general to give us confidence in the theory's universal applicability. Thus, for example, Axiom 2's limitation on the maximum size of a Darwinian clan is grounded in limits on energy consumption given in physical thermodynamics. Axiom 3 accords a level of fitness to every biological entity with respect to its physical environment. We know on physical grounds that nothing can be isolated from all the physical forces of its environment, and if we can accommodate "fitness" to physical theory, we shall have grounded this axiom as well. And we must so ground fitness for other reasons, as we shall see. Axioms 4 and 5 rest on the existence of physical substances and on the operation of physical laws that make for heredity. We know independently that these substances can be found and these laws operate throughout the universe. Although these remarks constitute nothing more than the intimation of a sketch of how to reduce evolutionary theory to physical theory, they are enough to underwrite its claims to the universality common among physical theories.

It will be recalled that the last chapter ended with the admission that the plausibility of the reduction there advocated for Mendelian genetics hinged on providing an account of fitness or adaptation that revealed its nonevolutionary, physical basis. It was this need that provided part of the motivation for exploring the theory of natural selection within which the notion of fitness figures. In an important respect, our exploration has not been provided the assurance demanded at the end of Chapter 4. For it has been held that with respect to the axiomatization provided, the term 'fitness' is a primitive, undefined one. So, accepting this account has merely postponed the question. Even if it is accepted that, with respect to the theory, this term is undefined, no one will accept this as the last word on the matter. Not only is there the overhanging question of the provenance of an undefined notion of fitness in the reductionistic explanations of molecular biology; there remains the threat, sketched above, that without a definition in some theory or other, the term 'fitness' will prove scientifically sterile, and the theory in which it figures will be subject to controversies about its cognitive content. Finally, the claim that this critical term for all of biology is a primitive one with respect to the theory in which it figures is in itself so novel, if not implausible, that it appears to be the Achilles' heel of our axiomatization. Objections to its adequacy may well begin with arguments that the term is not after all primitive in the theory and that therefore our axiomatization is inadequate. Therefore, our continuing exploration of evolutionary theory turns to a direct examination of the nature of 'fitness.'

Introduction to the Literature

T. Dobzhansky, F. J. Ayala, G. L. Stebbins, and J. W. Valentine, *Evolution* (San Francisco, Freeman, 1977) is an excellent introduction to the theory of natural selection.

On the nature of scientific theories, introductory students should begin with Hempel, *Philosophy of Natural Science,* chap. 6, and Nagel, *Structure of Science,* chaps.

5 and 6. These two works reflect postpositivist accounts of the nature of theories. Alternatives are described and defended by their proponents in the anthology *The Structure of Scientific Theories* (Urbana, University of Illinois Press, 1974), ed. F. Suppe. Suppe's introduction provides a useful introduction to and summary of these views. The applications of some of these ideas to the theory of natural selection is considered in P. Thompson, "The Structure of Evolutionary Theory: A Semantic Approach," *Studies in the History and Philosophy of Science*, 14(1982):215–30.

There is no substitute for reading Charles Darwin, *On the Origin of Species* (London, Murray, 1859). For an introduction to the intellectual context of the theory, see Michael Ruse, *The Darwinian Revolution* (Chicago, University of Chicago Press, 1979). Goudge, *Ascent of Life*, argues forcefully against the assimilation of the theory of natural selection to those of physical science, and G. L. Stebbins, "Modal Themes: A New Framework for Evolutionary Synthesis," in R. Milkman, ed., *Perspectives on Evolution* (Sunderland, Mass., Sinaur, 1982), pp. 1–15, is a very recent expression of reservations about a single distinctive theory of natural selection.

The most influential of many claims that the theory of natural selection is bereft of empirical content and unfalsifiable is that of Karl Popper, "Darwinism as a Metaphysical Research Programme," in P. A. Schilip, ed., *The Philosophy of Karl Popper* (La Salle, Ill., Open Court, 1974), pp. 134–7. Popper's views have been expounded and criticized in Michael Ruse, "Karl Popper and Evolutionary Biology," in *Is Science Sexist?* (Dordrecht, the Netherlands, Reidel, 1981), pp. 65–84.

The "synthetic theory of natural selection" was first advanced in Julian Huxley, *Evolution: The Modern Synthesis* (London, Allen and Unwin, 1942). Ruse's appeal to population genetics as the core of the theory of evolution appeared in his *Philosophy of Biology*, chaps. 3 and 4. The history of this synthesis of Mendel and Darwin is traced in W. Provine, *The Origins of Theoretical Population Biology* (Chicago, University of Chicago Press, 1971). For an influential recent statement, see R. Lewontin, *The Genetic Basis of Evolutionary Change* (New York, Columbia University Press, 1974), especially chap. 1.

Mary Williams's axiomatization of the theory of natural selection appears in "Deducing the Consequences of Evolution," *Journal of Theoretical Biology*, 29 (1970): 343–85. The discussion in this chapter follows her presentation as closely as possible. Readers should also see her discussion of the application and testing of the theory, "Falsifiable Predictions of Evolutionary Theory," *Philosophy of Science*, 40(1973):518–37, and "The Importance of Prediction Testing in Evolutionary Biology," *Erkenntnis*, 17(1982):291–306. E. C. Pielou, *Population and Community Ecology* (New York, Gordon and Beach, 1979), is an excellent introduction to mathematical and experimental applications of the theory of natural selection to problems in ecology. It summarizes the theoretical work and fieldwork that I have tried to systematize in Williams's axiomatization. Its references to further works in the field are particularly valuable. A more advanced treatment by the same author is *Introduction to Mathematical Ecology* (New York, Wiley, 1969, and later editions).

The theory of punctuated equilibrium is clearly expounded in S. J. Gould, "The Meaning of Punctuated Equilibrium," in R. Milkman, ed., *Perspectives on Evolution* (Sunderland, Mass., Sinaur, 1982), pp. 83–104. His original treatment of the subject appears in N. Eldredge and S. J. Gould, "Punctuated Equilibrium: An Alternative to Phyletic Gradualism," in T. J. M. Schopf, *Models in Paleobiology* (San Francisco, Freeman, 1972), pp. 82–115.

CHAPTER 6

Fitness

This chapter and the next are devoted to the notions of fitness and species, for a proper appreciation of these concepts answers almost every leading question that has been posed about the theory of evolution. These chapters show that biologists' uses of these notions are variants of standard practices in the physical sciences. Indeed, by the end of the next chapter much of the mystery surrounding the nature of biology as a natural science should disappear.

The first task is to provide reasons, independent of the axiomatization given in the last chapter, for supposing that fitness is a primitive term with respect to the theory of natural selection. The argument must be independent of the axiomatization if it is to win agreement, for the axiomatization's commitment to treating 'fitness' as a primitive term is apparently a vulnerable weakness. We shall see that, far from a weakness, this feature of the axiomatization is one of its most attractive features.

6.1. Fitness Is Measured by Its Effects

It will be recalled from the discussion in Section 5.2 that the chief threat to the cognitive significance of evolutionary theory is the difficulty of providing a definition of fitness that does not make the theory of natural selection circular, untestable, and therefore devoid of explanatory power. It is an important criterion of adequacy for any such definition that it come complete with an explanation of why biologists and nonbiologists have for over a century taxed the theory with this charge. For any definition that cannot be naturally applied to this task must fundamentally misrepresent the concept of fitness. If the definition cannot also explain the ease and persistence of the mistaken charge that 'fitness' is circularly defined, then it fails to properly express the concept as it is actually employed and understood. For it is a feature of the very meaning of evolutionary fitness that it be *easily* misunderstood.

This test, of simultaneously defining and explaining the naturalness of mistakes about it, is failed even by philosophers and biologists sensitive to the issue of circularity and eager to forestall it. Thus, Ruse writes,

A first priority for us must be to address ourselves to the question of whether or not natural selection is tautologous. But as soon as we do address ourselves to this question, immediately one thing becomes obvious. Whether or not some formulations of the principle of natural selection are tautologous, biologists in talking of natural selection are certainly pointing to a non-trivial possible way in which genetic change might be effected. (Ruse, 1973:39)

Ruse then gives a description of a hypothetical case of changes in gene frequencies due to differences in reproductive rates of the carriers of the genes. But this sort of hypothetical case will not solve the general problem. Changes in gene frequencies are not at issue. Nor is there any doubt that such changes will be the result of differences in rates of reproduction. The question is whether we can systematically identify the distinct and separate causes of differences in such rates. Ruse admits that natural selection is sometimes expressed in terms that make it vacuous, but he offers no assurance that it can be otherwise expressed. It is this assurance that is required if we are to credit the theory of natural selection with explanatory power.

Of course, biologists have noticed nontrivial ways in which genetic changes might possibly be effected. This is not at issue. What is up for dispute is whether the terms in which they describe these genetic changes and their alleged causes meet conditions on empirical content required of a scientific explanation. Short of a definition that frees fitness from reproduction rates, this cannot be assured.

Another philosopher attempting to come to grips with these issues, David Hull, holds that fitness is defined in terms of differential reproduction together with the causal mechanisms that generate it. In his view, the persistence of charges that the theory is vacuous rest on failures to recognize this second component of the definition. But he also recognizes that such causes are overwhelmingly heterogeneous and unmanageable in detail and complexity. So, "the principle of the survival of the fittest is officially a tautology in certain operationally oriented versions of evolutionary theory, and these versions suffer accordingly" (Hull, 1974:69).

But first of all, fitness could not be defined in terms that mention both the causes of differential reproduction and differential reproduction itself, or else fitness could not be cited to explain the latter as its cause: In this view, ascriptions of fitness both describe the phenomena to be explained and advert to the causes that explain them. As such, ascriptions of fitness would still consist in nothing more than redescriptions of what they are meant to explain together with an unspecific assertion that the differences did indeed have causes, ones left unmentioned in the explanation. Secondly, it is no solution to this problem to provide a definition of fitness within a theory that cannot be "operationally oriented." If this is the price of avoiding circularity, then it is either too high, or it is the tacit admission that a non-tautologous, operationally oriented theory of evolution would not appeal to fitness at all. But such a theory would not be the theory of natural selection.

Instead of attempting to answer this question directly, let us consider how the term 'fitness' is actually employed by evolutionary biologists. Instead of focusing on their self-conscious definitions (as illustrated in Section 5.2), let us examine the kinds of claims they make by the use of it, the ways they measure it, and the ways they correct their measurement of it. We shall soon discover that in some respects 'fitness' is a perfectly ordinary theoretical term, much like 'temperature,' and that the respects in which it differs from such an ordinary theoretical term explain the mysteries surrounding it.

Fitness is a relational property, reflecting the interaction of an organism and its environment. How we apply the theory of natural selection to explain adaptation, competition, and evolution depends on how we measure the fitness of, and especially the difference in fitness among, organisms. We need to be able to make comparisons between members of the same species in the same environment, between members of the same species in different environments, between members of different species in the same environment, and between members of different species in different environments.

An organism's fitness is relative to its environment, but an environment is not a spatiotemporal region. It is a region in more than four dimensions. Its dimensions are given by the set of forces that can interact with the organism's properties to determine its fitness level. Thus, two environments might be spatiotemporally identical and yet distinct from one another; for example, an open field may provide distinct overlapping environments for grouse, squirrels, insects, certain flora, etc. We identify environments along three spatial and one temporal dimension, but they lie in a space of a much larger set of dimensions, along each of which variations affect fitness levels. Exactly how large the dimensionality of an evolutionary environment is we cannot tell, because we know too little about the structure and behavior of organisms and about how external forces affect them. Even if we had adequate theories about the structure, behavior, and environment of organisms, the practical difficulties surrounding any attempt to explain fully why an organism had a particular fitness level would be staggering. The number and complexity of such theories would make reasonably complete explanation of an organism's fitness level unmanageable in length and incapable of actually predicting biologically interesting consequences of its fitness level.

Thus there are many different ways in which the same level of fitness may be possessed, and there is consequently no one-to-one relation between a given level of fitness and a manageable set of its causal conditions. We cannot identify levels of fitness by appeal to the properties of organisms and environments that determine it. Because of the one–many relation between fitness and its determinants, fitness must be measured *by its effects*.

Now measuring of a quantitative, theoretical property by measuring its effects is a common practice in physical science. Changes in temperature are measured by an alcohol thermometer that measures the effects of such changes on the linear expansion of alcohol in a narrow tube. The linear expansion in the thermometer does not define changes in temperature; it provides a way of measuring its magnitude. Because a thermometer measures temperature changes by measuring their effects, we can explain its function by appeal to the causal relation between temperature and linear expansion. If temperature were defined in terms of linear expansion, we could not do this. The operation of a thermometer is explained by citing the very phenomenon it measures. Moreover, the measurements thermometers give often need to be corrected; indeed, at extreme temperatures, these instruments do not give readings at all. Beyond the boiling and freezing points of alcohol, thermometers no longer respond to changes in temperature and cannot be employed to measure them. This fact is explained by appeal to thermodynamics and chemical theories of the ionic and vaporization properties of alcohol. Beyond the boiling and freezing points of alcohol, different measuring devices must be employed, and these instruments can also be compared with the alcohol thermometer's readings at intermediate temperatures. Differences between alternative devices' readings must themselves be explained, and some measuring instruments must be corrected in the light of others. To determine which are the most reliable and to effect the needed corrections, we employ theories about heat and about other nonthermodynamic properties of the instruments. Some of those theories may even show that it is possible to measure temperature changes by appeal to their causes instead of their effects. For instance, we can measure a change in temperature by measuring changes in electrical resistance, which causes it. But of course such a possibility is impractical for many purposes; measurement in terms of subsequent effects is preferable, even when measuring methods that rely on prior determinants are in principle available.

The measurement of fitness, as it figures in the theory of natural selection, differs from that of temperature in two respects. Fitness is a function of more complex relations among more variables than temperature is. And there is no theoretical edifice to stand behind fitness, to explain and correct measurements of it, as there is for temperature. Such theories may eventually be found, but they are not on the scientific horizon. Because of these two features of the conceptual position of fitness, measurement of it depends on its effects even more fully than the measurement of temperature does.

The fitness of an organism is measured by counting progeny – of the organism, of its ancestors and/or descendants, or of some subset of the progeny in a branching tree of descent. But although these demographic counts are the units in which fitness is measured and are the only common coin of evolutionary comparisons, it should be obvious that, as they stand, they are highly unsatisfactory measures of fitness. Consider, for example, identical twins that have reached reproductive maturity in the same environment. Presumably two such peas in a pod will have the same level of fitness. But if through some entirely random event one of them is destroyed, then the other will have vastly more offspring; if we measure fitness by number of offspring, without *correcting* for the intervention of short-run, nonselective forces, we will wrongly conclude that these two biologically identical organisms had different levels of fitness. The theory of natural selection assures us that, in the long run, persistent differences in the level of fitness, no matter how small, between competing members of a biosphere will lead to differences in reproductive levels. That is why we employ short-term differences in reproductive success to *estimate* fitness. But short-term trends, say ten or twenty generations, may not reliably reflect long-term ones, say a thousand or ten thousand, and therefore uncorrected fitness estimates may be wrong. Any actual level of fitness is consistent with any short-term level of reproduction, even zero reproduction, provided that the lack of descendants results from forces exceptional and/or indeterministic enough to be deemed random with respect to evolution.

We may argue that a particular property – like mimicry or disease resistance – makes for fitness by showing that it contributes to optimum or satisfactory design. To claim that a particular trait is suitable for attaining a certain end in a given environment, we need not enquire into reproductive success; instead, we may employ theoretical or engineering information to identify alternative strategies for meeting a design requirement and to grade the efficiency of these alternatives in the light of environmental and organic constraints. Pursuing this approach enables us to identify some of the determinants of fitness in particular cases and so grounds restricted judgments of comparative fitness on its causes instead of its effects. But the use of design criteria is unavoidably myopic. It can easily lead to misidentifying the more fit as the less, and vice versa; thus a parasite more efficient at wasting its host may be wrongly identified as fitter than one that survives at higher levels just because it is less efficient and therefore does not destroy its host. At best, employment of design criteria may enable us to make restricted, qualitative, non-fungible comparisons of fitness with respect to a small number of properties and a restricted class of environments. Optimum design cannot play the systematic, quantitative role required of a general measure of fitness that the theory requires. Fitness levels must be measured in a coin common to many biological systems if they are to play an explanatory function in the theory of natural selection.

Fitness levels are in fact related to actual reproduction rates in a way familiar to philosophers who have reflected on problems of probability. The hypothesis that an organism, in a given environment, has a given level of fitness is like the hypothesis that a given die is a fair one. This latter hypothesis is consistent with any finite sequence of outcomes of rolling the die, even 10^6 straight "2"s. No finite number of outcomes of rolling the die can refute the hypothesis. Still, the claim that the die is fair has empirical content, which is why it can satisfactorily explain a run of throws in which each face comes up the same number of times. This is true even though our only practical basis for believing that the die is fair is the finite number of times we have observed it to have been thrown in the past. There is in principle another independent way to assess the fairness of the die: by examining its material composition and structure. But although this method holds out the advantages of being independent of actually rolling the die, it has the disadvantage of being utterly impractical for anyone who might be practically interested in the hypothesis. If there were no possible way to determine whether a die was fair without actually rolling it, then the fact that one can roll any finite sequence whatever would deprive the fairness hypothesis of all its explanatory power with respect to actual sequences. The same must be true of fitness. If fitness levels could not even in principle be measured by anything but levels of reproduction, then differences in fitness could not explain differences in rates of reproduction and so could not explain evolution. However, the fact that biologists do correct values for fitness given by actual short-run reproductive rates shows that they do have access to alternative means; for example, they can appeal to optimal design for correcting comparative judgments in particular cases. The trouble is that, without general theories of organism structure, function, and environmental dependence, these corrections do not have the systematic basis that corrections of thermometers do or even the basis that the evaluation of the fairness of a die in principle has. And so these corrections can seem ad hoc. They are not in reality ad hoc, nor are they merely designed to preserve the theory from falsification. But they are based on practical and theoretical considerations that are heterogeneous, restricted in their application, and too piecemeal in their bearing on fitness to find a place in any canonical statement of the theory of natural selection.

At least one cause should now be clear for why the theory of natural selection is persistently charged with vacuity and circularity and denied the cognitive status of other scientific theories. Such charges are natural, given the features of the theory and its key theoretical term. If fitness were defined in terms of differential reproduction, then the theory would be guilty as charged. But if differential reproduction is a contingent, causal consequence of differences in fitness, and if it also provides the only practicable general means of measuring fitness, then anyone who supposes that measurements give the meaning of theoretical terms can be excused for mistakenly concluding that the theory is a grand tautology. This mistake is characteristic of *operationalism,* a doctrine still widely embraced among natural and social scientists. But even those not imbued with this discredited view may draw the same conclusion. Measurements of short-term rates are explained away, or corrected, in a manner that is theoretically piecemeal; piecemeal just because of biologists' ignorance of the general theories that can systematize their corrective tactics. But if piecemeal corrections are mistaken for ad hoc steps taken simply to preserve the theory, it is easy to infer that the theory is unfalsifiable and vacuous. Thus the operationalist deprives the theory of natural selection of any explanatory force by assimilating the key term of the explanans, 'fitness,' to the crucial terms of the explanandum, 'differential reproduction.' On the other hand, those who claim that the theory is unfalsifiable infer from the current want of an independent general theory for correcting demographic estimates of fitness that no such theory is possible at all and that biologists are therefore not entitled to make the piecemeal adjustments to fitness estimates required in the light of the theory. These two mistakes are often jumbled together. They are likely to be persistent as well, for *there is no prospect of finding theories that will provide manageable alternatives to measuring fitness in terms of its effects on reproduction, its explanandum phenomenon.*

But all this is just another way of saying that fitness is a primitive or undefined term with respect to the theory of natural selection. It cannot be defined by citing its effects, even though it is these effects that measure it; it cannot be defined by adding to the theory a manageably small number of statements describing the causes of particular levels of fitness. This will be so even though in a particular case we may be able to cite physical, physiological, anatomical, or behavioral traits of organisms that together with their particular environments make for fitness differences. If fitness cannot be defined within the theory of natural selection or by additions to it, then with respect to the theory it cannot be defined, period. It must be a primitive term.

That few have recognized this fact is really not surprising. Aside from the temptations of operationalism and the dissatisfaction with merely piecemeal treatments that have prevented biologists from seeing this fact about their key theoretical notion, there are other causes as well. One cause of the failure to recognize that the term is theoretically primitive is that 'fitness' has a meaning in ordinary language (like 'force'), and informal expositions of the theory of natural selection capitalize on this meaning. When fitness is explained or illustrated by piecemeal nonevolutionary means, it is not recognized that the interpretation is not part of the theory of natural selection but rests on biological theory and observation that are independent of evolutionary theory. Thus, according greater fitness to one organism than another on the basis of design criteria is useful in limited respects but is easily misrepresented as reflecting an implicit definition that interprets the term *within* the ambit of evolutionary theory. Because no general interpretation of 'fitness' can be given within the theory, evolutionary biologists who feel constrained to define their central theoretical term sometimes incautiously state that fitness is differential reproduction. This turns the theory into a tautology. Their error is to suppose that every term in a theory is defined within that theory. Because the only feature fitness has that could be employed in a general characterization of it is its effects on reproduction, biologists are apt to adopt this devastating definition wherever their exposition of the theory seems to require definitions of its key terms. To avoid this temptation, all that needs to be seen is that although 'fitness' is primitive with respect to one theory, it can be interpreted by other theories or even by nontheoretical, apparently ad hoc considerations about the items of which it is predicated.

Of course, these interpretations will not take the form of universal general statements about fitness that are both true and manageably short. The number and the heterogeneity of causal forces that determine fitness is very large, and our ignorance of the general theories that describe them and their interactions is very great. Therefore, we cannot expect an interpretation of 'fitness' that is anything like the interpretation of temperature, as mean kinetic energy, involved in the reduction of the ideal gas law to the kinetic theory. No doubt there is a vast though as yet unknown collection of nonevolutionary theories that together could generate a baroque and useless disjunctive interpretation of 'fitness.' But this is a reflection of the fact that, with respect to the theory of natural selection, this term is a primitive. This semantic vacuum is filled by misguided attempts to define fitness.

6.2. Fitness As a Statistical Propensity

Some biologists have been sensitive to the problem of defining fitness in a way that does not trivialize their theory, but they have nevertheless felt constrained to give it a definition within the theory of natural selection. They treat fitness as a propensity to reproduce, so that differences in fitness are to be understood as different *dispositions,* capacities, and abilities to reproduce.

Now a capacity to do something is not identical with doing it. Thus, brittleness is the capacity or disposition to break under specified circumstances, but objects may be brittle without ever manifesting this capacity, that is, without ever breaking. And because a disposition like brittleness is distinct from the manifest properties like breaking that it is properly said to cause, it follows that, even though its effects figure in the definition of a disposition, the ascription of the disposition still has real explanatory power. Thus, Molière's explanation of why opium causes sleep, that "it has a dormitive virtue [capacity]," is not necessarily illegitimate. What is required by a respectable dispositional term is that there be a uniform underlying mechanism by virtue of which objects have the disposition. Brittleness explains breaking because brittleness is supervenient on the molecular structure shared by all brittle things. Magnetism explains why magnets attract iron things, even though magnetism is commonly defined as the capacity to attract iron filings, just because magnetism is a dispositional property generated by the orientation of constituent iron atoms of a magnet. Having the capacity to attract iron filings is part of what we mean by magnetic, but *only* part. And part of what we mean by magnetic is having a nondispositional, "occurrent," persistent structure, which generates or constitutes the intermittently manifested disposition to attract iron objects.

This account of dispositionals may be applied to the notion of fitness, and it is sometimes held out both as an account of the biological meaning of this term and as a solution to the problem of circularity. In assessing this view, we must bear in mind that merely to say that fitness is a disposition is not to offer an alternative to the claim that it is a primitive term with respect to the theory. It is to make a comment about the term's role in the theory, to say that according to the theory the property of fitness has evolutionary effects only under certain conditions, just as a magnet has detectable magnetic effects only under certain conditions. The obvious truth that fitness is a disposition in this sense is not at issue between the view that it is a primitive term and the view that it can be defined in the theory. What is at issue is whether more can be said within the theory, besides the assertion that it is a disposition that has reproduction as one of its contingent effects; and whether what can be said falls short of trivializing the theory.

Although no biologists have formalized a definition of 'fitness' as the disposition, capacity, ability, or *propensity* to reproduce, this conception is certainly implicit in some of what they say: For instance, after giving a definition of fitness that trivializes the theory, Strickberger goes on to make a further statement that shows he views fitness in a light that does not trivialize the theory:

> Fitness has come to have a variety of applied meanings, such as fitness for sports, intellectual pursuits, business success — terms that generally refer to ability or aptitude. In its genetic sense, however, it is far more restricted, and refers only to *relative reproductive success*. As long as one genotype can produce more offspring than another in the same environment, its fitness is superior. In this sense genotypes can also be described as having *adaptive* or *selective* value, which are merely other terms for reproductive success. (Strickberger, 1968:720; emphasis in original)

The first sentence reflects a dispositional view of fitness common to ordinary thought and biology. The second asserts that fitness is reproductive rates. The third asserts that fitness is a term for what genotypes *can* do and thus is a disposition. The fifth sentence trivializes the theory again (to the extent that adaptive value or selective value are terminological variants on fitness).

The idea that fitness is a disposition, a capacity or ability, is an attractive one championed by several philosophers as well as biologists. The most developed argument for this view is that of S. Mills and J. Beatty. One of their aims is explicitly to undercut the idea that fitness is a primitive with respect to theory. They attempt to treat fitness as a statistical propensity, thereby making sense of its connection to actual measurements of fitness by reproductive differences while solving the circularity problem. The definition of fitness they advocate as reflecting and reconstructing evolutionary practice is as follows:

> The [individual] fitness of an organism x in environment E equals $n =_{df} n$ is the expected number of descendants which x will leave in E. (Mills and Beatty, 1979:274)

This characterization is qualified, and applied in several ways to demonstrate the justice with which it captures the features required of fitness by the theory of evolution. Mills and Beatty describe their definition as the "propensity interpretation of fitness" and view as its chief advantage the claim that it solves the circularity problem. Of course, fitness, survival, and reproduction are intimately connected in the theory. But, as they note, any direct definition of the first in terms of the other two generates the circularity problem and deprives evolutionary theory of its explanatory force. On the other hand, Mills and Beatty claim that "we cannot define fitness entirely independently of survival and reproductive success" (Mills and Beatty, 1979:269). Accordingly, they suggest that the correct definition will sever a *direct* connection between these three terms and *interpolate* a disposition or propensity to survive and reproduce between fitness and actual survival and reproduction. This suggestion appears to solve the circularity problem and leaves the theory's explanatory power intact: For the connection between having a disposition, an ability, to survive and reproduce and actually surviving and reproducing is a *causal,* indeed a probabilistic, one and not a logical or definitional connection.

The appearance of having provided a definition that circumvents the circularity problem turns out to be illusory. And the reason is that the dispositional connection between fitness and survival and reproduction is still too strong. For the disposition can itself only be understood in terms of an independent specification of fitness. And this specification must itself be free of just such connection, direct or indirect, to survival or reproduction. In other words, fitness must be a primitive with respect to theory.

Given their definition, determining fitness levels means determining the expected number of descendants, and determining differences in fitness means determining differences in such expected numbers. One apparent way to acquire estimates of these values is by measuring actual numbers of descendants. But, as they note, for reasons given in Section 6.1, raw measurements of actual numbers of offspring in successive generations are quite unsuitable as estimators of the expected numbers of descendants. When actual numbers of descendants are cited in attempts to measure the fitness of ancestors *retrospectively,* a variety of corrections must be made, corrections that take into account the varying ways in which, for instance, evolutionary theory, and particularly theories of heredity, allow for the persistence – and sometimes the expansion – of populations that are of comparatively lower fitness than some intra- and interspecies competitors (e.g., genetic drift, pleotropism). The same difficulty besets attempts to measure *prospective* fitness, or the current disposition to have an expected number of descendants, by measuring the reproductive rates of ancestors of organisms. Thus, recall the operational definition of fitness cited from Williams in Chapter 5. This operational definition is based on the actual reproduction of ancestors, suitably weighted to reflect the successively falling contributions that more distant ancestors make to the level of fitness of a given individual. Such a characterization may be a guide to current dispositions to reproduce, and therefore to current fitness levels, but only for Mendelian populations reproducing sexually in a stable environment and only *modulo* the absence of forces and changes known, on selectionist considerations, to have intervened in such a way as to vitiate these measures. For example, changed geological, meteorological, demographic, or ecological conditions will make irrelevant past reproductive levels, just as surely as they can isolate current dispositions from future reproductive rates.

Of course, the measurement of current levels of fitness, either by reference to their effects in the future or more usefully but more indirectly by reference to the effects of levels of fitness of predecessor organisms in the line of descent, is unavoidable. As we have seen, theoretical values in science are often measured indirectly by measurement of their effects. However, as with measuring instruments generally, the data provided by these measures must be corrected, and often corrected in the light of theories independent of those they are being used to apply. Now these corrections, which transform raw data into measures of fitness, reflect the relations between the theory of natural selection and the other theories employed to apply it. The piecemeal character and the invisibility of the theories in the elaboration and application of evolutionary theory is the source of its appearance of untestability and circularity. This invisibility remains undetected but undercuts a dispositional definition of fitness.

Mills and Beatty define fitness in terms of a probabilistic notion of expectation: Fitness equals *expected* number of offspring, which in turn equals the sums of the organisms' propensities to produce each of an indefinite disjunction of whole numbers of offspring, divided by the number of propensities to do this. How are these propensities to be identified and counted? Because the theory of natural selection says nothing more about fitness than that it is partially heritable, and that it affects the rate of reproduction of organisms, it is obvious that there are no resources within the theory either to identify or to count these propensities. Moreover, we have already seen that the process of identification moves in the reverse direction. That is, it is only by reflecting on evolutionary considerations that we can assess the bearing of raw data about actual reproduction on theoretical claims about propensities to reproduce.

Mills and Beatty's proposal is doubtless a true statement about fitness. But it is no solution to the problem of circularity. Rather, it is an "implicit definition." In geometry, a 'line' is often said to be implicitly defined as the shortest distance between two points. This is said to be an implicit definition of line because the statement is an axiom of the theory (and not really a definition) and, more important, the notion of 'point' that it employs is often implicitly defined as the intersection of two lines. Accordingly, point and line are circularly defined in geometry, if they are defined at all. The proper conclusion is that one or both terms are primitives with respect to the theory. The same must be said about fitness and the propensity to reproduce. Both figure in implicit definitions, general laws of theory. But one or both are primitives with respect to the theory.

It bears repeating that to insist on fitness as a primitive term with respect to the theory of natural selection is not to say that it cannot be characterized, theoretically or operationally, within the ambit of some other theory or theories. To see both why the term is a primitive one in the theory of natural selection and also how it can be characterized independently of the theory one need only reflect on what Mills and Beatty themselves say about propensities: "An object's *propensity* to manifest a certain property is a function of all the causally relevant features of the situation, independent of our knowledge or ignorance of these factors. The totality of causally relevant features determines the unique correct reference class, and thus the unique strength of the propensity to manifest the property in question" (1979:273). If fitness is a propensity, as they claim, then it is a function of "the causally relevant features" of the reproductive opportunities of the organism. But these features go completely unmentioned in the theory of natural selection. They are provided by all those other theories in geology, meteorology, anatomy, physiology, etc., that fitness causally supervenes on. This is why fitness is primitive in the theory of natural selection and is defined or at least characterized only by these other theories.

The theory of natural selection, however, neither can nor should take sides on which of the indefinitely many causally relevant features of a situation are the ones that determine the fitness of an organism in that situation. Another way of putting this point is to note that fitness is a disposition but unlike magnetism, for instance, it is one without a manageably specifiable base in occurrent properties. The theory of natural selection provides no such base. The only occurrent properties that figure in it are the very ones, survival and reproduction, that threaten the theory with vacuity when connected to fitness as its direct definition or occurrent base; and there is no acceptable, manageably describable base in occurrent properties of organisms and their environments specifiable elsewhere in the life sciences that can be annexed to the theory in a way that will define fitness.

Interpolating a disposition between fitness and actual survival and reproduction severs the direct logical connection between these three notions. But it does so only by introducing a fourth term, the disposition, which itself is unaccounted for in the theory, and accounted for in only piecemeal ways outside it. So this propensity opens up again the prospects for the circularity charges it was meant to forestall.

Within the theory, there are no more resources for providing a noncircular explication of the propensity to reproduce than Molière's physicians could provide for the dormitive virtue of opium. The propensity is said to be the cause of the differences in actual rates, is said to be transmitted from ancestors with the same propensity, and its strength differs within and between species. How can we know all this about the propensity? Through its causes and effects reflected in actual retrospective and prospective rates of reproduction. But this is the circularity problem all over again. Of course, we can widen the circle by insisting that 'the propensity to reproduce' is a primitive term with respect to the theory and is to be expounded by piecemeal appeals to other theories. But then the claim to have provided a definition for 'fitness' will be hollow triumph. True, 'fitness' will be definable in the theory, but its immediate definiens, 'expected propensity to reproduce,' will prove to be primitive. Under these circumstances, it would be less misleading to simply admit that 'fitness' is the primitive term and, corrected reproductive propensities are the units in which it is measured.

6.3 The Supervenience of Fitness

Fitness is a primitive in one theory. What are its relations to other theories in the rest of biology? The answer should be obvious given the relative independence of the theory of natural selection from the vast and heterogeneous body of theories and findings that provide piecemeal corrections of its fitness estimates, that specify the mechanism of the heredity it requires, and that describe the environmental forces that eventuate in evolution. The relationship evinced is identical to the relationship between functionally characterized terms and their disjunction of alternative underlying mechanisms; it is identical to the relationship between Mendelian properties and molecular genetics; and it is characteristic of the autonomy of biological theories and the processes they describe at varying levels of organization. The way to express this relationship is to note that the fitness of organisms is *supervenient*, in exactly the same sense as specified in Chapter 4. It is supervenient on the manifest properties of organisms – their anatomical, physiological, and behavioral properties – right down to their molecular constituents and their interaction with physical properties of the environment.

Recall the characterization of supervenience given in Section 4.8. Applied to the notion of fitness, it tells us first that if a given fitness level is supervenient on a given set of manifest properties of organisms – their anatomical, physiological, and behavioral properties – and on properties of their environment – the weather, geography, and other competing species, etc. – then no two distinct objects that share the same manifest properties can differ in fitness level. This is just the result that we want: Fitness must be a definite, objective feature of organisms, causally generated by other objective features of them and their environments. But, secondly, supervenience implies that it may be the case that no particular fitness level can be defined or connected to a manageably small number of basic nonevolutionary properties. This, too, is, as we have seen, a feature of fitness and one of the chief arguments for its status as a primitive term in our formalization. Proceeding more formally, we can establish the supervenient reducibility of fitness to manifest properties of organisms in just the way we did for Mendelian and molecular genes.

Consider the notion of fitness as a quantitative property, which can take on a continuum of numerical values. In effect, these values constitute a denumerable infinity of fitness predicates. Call this set of properties F (for fitness). Now consider the set of all the anatomical, physiological, and behavioral properties that it is physically possible for an organism and its environment to manifest. Call this set P (for physical). Let P' be the set of all properties constructible from P by any combination of conjunctions and disjunctions (and their negations) of finite or infinite length. Among the members of P' will be the P-maximal properties of P. An organism that manifests one of these exhaustive properties will be fully and completely characterized, both positively and negatively. Recall that such P-maximal, exhaustive members of P' will be mutually exclusive properties of organisms as well. Call this set of exhaustive, mutually exclusive, P-maximal predicates constructible from P, P^*. Now the set of properties F, the properties of having various fitness levels, is supervenient on the set of properties P if the following is true: When two organisms share the same member of P^*, the P-maximal properties constructible from the set of anatomical, physiological, behavioral, and environmental properties, P, the two organisms share the same property in F, that is, the same level of fitness. Thus an organism's level of fitness, $\emptyset_j$, is not dependent on or identical with its past or future actual, expected, or adjusted rates of reproduction but is a matter of the organism's and its environment's having a particular set of nonevolutionary properties in P. If an organism o has fitness level $\emptyset_j$, then there is a set of properties in P such that o and its environment have this set of properties, and anything else with the same set of properties has fitness level $\emptyset_j$. Although fitness is thus determined by features unmentioned in the theory of natural selection, it cannot yet be said to be reduced or reducible to these features. The members of P realized by o in a given environment are only jointly sufficient conditions for particular levels of fitness, and they may not be necessary and sufficient; but it is biconditionals, stating necessary and sufficient conditions, that are required for reduction. However, such statements of equivalence will be at least in principle constructible from members of P^*, the set of exhaustive and exclusive properties constructed from P, if the members of P are finite in number. To see this, suppose that organism o has fitness level $\emptyset_j$. Then, since fitness levels supervene on properties in P, there is some member of P^* that o manifests, say P_o^*, and if any other organism instantiates P_o^*, then it too has fitness level $\emptyset_j$. But as we saw in Section 4.8 this is tantamount to concluding that each member of P^*, the exhaustive and exclusive properties constructible from P, is sufficient for some level of fitness. If the set of members of P^* is finite, then a finitely long enumeration of the members of P^*, each of which is sufficient for some particular level of fitness, will provide both necessary as well as sufficient conditions for each level of fitness. Of course, if the number of members of P^* is very large or infinite, then such statements of equivalence will be either practically or logically impossible to construct. But the important point is that any particular level of fitness is a function solely of the manifest properties of organisms, and the function in question is that of supervenience. This explains how fitness can be *nothing more than* having a certain combination of anatomical, physiological properties in a certain environment; even though no set of such properties may be *statably* necessary and sufficient for a given level of fitness; even though differing organisms in different habitats, with differing prospective and retrospective reproduction rates, may have identical levels of fitness; and even though we may be unable to cash any particular level of fitness in for a complete and specific set of such manifest properties of organisms.

By appeal to the supervenience of fitness on physical, behavioral, and environmentally relative properties, we can account both for the employment of fitness to explain differences in reproduction rates and for the *intra-* and *interspecies* comparisons of fitness that biologists appeal to as the mechanism of successful competition, predator–prey relations, biogeography, niche occupancy, and other aspects of selection and evolution. A tighter connection to rates of reproduction deprives fitness of its explanatory force. On the other hand, a tighter connection than that of supervenience between fitness and the manifest traits of organisms would deprive the notion of its systematic employment. We will be unable to use it in the explanation of how differing organisms in similar or different environments can both survive and compete successfully against other organisms and of how differing organisms can supplant one another in the same environment. Only on the assumption of supervenience can different combinations of manifest and environmental properties constitute the very same level of fitness. Yet it remains entirely consistent with this flexibility in the relation between manifest properties of the organism and its level of fitness that any organism's *particular* level of fitness *at a given time* consists in, is identical to, nothing more than the organism's physiological, anatomical, and behavioral properties and the environment in which it finds itself.

Thus, although at the level of the theory of natural selection fitness is appealed to in order to explain differences in rates of reproduction. Levels of fitness of particular organisms are in turn explained by appeal to theories about their physical and behavioral properties and about the relation of these properties to the organism's environment. This enables us also to see more exactly the relation between the theory of evolution and those other biological, behavioral, and physical theories that, together with a particular account of heredity, *explain* the fitness levels of organisms.

The theory of natural selection rests on these theories because they can schematically explain each of its axioms. Because fitness consists in manifest properties of organisms regimented in present and future scientific theories describing these organisms, and because the hereditary transmission of (some of) these properties and the occurrence of variation among them also supervenes on these same theories, the leading principles of the theory of natural selection should be explainable on the basis of these theories. And indeed our axiomatization of these principles permits precisely this. Thus consider the axioms elaborated in Chapter 5.

Axiom 1. *Every Darwinian subclan is a subclan, that is, is composed of biological entities.*

This axiom is grounded in those biological theories that account for varying modes of reproduction – of genes, of organisms, and of populations or larger groups of organisms. These theories are to be found in physiology and in molecular biology. They explain why and how the members of a Darwinian subclan satisfy the parent-of relation and constitute a biocosm, in Williams's sense.

Axiom 2. *There is an upper limit to the number of organisms in any generation of a Darwinian clan.*

This "Malthusian" axiom finds its ultimate explanation in the physically determinable limits on the consumption of finite amounts of energy, for these limits determine extreme upper population limits.

Axiom 3. *For each organism, there is a positive real number that describes its fitness in a particular environment.*

Each organism has a certain proportion of those properties, dispositions, and abilities, which are the causal determinants of its reproductive opportunities and thus of the number of its offspring. Because fitness is supervenient on these properties, this axiom can in principle be explained by theories that account for why particular organisms have certain proportions of these properties and how they eventuate in reproductive differences. But because fitness is supervenient on these properties, different organisms can have the same level of fitness in a given environment, and the same organisms can have different levels of fitness in different environments.

Because fitness is determined by those properties causally responsible for its effects in reproductive opportunities and successes, and because genetic theory assures us that some of these properties are hereditary, it should be expected that:

Axiom 4. *If (a) any Darwinian subclan, D, has a subcland D_1, and (b) D_1 is superior in fitness to the rest of D for sufficiently many generations, then the proportion of D_1 in D will increase.*

The fifth axiom asserts that the antecedent of Axiom 4 is fulfilled. It is the assertion that fitness is sufficiently hereditary to make for this secular increase in the proportion of D_1 within the entire population.

Axiom 5. *In every generation of a Darwinian subclan D (that is not on the verge of extinction), there is a subcland, D_1, and D_1 is superior in fitness to the rest of D for long enough to ensure that D_1 will increase relative to D; and will retain sufficient superiority to continue to increase, as long as it is not fixed as the whole of D.*

This axiom is to be explained by theories that assert differences between the features of biological entities and their environments, theories that establish the heritability of these differences, and ones that establish their effects on the differential reproduction.

Our discussion of the supervenience of fitness has proceeded at exclusively one level of interpretation of the theory of natural selection and of the term fitness: We have examined the repercussions of the thesis that the fitness of organisms is supervenient on their physical properties and that of their environments. But the theory is susceptible of alternative interpretations, and we have already instanced two in the previous chapter (it governs the evolution of genes and of populations of organisms), just in case these items satisfy the constraints on being biological entities. Thus, the fitness of a gene is supervenient on its properties: Its molecular structure, the particular sequence of base pairs out of which it is composed; and its environment – its molecular milieu, its pH, the concentration of other chemical constituents in the milieu, the presence of other genes whose products regulate its activities. For all these forces and systems work together with it to produce and operate the organism that houses the gene and through which it survives to replicate and transcribe further.

Thus, looking back at our account in Section 3.1, of why DNA contains thymine whereas RNA contains uracil, we can complete the evolutionary explanation of the greater fitness accorded by thymine in DNA and uracil in RNA. In particular, the difference in fitness of the thymine-based DNA over the uracil-based DNA is supervenient on the following facts: Cytosine can spontaneously deaminate, such deamination when unchecked produces mutations that interrupt function in subsequent replications; but there is a uracil-DNA glycosidase that removes uracil bases from DNA while leaving thymine bases undisturbed. Similarly, the presence of the uracil-DNA glycosidase as a DNA-product can itself be explained by noting that it confers greater fitness. This greater fitness is supervenient on several of the same molecular conditions that explain the fitness of thymine in DNA. But because the relationship between the adaptive properties of genes and their molecular ones is only supervenient, we cannot expect that this particular account will be generalized as a model for the explanation of all evolutionary claims about macromolecules.

On the other hand, we can now complete the reductionist's argument by discharging its allegedly irreducible commitment to the fitness of certain molecular primary sequences. Recall the account of the blood's function – with respect to the transport of oxygen, hydrogen ions, and CO_2 – that appeals to the primary structure of hemoglobin. It was noted that 9 out of the 140 or so amino acids of the primary sequence were conserved across species because of their crucial function in determining secondary structure. But this functional explanation for their persistence is underwritten by evolutionary considerations and in particular by attributing to these 9 amino acids adaptive value, greater fitness. Now the axiomatized theory of evolution explains why this fitness difference should lead to the conservation of these amino acids in the sequences of hemoglobin across all the mammalian species thus far examined. But these fitness differences are in turn supervenient on the chemical facts about the amino acids that can be polymerized into a polypeptide – and on the molecular consequences of such polymerization. The adaptive differences are also supervenient on the differences in the primary sequences of the hemoglobin genes that code for these polypeptides and on the regulatory genes that enable them to be transcribed and translated into the hemoglobin molecules. Thus, the whole story of why some amino acids in the hemoglobin molecule are conserved can ultimately be told by appeal to an evolutionary theory that is itself nonteleological in form and concept. It can be told by appeal to a notion of fitness that is reducible at least in terms of supervenience to purely physical properties of the molecular constituents of biological systems.

Appeals to fitness may in fact recur at other levels of biological organization. Thus an evolutionary account of species selection, if there is such a thing, may involve attributing differences in fitness between whole species. Such attributions will be supervenient, in part, on claims about fitness differences between organisms within these species. But the full account of these differences may hinge on attributing differences in fitness to the supervening components of these organisms, and eventually on the attribution of fitness differences at the level of the genome. These differences will in turn supervene on biochemical differences at the level of the hereditary material.

It appears to be the case that the polynucleotide is the universal carrier of hereditary information. Apparently, among all the molecular alternatives available in the primal soup at the beginning of terrestrial evolution, it proved to be the fittest and so evolved in accordance with evolutionary theory. Now if it did so, then there must have been a hereditary mechanism for the transmission of whatever properties its superior fitness supervened upon. For evolution requires hereditary mechanisms. But this is puzzling, for I have argued that although the existence of hereditary mechanisms is required by the theory of natural selection, their particular character is itself determined by selective forces governed by this very theory. So, it would appear, that in the chicken-and-egg sequence of hereditary and selection, selection comes first after all. For it appears that we will need an evolutionary explanation for the hereditary mechanism of the initial biological entities on the earth, or on any other evolving biosphere for that matter. This conclusion, however, does not follow.

If, as is reasonable to believe, there is only one physically possible initial replicating scheme physically consistent with the constituents and conditions of the primal soup, then there is no range of alternative hereditary mechanisms from which natural selection can select. There is no choice. There is only one possible hereditary mechanism at this point of evolution. There is no scope for an adaptational explanation of its appearance in terms of competition with other mechanisms. There must be a nonevolutionary explanation for its appearance, of course, but the evolutionary buck really does stop here. At this fundamental level of biological phenomena, the disjunction of a vast and heterogeneous set of alternatives gives way, and it finally disappears, leaving only one way that the cat is skinned: the way determined by the operation of the laws of physics and chemistry on the boundary conditions of the origins of life.

6.4. The Evidence for Evolution

Let us turn from the expression of the theory of natural selection to the question of its truth and the evidence that can be offered in its favor. We shall see that the treatment of the theory as an axiomatized system whose key explanatory term is undefined, but supervenient on findings in other theories, helps us to understand the difficulties involved in marshaling general evidence for the theory. These difficulties are not belabored with a view to undermining the theory. Its evidential warrant is in my view beyond question. Moreover, the very question of whether it is well confirmed is a matter best left to biologists. The point of this section is rather to strengthen further the case for the view of evolutionary theory hitherto developed, not to discredit that theory.

A serious obstacle to finding evidence that bears directly on the general claim that evolution results from differential selection over heritable variations is itself biological. That is, when the temporal rate of reproduction of organisms under observation is close to our own, and when environmental conditions are subject to changes of geological slowness, the individual *Homo sapiens* will simply not live long enough to detect many changes that confirm the theory. These facts have two related consequences for the theory. They permit its exponents to explain away the absence of confirming evidence from the evolution of middle-sized mammals, whose evolution interests us most and would provide the psychologically most striking and forceful confirmation of the theory and of its explanatory range. On the other hand, evolutionists' temptation to cite these facts as excusing the absence of certain sorts of evidence can be turned to their disadvantage by an opponent of the theory who wishes to charge it with unfalsifiability. Nevertheless, there have been detectible changes in wild populations that strongly confirm the limited claim that evolution at least sometimes proceeds in accordance with Darwinian mechanisms. Standard examples include industrial melanism in moths and evolved resistance to myxomatosis among Australian rabbits. But in these cases the evolutionary change is very small and exclusively intraspecific. They provide no evidence for interspecific evolution. They cannot decide between a theory that allows for intraspecific evolution through natural selection while providing a nonselective explanation for species diversity. Yet it is unambiguous evidence for the evolution of distinct species from one another that the sternest critics of the theory demand. This demand, as I shall show in Chapter 7, is misconceived. Nevertheless, the absence of evidence from wild populations for anything more than the existential claim – that natural selection over heritable variation sometimes obtains – is serious. For the contemporary Darwinian theory asserts not the mere occurrence of this sort of evolution, it claims ubiquity for it; moreover, it is widely held to claim that all evolution proceeds not only by selection but by selection over *small* variations at a constant rate. Because of these features of the theory, and because of the facts about the biology and environment of the agents that test the theory, recourse to other arenas of confirmation is essential.

For Darwin himself, one of the most crucial of these further arenas of testing is to be found in the phenomenon of artificial selection. The appeal of Darwin and his successors to artificial selection as evidence for the theory raises three questions: Does agricultural artificial selection confirm evolutionary theory any more strongly than studies of wild populations? Can we employ the horticulturalists' and animal breeders' techniques to design experiments that will confirm the theory? Can we simulate natural selection in the laboratory and thereby confirm the theory?

Biologists have traditionally viewed the connection between artificial and natural selection as at best an analogical one, a suggestive comparison, useful for expounding evolutionary theory but not for justifying it. The two sorts of selection bear similarities (both manifest differential reproduction with cumulative effect involving relatively slight intergenerational variation). But it has been held that there are significant disanalogies. Moreover, from the fact of artificial selection, it is argued, one cannot conclude that *natural selection* ever does in fact obtain, nor that it selects the same kinds of variations sought and attained by breeders. Breeders of dairy cattle select for short horns and high butterfat, but this does not show that either of these traits is subject to natural selection. The most such results show is that some types of selection *can* have effects. They neither show the existence nor the direction of natural selection. So, at any rate, it is argued.

These arguments are too strong. It is quite true that artificial selection cannot reveal the direction of natural selection, but it certainly does more than confirm its bare possibility.

It is clear that in advancing the theory of natural selection Darwin argued analogically from the existence of artificial selection to the possibility of natural selection. However, on the assumption that Darwin's theory is correct, the distinction between these two sorts of selection is not that of exhaustive and exclusive difference. Rather, the distinction is that of general case to special case. Artificial selection is not another form of selection, different from natural selection; it is a species of natural selection. The forces operating in artificial selection are not just similar to those acting in natural selection, they are a special subset of those forces. The operative force in artificial selection is a systematic set of behavior of members of one species (*Homo sapiens*) that determines the reproductive success of members of other species (e.g., *Canis familiaris, Felis domesticus,* or any other domesticated species). Regularities in the behavior of a predatory species determine the differential reproductive successes of members of its prey species, and this behavior is rightly deemed a natural-selective force. It differs from other selective forces, like climate, geological stability, available food supply for its prey, etc., but it is undeniably a selective force. How are we to distinguish the literal selection for fitness of the human plant and animal breeder from the metaphorical selection for fitness of the nonhuman predator? Surely the fact that in the former case the causal variables include the conscious agricultural intentions of the *Homo sapiens* and in the latter case the causal variables do not include such conscious purposes is irrelevant. It is no reason to distinguish artificial and natural selection as mutually exclusive mechanisms of evolution. To suppose otherwise is tantamount to erecting that barrier between *Homo sapiens* and the other species that Darwin's theory did the most to bring down.

But if artificial selection is a type of natural selection, then the claim that it has only analogical bearing on evolutionary theory is either too weak or false. It will clearly turn out to be false that the existence of artificial selection does not imply the existence of natural selection. For if the former is a species of the latter, whose existence is obviously assumed by all parties in the present dispute, natural selection occurs if artificial selection does. Similarly, it is wrong to say that the selection of traits that breeders make reveals nothing about the selection for traits that nature makes. Because artificial selection is a form of natural selection for traits that nature makes, it will include the ones that humans make. The claim that artificial selection shows only that natural selection *can* have big effects while operating on small variations is clearly too weak. For artificial selection reveals more than a bare possibility. It demonstrates the actuality of large changes arising through natural selection on small variations. Finally, insofar as we treat artificial-selection experiments as a source of data *prima facie* different from the observations of wild populations, they clearly strengthen Darwinian theory's claims for the ubiquity as well as the existence of the mechanism of natural selection.

Darwin's reasoning by analogy from artificial selection to natural selection in the initial chapters of *On the Origin of Species* (1859:chap. 1) was of course perfectly in order, and it represented a heuristically invaluable means of introducing an initially implausible theory of evolution. Indeed, had Darwin argued from the outset that artificial selection was but a special case of natural selection, he would quite plausibly have been accused of begging the question of how the former worked. But we must distinguish the introduction or presentation of the theory from its content; especially when we set out to test it. And it seems a clear consequence of the content of the theory that artificial selection is but a form of natural selection. This strongly augments the bearing of artificial selection as evidence for the hypothesis of natural selection. (For an excellent account of the introduction of Darwin's theory and the analogical role played therein by artificial selection, see Ruse, 1979.)

The unwarranted distinction erected between natural and artificial selection also generates unwarranted skepticism about whether one can devise experiments that will support the theory of natural selection. Biologists often answer this question with a cautious yes. But the caution is more than that invariably invoked against closed laboratory investigations of phenomena difficult to monitor unambiguously in nature. Any laboratory experiment in even so well tested and relatively unproblematical a domain as that of Newtonian mechanics is attended by cautions. Inferences from its results may seriously underestimate the complexity of the "same" type of phenomenon beyond the laboratory. But cautions in biology often go further. And again the caution is based on the supposition that inferences from laboratory selection to natural selection in wild populations are analogies and not inferences from particular cases to general hypotheses. For instance, we can establish geographic isolation in the laboratory and thus produce reproductive isolation, or regulate food, space, air temperature and pressure, or wind velocity, and detect changes in the distribution of phenotypes among, say, *Drosophila*. From such experiments we may learn how natural selection *could* proceed, at what rate it might do so, and with what potential consequences. But, it may be held, nothing follows from these experiments about *how it does* in fact operate. Again, such conclusions are unreasonable. It must be admitted that biological experiments are more complex and more removed from natural settings than many physical ones. But surely the differences between selection experiments and, for instance, laboratory experiments to test the constancy of gravitational attraction are matters of degree and not of kind. If this is so, then either we must say that terrestrial experiments merely show that gravitational attraction *could* be a universal constant, or we must say that selection experiments in the laboratory show the direction that natural selection does take. For differences of degree in complexity of experiments cannot make for modal differences in the form of conclusions about possibilities instead of actualities drawn from them. They can only make differences in the strength of our rational beliefs about the conclusion.

The special importance of laboratory selection experiments is that they increase our evidence for the extent of evolution by natural selection. Observation of wild populations establishes the occurrence of Darwinian selection. But there are natural obstacles to the human accumulation of evidence as to its extent or ubiquity. By replicating in the laboratory various naturally occurring phenomena hypothesized to act as selective forces, we can provide evidence for the extent as well as the existence of natural selection.

Nevertheless, behind the doubtful arguments for treating laboratory tests of the theory of natural selection with caution stands the implicit recognition of a real problem facing the confirmation of the theory. For in the laboratory experiments that provide the most detailed and precise confirmation of the theory of natural selection, that theory fills no explanatory need. And its key notion, fitness, turns out to be superfluous to the best descriptions of the experimental phenomena. To see this we must return to our comparison of fitness and temperature.

Temperature is a property simple and assessable enough to be entrenched in scientific theories that are now two hundred years old and more. Once temperature was connected to the mean kinetic energy of the molecules of a gas, thermodynamics was freed from theoretical dependence on such temperature-measuring devices as the skin and the alcohol thermometer. Indeed, the situation was reversed, and the operation of at least some of these measuring devices became explicable. More important, the theory could be employed together with temperature measurements to make predictions about how a system responds to thermodynamic disturbances of varying kinds. And of course where the theory cum measurement was disconfirmed, the measurement was more often than not rejected or corrected. By contrast, Darwin could make no particular predictions or for that matter retrodictions on the basis of evolutionary theory *alone,* or at least none *specific enough* to confirm the theory in the opinion of those not already wedded to it. The importance of Darwinian theory does not lie in its predictive strength, for, as we have seen, the nature of its key explanatory concept, fitness, precludes such strength. The importance of the theory lies in the freedom it provides biologists to view natural phenomena as just that, as natural, and not as the creation of an artificer with designs for natural phenomena. It was only with the advent of other nonevolutionary theories that the theory of natural selection could generate specific predictions substantial enough to confirm it. Part of this predictive power is provided by specifying a mechanism for hereditary transmission, like Mendelian genetics. For this theory provides an independent means of identifying characteristics subject to inheritance. What accounts for evolution is the fitness or lack of it these traits confer.

And differences in fitness can be identified on a case-by-case basis through appeal to piecemeal theory and other unsystematic means. For example, an *Escherichia coli* bacterium resistant to tetracycline is fitter in a tetracycline-rich environment than one lacking this trait. So we may use nonevolutionary, design considerations to predict that one is more likely to survive and reproduce than the other. But these case-by-case determinations will not enable evolutionary biologists to make many of the comparisons they want to make. They cannot say whether a tetracycline-resistant bacterium is fitter than an ampicillin-resistant one in the ambient environment when the respective environmental level of each of these drugs is unknown. To answer this question, they must examine growth rates of the bacterial cultures; that is, they must measure population changes. But this involves them in the circle of evidence and explanation that the notion of fitness engenders. This circle deprives the theory of tests that employ measuring "devices" independent of the theory. Only in simple laboratory settings, involving huge numbers of very simple organisms, whose genomes are relatively small and well known, in environments subject to complete control, can the theory of evolution be applied, tested, and confirmed *with all the precision* that is demanded by its critics and desired by its defenders.

But everything that the theory of natural selection can explain about what is happening in a well-controlled laboratory experiment can be explained more deeply, more directly, and in greater detail by physiological and biochemical principles that *do not mention the supervenient evolutionary concept of fitness.* When enough theoretical and experimental detail has been gathered to make a prediction that specifically confirms the claims of evolution about the maximization of fitness, the theory of natural selection and the notion of fitness become *superfluous:* They are no longer required to effect the prediction or to explain the occurrence of the predicted phenomenon. The prediction that can be extracted from the theory in such cases is at best generic, and the explanation it provides will be qualitative at most. The theory of natural selection and any appeal to evolutionary fitness are superfluous in these settings because in them biologists are already able to identify and measure directly the determinants of differential reproduction. They can do so without making a detour through estimates of fitness based on these reproductive differences. Although it is superfluous in these cases, the theory is also strongly confirmed by them just because in these laboratory experiments the biological processes are so rapid and so protected from the intervention of unknown or unexpected forces, and the number of organisms is so huge. Therefore, what evolutionary theory tells us will happen in the long term, happens in the short term, and happens invariably (provided the experiment is well designed). Outside the laboratory, biological processes are slow, intervention of unknown forces is the rule, the number of organisms is small, and we must have recourse to fitness if we are to explain evolution at all.

The superfluousness of the theory of natural selection for explaining and predicting evolution in the biochemist's laboratory is reflected in the reluctance many show to accept laboratory experiments and simulations as tests of the theory of natural selection. Experiments are of course tests of it, but tests in which the predicted events can be explained in greater detail by nonevolutionary theories of biochemistry, cell physiology, bacteriology, etc. Indeed, the theory of natural selection gets its best confirmation in these laboratory settings where it is not actually needed at all. Its confirmation in contexts like paleobiology, ethology, and sociobiology, where it is required, is much less precise and detailed.

It is worth noting that, if the theory of natural selection can only be decisively confirmed in the artificial setting of the laboratory, then it is condemned to perpetual dispute by its opponents. For they want claims about the evolution of middle-sized organisms mentioned in the book of Genesis to be tested with an allowable range of error that would do credit to astronomy; so of course they will remain skeptical. Proponents of the theory freely employ it and are at least content that the theory has been tested and never yet disconfirmed. It is no defect in evolutionary theory that it can issue in no more than generic predictions and post facto explanations. For this results from the contingent fact that the determinants of fitness are too diverse and complex to permit us to improve the powers of the theory. It would, however, be a mistake to conclude that no improvements are possible in our explanations and predictions of evolutionary phenomena. But such improvements are forthcoming only by passing beyond evolutionary theory and focusing on the discovery of theories about particular members of the vast and heterogeneous class of determinants of fitness: theories in nonevolutionary, functional biology.

6.5. The Scientific Context of Evolutionary Theory

Theories are not assessed in isolation. Successful ones are survivors in a competition with other alternatives. And the theory of natural selection is widely held to be strongly confirmed because it fares better than its competitors in the light of certain biological findings. The theory of natural selection held to be thus confirmed is not our axiomatized one, but the synthetic theory, harnessing selection together with Mendelian genetics. Among these competitors, at least three have had responsible biological proponents in the post-Darwinian period. The three theories are Lamarck's hypothesis of the hereditary transmission of acquired characteristics; the saltationist theory, according to which evolution proceeds not exclusively or even mainly through selection on small variations, but at least sometimes through the appearance of large heritable changes, so-called macromutations; and the orthogenic theory, which has it that evolution is the result of changes in heritable variations that invariably take them beyond adaptiveness toward extinction-producing maladaptation. In our assessment of claims about the greater confirmation of the synthetic theory by comparison to each of these, it must be remembered that each of these theories may be held to assert the existence of phenotypes, of heritable properties of organisms, and each will require at least the existence of a mechanism to provide for the invariable transmission of hereditary properties in the absence of forces of evolutionary change. It seems safe to say that biologists generally reject these alternatives to the synthetic theory because they all fail to account for the evidence of genetics and cytology. But if evolutionary theory must be distinguished from a theory of heredity, then this is not a good reason to prefer the theory of natural selection. For by themselves cytology and genetics only bear on heredity, and not on selection. Indeed, this approach obscures the theory's relations to other biological findings that do confirm it. Distinguishing evolution from heredity in the way the axiomatization of Chapter 5 does should help us better understand the evidence for evolution.

The cytological evidence in question is the microscopically observed cellular phenomena of mitosis and meiosis. The genetic evidence is taken to be the transmission and distribution of phenotypes in sexual species in rough accord with the Mendelian laws of segregation and independent assortment. Now it is clear that the relation between the existence of mitosis, and meiosis, and the heritability of traits in accordance with Mendelian ratios is not logically incompatible with any of the three competitors to Darwin's theory. Indeed, all of them require the existence of traits with varying but considerable degrees of heritability. Therefore, they all require some theory of heredity and some account of the cellular mechanism of hereditary transmission that meiosis provides and of the developmental ontogeny that mitosis affords. Moreover, it would take little ingenuity to create bridge principles that establish connections between these non-Darwinian theories and descriptions of cytological and genetic phenomena; in fact, Section 5.3 did so for Lamarckian theory. In what sense, therefore, does the theory of natural selection account for these cytological phenomena when its competitors do not?

The phenomena of meiosis and mitosis, the Mendelian laws, and the theory of natural selection are interconnected in a vast network of biological findings and theories, as well as auxiliary hypotheses and theories drawn from chemistry, optics, x-ray crystallography, ecology, etc. Relevant parts of the network may be sketched as follows: The existence of meiosis and mitosis is inferred from indirect observations of theoretically identified organic material, which has usually been stained by mixture with a dye, prepared on a slide, mounted on a microscope, and then manipulated until a theoretically calculated degree of resolution is attained. Conclusions about the existence of the cytological phenomena occurring in the nuclei of the cells on the slide turn on independently substantiated assumptions about physical optics, on generalizations about the effects of dyes on organic material, on our observation of this material, and on principles of zoological identification and cellular anatomy. Once an agreed description of the topography of these cellular occurrences (the stages of meiosis and mitosis) is in hand, the principles of Mendelian genetics may be appealed to in order to functionally explain the topographical order of events in meiosis and mitosis. They may also explain the consequencés of varying sorts of breakdowns (like nondisjunction) in these processes. But the complete explanatory relation between Mendelian genetics and the cytological phenomena is extremely complex and involves appeal to at least the following further assumptions: The findings and methods of molecular and fine-structure genetics, the physical regularities governing electron microscopy, x-ray crystallography, electrophoresis, DNA hybridization, and autoradiography. It is only together with appeal to the vast theoretical edifice of physics and chemistry that Mendelian genetics "accounts for" the cytological phenomena. Moreover, as we have seen in Chapter 5, although Mendelian laws may functionally explain the cytological phenomena, these in turn provide a structural explanation for the Mendelian laws! Of course there is further independent evidence for the Mendelian laws to be found, for instance, in agricultural experiments of the type that led Mendel to frame these laws. And without such further evidence, of course, the mere fact that, together with other assumptions, Mendel's principles functionally explain the topography of cytological phenomena is little reason to embrace these principles.

Moreover, if the synthetic theory of evolution is but the conjunction of a theory of natural selection and an independent account of Mendelian inheritance, then the confirmation of the Mendelian principles provides only the weakest evidence for the conjoined theory of natural selection. On the other hand, if the laws of independent assortment and segregation must be explained by the theory of natural selection, then it will share some of the confirmation that the phenomena of meiosis and mitosis indirectly provide these "laws." The relations between these "laws" and natural selection is very far from being direct. It is mediated by further auxiliary assumptions over and above those already invoked to link Mendel's principles and cytological phenomena. As noted several times already, the theory of natural selection is neither presupposed by nor presupposes Mendel's laws. The undisturbed behavior of genetic material in accordance with Mendel's laws not only does not produce evolutionary change, it *precludes* it. The Mendelian law of segregation implies the Hardy-Weinberg law, according to which gene ratios must remain constant across indefinitely many generations, ceteris paribus. Under what *additional conditions* will the operation of Mendel's laws provide support for the theory of natural selection?

To arrive at the phenomena of natural selection, we must add practically the rest of biological, geological, and meteorological theory, known and unknown. For natural selection involves the survival of the fittest and fitness is supervenient on this vast body of theory. Differential fitness is given by appeal to the consequences of causal variables of all these theories working together to determine the ability of the subject of selection to survive and reproduce. It is only at this point that heredity enters, and with it, for sexually reproducing species, Mendel's laws. If differences in fitness are hereditary, then in the course of successive generations, small variations in hereditary properties (caused by biochemical changes in nucleic acids, and their aggregates, that are random with respect to selection pressures — another auxiliary assumption) will result in natural selection. Mendel's laws give the mechanism of heredity required to account for gradual evolution among sexual species, but only when they are added to the rest of this edifice so vast that we will never see its details filled out. We believe that levels of fitness are determined by nonevolutionary facts about the subjects of selection — facts that have been, will be, or can be provided by all the rest of the theories describing the causal forces acting on individual organisms, species, or genomes. But we have no reason to suppose that this plainly unattained possibility can be converted to actuality than we have reason to believe in the uniformity and the expressible finitude of nature and its regularities. Because biologists nevertheless hold the relevant beliefs, they hold that, together with this perpetually unfinished body of theories, Mendel's theory does account for the operation of natural selection among sexually reproducing systems.

This conclusion may shed some light on the paleontological evidence cited against "Darwinian gradualism" by proponents of punctuated equilibrium theories. Because "gradualism" — evolution at a constant rate by small steps — is introduced by the Mendelian component of the synthetic theory, fossil evidence bears on this independent component. It does not undermine the evolutionary component of the theory of natural selection. The theory of natural selection seems to be established beyond reasonable doubt. The reason is that it is sustained by the joint operation of a larger number of independently confirmed theories, findings, and principles than almost any other theory available. Its evidential warrant is the product of what philosophers call a consilience of inductions. This is no surprise considering the fact that it governs the behavior of entities that are also subject to all the laws of physics, chemistry, and the rest of biology, whereas the converse is not true. Of course, had one or more of the findings and theories of the rest of natural science been different from what they in fact are, some other evolutionary theory might have been as well as or more strongly confirmed than Darwin's. Exploring what sorts of differences in the findings and theories of other areas of natural science would make for such conclusions is probably the best means of manageably detailing the evidence that confirms evolutionary theory more strongly than any other. Explorations that focus on the cytological evidence or on Mendel's laws are no more likely to reveal this evidence than the examination of other areas of science.

If the interrelations among the theory of natural selection and the other components of biological science are as complex as I have drawn them, the testing of the theory is a highly indirect matter, one that involves many other theoretical claims. It involves almost no direct assessment in terms of the observable diversity of past and present terrestrial life. Testing any theory is often and rightly said to be indirect, because it requires auxiliary hypotheses to ascertain that the boundary conditions required by the theory obtain, before we allow the drawing of a conclusion from the theory to count as a test. Thus, no test of Newtonian mechanics involving the motion of steel balls in a strong magnetic field would be acceptable, because such settings are outside the intended domain of Newtonian mechanics: It is not an account of magnetic phenomena. But to assure the absence of magnetic forces we need some means of detecting magnetic forces. If a test of Newtonian mechanics appears to disconfirm it, we will want to be sure that no undetected magnetic forces interfered with the mechanical behavior of the test system. Only if we accept at least an implicit theory about such forces and how to detect them can we affirm their absence. But in this case a disconfirmation may reflect mistaken assumptions about such auxiliary theories required to test the Newtonian hypothesis. Perhaps the hypothesis under test is not undermined by an apparent disconfirmation, because of the falsity of an auxiliary theory's assertion that magnetic forces are absent. Although in most practical contexts such alternatives are reasonably ruled out, it is important to be aware of the role of auxiliary theories in tests of a hypothesis, and of the fact that therefore such tests are "indirect" and involve inferences as well as straightforward observation. In the case of the theory of natural selection, any test of the theory, whether in the laboratory or the field, is indirect in that it presumes the test subjects are biological entities and satisfy the conditions of bearing the 'parent-of' relation. This is a relatively safe, but nontrivial, assumption.

However, there is a much stronger sense in which any test of the theory of natural selection must be indirect. For many organisms, especially the ones that interest us most, the predictive weakness of the theory is unavoidable and cannot be lessened by any practical improvements. So even the (indirect) tests by deduction of consequences for observation about particular organisms are seriously restricted, except in the artificial confines of the laboratory. The theory's evidence is therefore to be found not in its consequences, but in the other parts of biology and of physical science that "feed in to it," that explain its principles. Its confirmation is to be found in the degree to which it synthesizes these other bodies of biological knowledge and non-biological theory into the best available description of a mechanism for evolution. But this sort of confirmation is indirect because it accords warrant to the theory indirectly: by the influence of the evidence for the other theoretical bodies that explain it and that it brings into connection with one another. This kind of confirmation is obviously even more indirect than the deduction of consequences that can be tested and is also much weaker in one respect: No one piece of evidence "points" directly at the theory of natural selection. On the other hand, this sort of confirmation is stronger, for so many different pieces of evidence "point" in the general direction of this theory that surrendering it will have untoward consequences across the whole fabric of biology and indeed within the epistemological and metaphysical convictions of the rest of the sciences as well. For this reason, it is hard to imagine what a single piece of evidence that could seriously undermine the theory would be like. And by the same token it is easy to view the theory as untestable.

Because of the nature of its indirect confirmation, there is a lot of slack in the theory of natural selection. There is a great deal of scope for controversy that appears to be quite fundamental and immediate, like that between gradualists and punctuated evolutionists. There is also scope for controversies that are less immediate though apparently just as fundamental, like that between neutralist and adaptationalist hypotheses about genetic evolution. But once we see how complex and multifarious are the evidential supports of the theory, it becomes clear both how such lively controversies are possible and why they do not affect our general confidence in the theory no matter how they come out. Thus, a decision between neutralist and adaptationalist views of genetic evolution will still point in the general direction of natural selection for some genetic changes and for all organisms; a decision about the mode and tempo of evolution will settle questions about the values of parameters and the magnitude of variables, as well as the units of natural selection. But it will not affect the theory itself.

Still, for all these philosophical rationalizations of the theory's differences from theories in physical science, a nagging doubt must remain among biologists. Many reasons have been given for lengthening the distance between the theory of natural selection and the facts that it is actually called upon in biology to systematize. All this talk of Darwinian subclans and subclands, of fitness as a notion that cannot both be operational and predictive, of the specification of a very general abstract version of the theory, leaves open the question of exactly how the theory does link up with real evolutionary biology, with paleontology, with systematics, with ecology. If our account of the theory of natural selection accords it universality and generality at the cost of making it a stranger to biology, irrelevant to the agenda of this subject, then our accomplishment will be irrelevant to biology. At best, it will simply be a misconceived piece of merely philosophical speculation. It is to allaying such doubts that the next chapter is devoted.

Introduction to the Literature

An excellent discussion of the nature of scientific definitions is provided by Hempel, *Philosophy of Natural Science,* chap. 7. A deeper discussion of these matters is Hempel's *Fundamentals of Concept Formation* (Chicago, University of Chicago Press, 1952).

A lively discussion of the charge that fitness is a vacuous concept appears in S. J. Gould, *Ever Since Darwin* (New York, Norton, 1977), pp. 39–45. The difficulties of defining fitness in a nontautological way are illustrated in Hull, *Philosophy of Biological Science,* chap. 2, and Ruse, *Philosophy of Biology,* chap. 2.

The best exposition of the analysis of fitness as a probabilistic disposition is S. Mills and J. Beatty, "The Propensity Interpretation of Fitness," *Philosophy of Science,* 46(1979):263–8. R. Burian, "Adaptation," in M. Grene, ed., *Dimensions of Darwinism* (Cambridge, Cambridge University Press, 1984), analyzes fitness into three notions: fitness tautologically defined, fitness as optimal design, and fitness as a propensity. His paper illustrates all three of these meanings in the writing of biologists from Darwin to the present.

An influential discussion of fitness in the context of theoretical biology is R. Levins, *Evolution in Changing Environments* (Princeton, N.J., Princeton University Press, 1968).

A contemporary discussion of practical problems in the measurement of fitness can be found T. Prout, "The Relation Between Fitness Components and Population Prediction in *Drosophila,"* *Genetics,* 68(1971):127–49, 151–67, and "The Estimation of Fitness from Population Data," *Genetics,* 63(1969):949–67.

For discussions of supervenience in general, see J. Kim, "Supervenience and Nomological Incommensurables." This notion is applied to the analysis of fitness in A. Rosenberg, "The Supervenience of Biological Concepts," *Philosophy of Science,* 45(1978):368–86.

Ruse, *Philosophy of Biology,* most clearly represents the account of the evidence for evolution attacked in this chapter. But most of the works cited in the introduction to the literature for Chapter 5 are relevant to this subject.

CHAPTER 7

Species

The theory of natural selection is a theory about the evolution of species. This claim seems beyond dispute, but it suffers from a serious embarrassment, evident to the leading figures in contemporary biological science:

Darwin's choice of title for his great evolutionary classic, *On the Origin of Species,* was no accident. The origin of new "varieties" within species had been taken for granted since the time of the Greeks. Likewise the occurrence of gradations, of "scales of perfection" among "higher" and "lower" organisms, was a familiar concept, though usually interpreted in a strictly static manner. The species remained the great fortress of stability and this stability was the crux of the antievolutionist argument. "Descent with modification," true biological evolution, could be proved only by demonstrating that one species could originate from another. It is a familiar and often-told story how Darwin succeeded in convincing the world of the occurrence of evolution and how – in natural selection – he found the mechanism that is responsible for evolutionary change and adaptation. It is not nearly so widely recognized that Darwin failed to solve the problem indicated by the title of his work. Although he demonstrated the modification of species in the time dimension, he never seriously attempted a rigorous analysis of the problem of the multiplication of species, of the splitting of one species into two . . . [F]oremost among [his reasons for this failure] was Darwin's uncertainty about the nature of species. (Mayr, 1970:10)

The seriousness of this problem should not be underestimated. After all, it has sometimes been said that without a solution to the problem of species diversity there is no distinctive Darwinian theory of natural selection at all. Recall the six-part presentation of the theory of natural selection in Section 5.1. There, it was noted that these claims are so obvious that Darwin did not require a five-year voyage around the world to uncover them. Any rural setting in England would have vividly presented their operation. What is more, it has sometimes been claimed that any of Darwin's own most vigorous opponents could have or did subscribe to them, while still rejecting Darwin's theory of the evolution of species. Accordingly, either these six statements (and the axiomatization based on them) do not adequately capture Darwin's theory, and that of his successors, or Darwin's theory fails to solve the key problem that he claimed to have done: the problem of the diversity of species and its causal origin.

As Mayr notes, the source of this problem is the difficulty of providing a correct account of the nature of species. As we shall see, the best account of the matter dissolves, rather than solves, the problem of explaining the origin and diversity of

180

species. At the same time, it clarifies several remaining puzzles about the structure, explanatory power, and empirical confirmation of the theory of natural selection.

Darwin's uncertainty about the nature of species has persisted among his successors, and there is less agreement now about what a species is than there was in Darwin's time and before. This of course reflects a substantial increase in our knowledge about matters described by the term 'species.' Nevertheless, there can be no more serious cause for concern about the foundations of a discipline, its future prospects, and its current claims to knowledge than the admission that its key notion remains without an agreed theoretical significance over a century after the central theory in the discipline was framed. The point is not that the ordinary term 'species' has no commonly accepted meaning (quite to the contrary, it does), nor is it that evolutionary theory's claims about species are subject to popular misunderstanding or misrepresentation. Still less is it the case that the theory gives a precise technical sense to the term foreign to its ordinary meaning. No, the problem runs much deeper. In biology there is at present no general agreement on an explicit definition of the term 'species,' or on what counts as a logically or causally sufficient condition for a set of organisms to constitute a species, or for that matter on a necessary condition for species membership. Nor is there any agreed-upon operational mark of species membership. Most of the available definitions of species and the theories of speciation in which they figure are incompatible and, what is worse, inadjudicable.

The absence of any agreement on this notion is crucial because the biological discipline within which it does the most work, systematics, is widely thought to be the most fundamental department of the subject. Systematics (or taxonomy), the study of the diversity of organisms, the specification of their fundamental kinds or units, and the relations among them, seems central to all biology. Systematics is not only a necessary condition for scientific description of the very subject matter of the discipline; it is also the most encompassing, because of its employment of the findings and theories of every other compartment of biological science in the establishment of the types, kinds, and units required for scientific description. The presuppositions of traditional systematics are twofold:

1. There is a single correct description of the basic types of flora and fauna in the world.
2. This single correct description is not merely compatible with the rest of science, but actively coheres with it, as reflected in the conviction that the explanatory power of the uniquely correct taxonomy of terrestrial flora and fauna can be grounded in the rest of biology.

These two presuppositions parallel those of the periodic table of the chemical elements. This single correct systematization of the elements is crucial to all aspects of chemistry just because it can be explained by appeal to the fundamental underlying theory of chemistry: atomic theory. Systematics aims at providing the biologist with a "periodic table" at least as suggestive and exact as the one Mendeleev provided chemistry. That is why systematics embraces these two presuppositions. But these very presuppositions make the absence of agreement on the meaning of the notion of 'species' particularly acute. If the only suitable basic category of taxonomy is the species, and there is no theoretically grounded definition or characterization, operational or otherwise, of this notion, then there is no basic category for — no single correct description of – the basic types of flora and fauna. But this conclusion is tantamount to denying the possibility of scientific systematics, denying the possibility of manageable explanation of the most salient facts about diversity in any terms familiar to contemporary biology. One basis for such a conclusion would be the decision that presupposition 2 is wrong, that there is no single taxonomy that can be manageably related to morphology, physiology, behavior, or any more fundamental level of biological description, for example, genetic or biochemical.

Although some few biologists might interpret such a conclusion as still another argument in favor of biological autonomy, for the existence of irreducible levels of biological organization, most would consider a basic feature of biology jeopardized. Indeed, most biologists would hold that presuppositions 1 and 2 are so intimately linked that 1 cannot be embraced without holding a minimal version of 2.

Suppose that presupposition 1 is incorrect, because, among other reasons, presupposition 2 is false: In particular, there is no unique system of species, because there is no single or small number of causes of speciation. This will not obviate the problem set for Darwin in the extract above. If the notion of 'species' cannot be satisfactorily explicated, the problem of explaining 'speciation' – the origins of species – will disappear as such. The problem of explaining diversity will remain, but the explanation will not be required to explain biological diversification in terms of the notions of species and 'speciation.' Accordingly, it will be no defect of Darwin's theory that it cannot after all provide a univocal explanation for species diversity. For there is no single phenomenon of speciation to be univocally explained.

If presuppositions 1 and 2 are false, then the evolutionary biologist will be able to view diversity not as the consequence of one or a small number of factors or forces but as the outcome of a large number of very different ones. These factors may divide nature just as firmly as traditional species notions, but they may do so by acting along many different dimensions, dimensions that reflect the characteristic disjunctive heterogeneity of the biological realm. This conclusion would come as no surprise in view of its appearance at other junctures in this account of biology. In this chapter, we shall examine the various accounts of 'species' with a view to assessing these two presuppositions. Our conclusion should finally make clear the nature of biology's limits and its differences from physical science, differences that sustain neither autonomy nor provincialism.

7.1. Operationalism and Theory in Taxonomy

The notion of species seems to be ambiguously between a theoretical term in the theory of natural selection and a theory-neutral term employed to describe the most important or most obvious classes of individuals. In the latter sense, the term is employed to describe phenomena that evolutionary theory can be expected to explain and that can be expected to test the theory. In this usage, it is theory-neutral in that its definition should not be expected to presume that one or another theory of evolution or, more particularly, of speciation is correct. A mark of this neutrality is the fact that the term long antedates Darwin's or any other competing evolutionary theory. Indeed, it was once used to describe a diversity of phenomena no one supposed to have been the product of evolution at all.

It may appear desirable that any contemporary definition of the notion preserve these features. For only a theory-neutral notion of species could provide a description of biological phenomena that did not beg the question of theory confirmation in favor of one theory or another. Moreover, such a definition would make sense of the history of biological enquiry – stretching back past Darwin and Lamarck to Linnaeus and beyond all the way to Aristotle – by showing that its subject matter remained commensurable among the succession of theories. Finally, insofar as systematics has other functions besides testing or expounding the consequences of evolution for biological diversity, its key term should not be too closely tied to evolution. For many purposes, it is important to have a canonical description of the different sorts of organisms to be found in various regions, together with reliable means of identifying them. These needs are apparently independent of any explanation of the causal origins of the sorts identified, or even of the theoretical importance of the particular features of organisms hit upon for effecting these identifications. In consequence, it is sometimes held that a theory-neutral account of species is at least possible. When we add the fact that there is much disagreement about the role and meaning of the term in the theory of natural selection, the stronger conclusion that we need such a theory-neutral notion becomes plausible.

For a time in recent years, some biologists took this sort of argument seriously and attempted to construct such a theory-free notion of species: an operational definition of the notion. It was hoped that this would provide objective, replicable, indeed computationally programmable specifications of particular species, specifications that would enable us to predict the properties of organisms from their species membership and vice versa. This approach to taxonomy, known as phenetic or numerical taxonomy, is now without many adherents, but its initial appeal and eventual fate reveal important things about the notion of species.

First, we must distinguish the problem of definition for the general category or taxon of species from that of defining or identifying particular species like *Cygnus olor,* the conventional swan, or *Didus ineptus,* the dodo bird. A definition of *Didus ineptus* can be expected to list the salient features of members of this species – size, color, habitat, means of locomotion, predators or prey, number of young, social structure, etc.; or it may be a more refined characterization in terms of physiologically identifiable traits or features of cellular fine structure, say, the number of chromosomes in a cell at some accessible stage of development, etc. Such a list is usually accompanied by a specimen. A similar list is to be provided for *Drosophila melanogaster,* for *Escherichia coli,* or for *Saccharomyces cerevisae* (yeast), etc. The ultimate aim of systematics is the provision of lists of species and of higher taxa, the genus, family, order, class, and phylum. But it must begin with a schema for such lists – an implicit or explicit account of what it is to be a species at all, any species, and of what kinds of traits make for species membership or species difference. It must begin with a grasp of the general taxonomic notion of a species and then, employing this definition of the schema "x is a member of species S just in case x has properties of the following kind . . . ," fill in the S and the ellipses. The problem is specifying what kinds of properties go into the ellipses. This is the problem of defining species.

Pheneticists diagnosed the lack of agreement on how the ellipses are to be filled in as reflecting disagreements about theoretical issues in evolutionary theory. Furthermore they held that, even if these disagreements were settled, applying the theory would result in "subjective," unreplicable, and predictively empty taxonomies in any case. They urged that species be defined in terms of the sharing of observably detectable properties, marshaled together by strictly specified mathematical operations. The result is a recipe for taxonomy, a definition of species as sets of organisms sharing a set of observable or objectively detectable properties that satisfy a stipulated mathematical relation to one another. Part of the rationale for this approach to taxonomy is philosophical: It reflects the exigencies of operationalism, a doctrine according to which the meaning of a term is given by operations for deciding whether it correctly applies, and according to which a term with no explicit list of such operations is without meaning altogether. As proponents of pheneticism have urged: "Operationalism in taxonomy (as in other sciences) demands that statements and hypotheses about nature be subject to meaningful questions. . . . If we cannot establish objective criteria for defining the categories and operations with which we are concerned, it is impossible to engage in meaningful scientific dialogue about them" (Sokal and Camin, 1965:175).

Numerical taxonomy begins with the organisms to be classified and enumerates the various types of properties they exhibit to varying degrees. A table or matrix of the organisms and types of properties listed will reveal differences and similarities among the organisms, based on the number of properties any two organisms share. Statistical methods can quantify these correlations, enabling us to attach weights to differences and similarities or at least to order the degree to which any of the properties is shared by pairs of organisms. Some properties will be shared by all the organisms under classification and will not help separate species within the set. Other properties may covary perfectly or very closely among all pairs of organisms and be always present or absent together. These should be treated as single properties in order to circumvent any bias toward the choice of taxonomically insignificant properties imported from presystematic common sense descriptions of organisms. In fact, the counting of distinct properties will involve other sorts of adjustments. For example, certain properties can be measured and quantitatively ranked. Therefore organisms bearing distinct properties within this ranking may be deemed in addition to share a higher order-property, mathematically constructed from the observed properties that can be directly measured and ranked. If organisms really can be systematized in this way, then the more properties correlated with pairs of organisms that share or do not share them, the more a given taxonomy should stand out as increasingly well confirmed by the data. Beyond a certain point, further evidence will only increase our confidence in the categorization without increasing its refinement. Once these data are collected, proportions of properties shared and unshared among organisms can be mathematically manipulated in accordance with any consistently applied algorithm or rule to generate quantitative measures of similarity between organisms. These are then employed to generate the smallest units of association, the species, and, by lowering the quantitative degrees of association required, the more and more general taxa of the phylogenetic tree, genera, families, orders, etc. Of course, the results may or may not agree with the subjective, nonreplicable, theory-loaded taxonomy of nonnumerical taxonomy, but this is no defect in the latter's systematization.

It is easy to show that in general a strict operational approach to species is untenable, because of the untenability of operationalism throughout natural science in general. Waiving the question of whether the traits selected can themselves be operationally identified, any strict and exclusive definition of species in operational terms will sever any possible connection between taxonomic features and theoretically described properties of organisms. But these theoretical properties often provide a deeper and more accurate mark of species membership. They can explain why the operationally specified properties are biologically relevant to membership in the species, instead of being merely connected to it by definitional fiat. Additionally, the discovery of new correlations between previously unnoticed traits may produce a numerical taxonomy compatible with or inconsistent with a previously constructed one. Either way, the results will be untoward. For in the former case we will not be able to say that the new findings support the old ones, in that the new ones provide different operational meaning to the same species concepts, generating ambiguity and equivocation instead of confirmation. Should a new systematization provide apparently inconsistent taxonomic conclusions, they will not disconfirm the prior division. For again, the species terms associated with a new body of operational properties define a new and different set of terms, homonyms at best.

Finally, we return to the traits originally enumerated. It is easy to say that they are directly accessible to observation and mathematical manipulation, but it is harder to show this. To begin with, because every organism has an indefinitely large number of traits, any selection among them as irrelevant or significant must be either arbitrary or theoretically motivated. If the former, there is no assurance that they will group together in any useful way; indeed, the selection can never be entirely random, but will at least reflect a priority of selection encoded in our own sensory equipment; as such, it will be open to the very charge of subjectivity and nonreplicability that phenetic taxonomists hope to avoid. If, however, the selection is based on an articulated theory of what traits are important and what are not, then the operationalist proscription of theory is violated at the outset. The operationalist rationale for the whole procedure would be revealed as so much philosophical window dressing. It would turn out to be merely a methodological rationalization for some preferred theory of the properties biologically significant for species differences. As such, pheneticism is no more theory-free or theory-neutral a classification system than the ones it seeks to supplant.

Of course, numerical taxonomists can reformulate their doctrine so that it is free from the special philosophical defects of operationalism: They can admit that their methods are not strictly operationalist. They can allow that terms characterized by different operations can have the same meaning and refer to the same range of objects, permit the introduction of theoretically motivated character traits as relevant to species differences, etc. Their doctrine will no longer be justified by operationalist philosophy. But no doubt a justification can be constructed out of the scientist's general desire to stay as close to the facts as possible in laying out a descriptive typology. Like all scientific research programs, numerical taxonomy will not allow itself to be detained by philosophy, and it ultimately finds its best argument in the scientific successes it can produce. Thus, its chief exponents reflect their attitude toward conceptual objections and practical success as follows:

The philosopher may argue that it is not possible to make absolute measures of resemblance, because such measures would involve an arbitrary selection among the endless array of attributes which could in some sense be called characters of the organisms. Nevertheless, meaningful estimates of resemblance can be made once there is agreement on what characters are to be admitted as relevant in taxonomy. (Sokal and Sneath, 1963:90)

Although the question at issue is whether we can agree on what characters are relevant for taxonomy, in a sense Sokal and Sneath are correct. Every successful scientific typology is a miracle of question begging and the result of pulling oneself up by one's own bootstraps. One starts somewhere, anywhere, with a special theory, or with the presumptions embedded in ordinary descriptions of the phenomena, and attempts to construct a taxonomy. This is roughly how Mendeleev did it for the periodic table, arranging elements according to similarities and differences between well-known properties of chemicals, and without any knowledge of the underlying atomic structure. It was only in the next generation that such knowledge vindicated his typology as *the* correct one. Doubtless, a philosopher could have hamstrung Mendeleev's "operationalism," and it was probably fortunate that he did not offer any philosophical justification for his procedure. But the crucial difference between Mendeleev and numerical taxonomists is that he succeeded, and they did not. Within a finite amount of time he provided a natural, apparently nonarbitrary, unique systematization of the elements. Phenetic taxonomy has no equivalent accomplishment. Of course, the range of objects to be sorted is many orders of magnitude larger in systematics than in chemistry. Numerical taxonomists have only devoted the better part of two decades to the work, so any final judgment on the work would be premature. Nevertheless, just as the real argument in favor of this approach is not philosophical, the real argument against it is not philosophical either; it is factual. These methods have simply not generated a workable taxonomy, one that will meet all or most of the needs for which biologists appeal to this discipline. The facts of the matter simply cannot be carved up and rearranged in accordance with numerical taxonomists' hopes.

The moral to be drawn from the failures of phenetic taxonomy is that systematics is a thoroughly theoretical and factual undertaking. It is not a matter of finding out how nature has divided things up into kinds by just looking, or even by just measuring and manipulating. Success here will require theory, both to underwrite the traits identified as significant for taxonomic purposes and to assess the adequacy of the systematic result. That is, whether a particular systematization is correct will be in part a matter of whether it can be explained by a theory of diversity and whether it serves the descriptive needs of tests of this and other theories. Again, the comparison with the periodic table of the elements is enlightening. Mendeleev's taxonomy of the elements is correct because it can be explained by atomic theory, and it is useful because it can be employed both to extend atomic theory and for other physical and chemical purposes independent of the theory.

Similarly, a taxonomy of organisms must have the same role. Numerical taxonomists are right to demand a classification that is objective, replicable, and predictively useful. They are wrong to think that such a classification either can or must be theory-free. On the other hand, if no theory-based classification meets the criteria we set up for taxonomic adequacy, then, short of jettisoning all relevant theory, it is the aim of providing such a taxonomy at the level of organisms that we must surrender. But to surrender this aim is to surrender factual beliefs, namely, that biological diversity really does reveal a small number of significant general differences; that these differences between organisms and groups themselves have some explanatory import, one that is relatively uniform.

Classification, it turns out, is not a purely formal or conventional matter of stipulation. Although it has been only rarely true in biology, outside of it philosophical doctrines have long fostered the view that the scheme of categories we employ to describe phenomena must ultimately be conventional, and the preferability of any one must reflect considerations of convenience and economy of description. This view is an inevitable consequence of the notion that the descriptions of such categories must be definitions, analytic truths, and therefore without content. In this view, different classification systems are like different codes for writing numbers: There is the binary code, the decimal code, etc. Each of them can equally well express all the truths of arithmetic. Which one we employ is solely a matter of convenience: binary for computers, decimal for organisms with ten fingers. Like the descriptions of such codes, alternative taxonomic definitions make no claims about the world, about matters of fact; they are compatible with whatever happens. Because their definitions permit us to translate between them, we may describe any fact in any rich-enough system of concepts, provided only that we are willing to tolerate circumlocution, at some points at any rate. Thus, the choice among descriptive schemes and typologies is dictated by our desire to avoid tedious circumlocution where possible. Economy and convenience are the ultimate grounds of choice. Though widely and unreflectively held, this doctrine cannot be brought into line with the nonconventional factual warrant of atomic theory that stands behind Mendeleev's particular choice of typology for the elements. His typology is not just the most convenient among alternatives. It divides up the world in the *right* way. If atomic theory were to collapse, so would the periodic table. This shows that there is no firm line between logical, analytic, definitional truths of taxonomy on the one hand, and contingent, factual, empirically assessable statements of theory on the other. Once we see that the facts may make us *give up* a typology to preserve an empirical finding of supreme importance, we must surrender the notion that a different standard governs truths of typology. We must face the fact that their choice is constrained by the same forces that decide between factual theories. This is why there is no theory-free taxonomy and why all taxonomic decisions are about factual matters. Had numerical taxonomy been a success, it would have constituted a large set of important empirical generalizations about the properties of species, generalizations that would have to be further systematized and explained. And both these tasks would have had to be undertaken by evolutionary theory. In this respect, too, phenetic taxonomists who preached the independence of their approach from evolution, or its priority with respect to evolutionary enquiries about the taxonomies it proposes, were quite mistaken.

In the light of this examination of numerical taxonomy, we may conclude that an adequate taxonomy will reflect a sound theory about the nature of species and speciation. Reflecting such a theory will be a necessary though not a sufficient condition of adequacy. Failing it, there will be no classification of terrestrial diversity at the level of the species. There may be a classification at some other level, or in some other terms. And this level or these terms will have to be adequate to describe the stability among types throughout evolution that the term 'species' is in fact employed to refer to. Failure to define species adequately will not do away with the problem of diversity, and we must keep this fact firmly in mind.

7.2. Essentialism – For and Against

There are at present three prominent definitions of species and associated accounts of speciation. Each reflects insights that weaken the appeal of the others, but each is also bedeviled by difficulties of its own. They are: the so-called biological species notion, which hinges on reproductive relationships among conspecifics; the evolutionary species concept, which stresses the unitary response of a lineage to selection; and the ecological treatment of species, which identifies them indirectly through the specification of their position in an ecology. All three of these theories of species are self-avowedly "antiessentialistic": They reflect post-Darwinian rejection of the traditional species concept, because it is said to be vitiated by a commitment described as "biological essentialism." Although the term has become an epithet of abuse, there is nothing mysterious or pseudoscientific about essentialism. In fact, the thesis of essentialism turns out to reflect a continuing constraint on scientific adequacy that even "antiessentialist" theories must satisfy.

Essentialism with regard to species is the claim that for each species there is a nontrivial set of properties of individual organisms that is central to and distinctive of them or even individually necessary and jointly sufficient for membership in that species. Thus, essentialism is nothing more than the doctrine that a particular species name, say, *Didus ineptus* (the dodo), can be given an explicit definition, or at least as close to one as we can get in science. Essentialism with respect to species is no different than essentialism with respect to the fundamental particles of microphysics or the elements in the periodic table. Once Mendeleev set up the table, physicists and chemists were able to search for the underlying features of various elements that explained their position in the table. Eventually, atomic theory provided this explanation by connecting the chemical and physical properties of each element to the atomic structure of the constituent atoms of elements: their respective numbers of protons, neutrons, and electrons, and the arrangement of electrons in the "shells" surrounding the nucleus. Thus, the chemical and physical properties that were the original basis of classification gave way to the atomic properties of elements, and their privileged position was reinforced by further explanatory accomplishments: the prediction of the properties of as-yet-undiscovered elements, the accommodation of isotopes, etc. The atomic structure of an element thus is said to provide its essence, to convey its essential properties.

But this claim should not be understood in any very strong metaphysical sense (at least for the scientist's purposes). The claim is not intended to suggest that atomic structure is logically necessary for the nonatomic properties of a particular quantity of an element, or that atomic structure can be uncovered by reflecting on the concept of an element. Essentialism is not here a claim about noncontingent connections that bestow identity, in the way academic philosophy employs the term. In this regard, the term is an unfortunate one, for it tars with the brush of medieval philosophy and rational theology a perfectly respectable scientific research program: that of discovering the underlying uniformities governing the constituents of more accessible phenomena, uniformities that more adequately systematize and explain them. Essentialism in chemistry is just the demand that we undertake an experimental enquiry to discover the causes of the order Mendeleev uncovered. Had twentieth-century physics failed to uncover such an order, Mendeleev's taxonomy of the elements would have been thrown into doubt, and rightly so. Similarly, had twentieth-century physics uncovered an underlying structure incompatible with Mendeleev's table, the table would have been adjusted to accommodate the more fundamental atomic theory. Because atomic theory is held to provide a basic account of the nature of chemical elements, it provides an "essentialist" definition of chemical elements. In the physical sciences, essentialism is just another term for our belief in the unity and simplicity of nature, and theories that violate essentialist strictures are ipso facto suspect.

In biology, however, this doctrine has fallen into disrepute. For "essentialism" about species has been held to be the chief intellectual obstacle to the idea of evolution and to a proper understanding of biological variability. It is certainly true that the conception of 'species' from before Plato was essentialist. It was indeed held, right through the nineteenth century, that species were like elements. Like the quantities of an element that have something in common, the members of a species had something in common as well. Until the late nineteenth century what quantities of an element had in common was not known, though it was reasonable to suppose there was something they did. Similarly, in the absence of evidence for evolution, there was reason to suppose that there was something that members of a species have in common, that this shared property causes them to be members of the species and that it explains their distinctness from members of other species.

The entrenchment of this essentialist conviction about species was an impediment to evolutionary thinking. It made it difficult to talk of the evolution of species, of change within and between species, just as it is difficult to talk of change among elements. It is not impossible, of course, to say that one quantity of matter has changed from being radium to being radon by decay, but there is no such thing as the element radium changing into the element radon. At most, all quantities of radium may decay into radon, and then there will be no samples of radium left. But its place in the periodic table will not be expunged. There is no difficulty in the notion that through transmutations or decay, quantities of another heavier element may become samples of radium. But the *kind* radium cannot change into the *kind* radon. The notion that species evolve is, in an essentialist view of the matter, to be understood in this latter sense, as the change not of organisms but of the kinds they belong to. As such, species can evolve no more than the kind radium can change into the kind radon. Of course, we can give an interpretation of species that allows us to talk about their evolution in the same way we talk about the change in quantities of matter from one element to another. We must say that lines of descent may enter into and leave particular species taxa. But although this is a possible way of speaking, it is unnatural, it is difficult to reconcile with what we say about species extinction, and it does reflect something of an obstacle to thinking about evolution. To this extent essentialism is at odds with evolutionary thinking.

Some contemporary biologists have made more out of this obstacle than it warrants, however. They have held that essentialism is incompatible with evolutionary thinking. This conclusion is too strong. At most, we must admit that treating particular species as types with fixed necessary and sufficient conditions or at least central and distinctive explanatory properties will make us rephrase many of the things we wish to say about the evolution of species. Of course, if we find it inconvenient to do so, we should —while retaining the theory of natural selection — surrender the essentialist approach to species in favor of one that does not require them to have fixed explanatory membership conditions.

Although this looks like the same conclusion that antiessentialists draw, it retains the right to call upon essentialism elsewhere and to continue to employ it as a methodological stricture. What we surrender is the view that species notions, like *Homo sapiens,* are like the notion of oxygen or other chemical elements. We surrender the hope that they can be accommodated in a theory in the way the elements can be accommodated in atomic theory. From this conclusion we may infer that essentialism, a method common to the physical sciences, is inappropriate in the life sciences. Or we may infer that there is something seriously wrong with treating particular species names as general kind terms that can be entrenched in scientific theories, as terms whose characterization will be useful in providing an adequate taxonomy and will describe divisions of the biosphere open to systematic explanation.

The vigorous rejection of essentialism with respect to the species notion and the banishment of all essentialist thinking from biology are characteristic of autonomists, for they see biology's unsuitability to essentialist strictures as one of the strongest potential arguments in the provincialists' armory. Provincialists of course do not label their requirement for physically coherent theory in biology "essentialism" just because of its undesirable scholastic associations. But their demand that biology uncover regularities that can be systematically linked to the remainder of science is just essentialism without the name. For such linkage requires the existence of a small number of causal connections between the categories ordering biological phenomena and those that describe nonbiological phenomena. The statement of such connections is tantamount to a vindication of essentialism, construed as the doctrine that for any biological category there are central and distinctive properties to be explained by a theory that actively coheres with the rest of science. Failing to vindicate essentialism is thus another part of provincialism's indictment of biological practice.

The trouble with essentialism about species is not philosophical; it is factual. As a hypothesis about *Didus ineptus, Homo sapiens, Canis familiaris,* or *Escherichia coli* it seems false. There does not seem to be any trait or characteristic common and peculiar to all members of any of these species, none whose manifestation by an organism causally determines its species membership. As Darwin wrote, the most salient characteristic about species is the substantial *variation* among the members of one. As Sober notes (Sober, 1981), if essentialism were correct then this variation would have to be viewed as a deviation, disturbance, noise, or interference of the environment that deflects organisms away from a uniform, undisturbed "natural" state. Consider a quantifiable trait of a species, like clutch size. Essentialism is the tactic of treating the mean value of the trait as privileged, as more representative and more crucial to species membership than other values, which reflect accidents of the environment. Essentialism may be adopted with respect to phenotypic traits or genotypic ones. In either case, antiessentialists maintain, evolutionary theory and population biology both reveal that this view is simply mistaken. There are no privileged traits, and variation around a mean is not just random error produced by environmental disturbance. Genetic variability, is, dare one say, essential to the biosphere, and no genetic characteristic can be singled out as a baseline value around which others vary. Doubtless, if matters were much simpler, essentialism might have been right. For this reason the doctrine is not to be rejected on philosophical grounds. It is by no means incoherent. It is just false. And it is false for the same factual reasons that numerical taxonomy fails to systematize biological diversity.

There are no common and peculiar traits of the sort either requires to establish a satisfactory taxonomy.

Contemporary systematists argue that evolutionary and population biology are logically incompatible with essentialism because the latter views variation as a disturbance whereas the former views it as representative:

The assumptions of population thinking are diametrically opposed to those of the typologist [the essentialist]. . . . Individuals, or any kind of organic entities, form populations of which we can determine the arithmetic mean and the statistics of variation. Averages are merely statistical abstractions; only the individuals of which the populations are composed have reality. The ultimate conclusion of the population thinker and of the typologist are precisely the opposite. For the typologist the type . . . is real and the variation an illusion, while for the populationist the type (the average) is an abstraction and only the variation is real. No two ways of looking at nature could be more different. (Mayr, 1976:28–9)

Because population thinking is the kernel of the synthetic theory of evolution, essentialism turns out not to be a mere inconvenience, nor a false hypothesis disconfirmed in the light of evolutionary findings, but rather something approaching a conceptual mistake to be avoided at all costs. We shall see what the costs of avoiding it really are.

7.3. The Biological Species Notion

The sternest critic of essentialism has been Ernest Mayr; at the same time, Mayr has offered the most widely adopted characterization of species and of speciation, the so-called biological species notion. (My discussion of this and competing definitions in the next section are heavily indebted to Kitcher, 1985.) Since he first propounded his definition in 1942, Mayr has provided some significant qualifications on it. It originally took the form:

Species are groups of actually or potentially interbreeding natural populations which are reproductively isolated from other such groups. (Mayr, 1942:120)

The first thing to note about the statement is that it is very far from an operational definition of the notion of species. This is in itself no defect. Failing or succeeding to interbreed is of course a detectable, observable relation among organisms, so at least sometimes we can tell if two organisms are conspecific. But more often than not we can perform no operation that will decide the question of whether interbreeding is actual or possible. As with most theoretical notions, there is no operational definition of species. But for the notion to have any explanatory role it must be related directly or indirectly to operations that can at least test a claim about species differences. To be practically useful in taxonomy, the notion will have to be related to a relatively small disjunction of such tests. These tests must themselves be theoretically grounded or related if they are viewed as testing for the same thing. Such theoretical grounding will include not only a theory of speciation but additional biological theory. Failure to provide such a manageable set of operational tests, tests that are related to a definition of 'species' in a non–ad hoc way, will cast doubt on the definition as reflecting real distinctions among organisms.

The two crucial notions of this definition are 'interbreeding population' and 'reproductive isolation.' The former notion is to be understood as a set of organisms that are panmictic — that is, randomly interbreeding: "a group of individuals so situated that any two of them have equiprobability of mating with each other . . . A species in time and space is composed of numerous such local populations, each one intercommunicating and intergrading with the others" (Mayr, 1963:136). How are we to judge "equiprobability of mating"? The judgment that such equiprobability obtains for any population is a highly theoretical conclusion, one that hinges on statistical methods and substantial biological assumptions. If the methods are inappropriate or the assumptions unjustified, the conclusion will be undermined. Moreover, among biologists there are differences on which methods and which assumptions to employ in determining "equiprobability of mating." Thus the biological species notion will not generate a single classification if uniformity cannot be imposed on such judgments.

The notion of reproductive isolation is equally indeterminate from a theoretical point of view. There are of course many different isolating mechanisms, which can intervene at any point in the causal chain from initial encounter to viable birth of nonsterile offspring. Moreover, populations sometimes identified as differing species can hybridize, and this can occur with the same frequency as interbreeding by migration between widely separated populations of the same species. Where this happens, the biological species notion has broken down. Moreover, the rationale for interest in reproductive isolation is to be found in the belief that species are constituted by channels of gene exchange. Accordingly, we require a reason why, among all the phenotypical consequences of heredity, reproductive isolation should be privileged as the criterion of species distinctness. The response given is that without it gene exchange would result in an utterly different pattern of diversity. Each distinctive region would come to have a single homogenized type of flora and fauna as a result of gene exchange unlimited by barriers at the points of reproduction. This response requires that gene exchange, even in small quantities, can lead to the disappearance of the diversity that characterizes any locale. Where this presumption is false, notably among plants, the theoretical basis for focusing on reproductive isolation breaks down. It is no solution to these problems to find the explanation of breakdown in genetic considerations, for such appeal is a covert surrender of the biological species notion in favor of some other, genetic, principle of demarcation.

The problems so far isolated do not by any means undermine this definition beyond usefulness, but they show, first of all, that any definition of species must come to grips with theoretical and empirical problems and is in fact judged not as a conventional stipulation, but as a contingent hypothesis. Second, they show that at some points the notion breaks down and either lumps distinct species together or fails to explain why some isolated populations are members of the same species and others are not. But there are more crucial problems facing the definition. One problem is distinctly theoretical, and the other is operational.

The theoretical problem is that of reconciling the notion of species as reproductively isolated interbreeding populations with the existence of asexual species. Members of such species — which reproduce without interbreeding, without exchanging genes — are, with respect to reproduction, as fully isolated from one another as they are from members of any other species. This is a problem immediately recognized by exponents of the biological species notion, but it is played down as little more than a fly in the ointment. In his initial exposition of the definition of species, Mayr noted that there is some question as to whether this species definition can also be applied to aberrant cases, such as the mating types of protozoa, the self-fertilizing hermaphrodites, animals with obligatory parthenogenesis, and certain groups of parasites and host specialists . . . The known number in which the above species definition may be inapplicable is very small, and there seems to be no reason at the present time for "watering" down our species definition to include these exceptions. It will always be possible to add supplementary clauses, should a need for them arise. (Mayr, 1942:121–2)

But the problem of asexual species has proved to be more serious than Mayr first supposed, and in the long run his definition has been modified in the hope that it can accommodate this phenomenon. Unless a relatively simple, theoretically motivated qualification can be added to the definition, it must be concluded either that the definition does not have the generality required for any explanatory account of species diversity and its causes or that talk of asexual "species" is the employment of a homonym to describe facts utterly different from those described in terms of biological species. This latter conclusion is intolerable for anyone who, like Mayr, finds the notions of species and of evolution inseparable, and who holds that sexual species evolved from asexual ones.

At first, Mayr adapted the definition to accommodate asexuality by importing morphology as a mark of species along with interbreeding, unifying them both by appeal to genetic mechanisms: In effect, he adopted a double standard, though he disguised it:

To draw conclusions from the degree of morphological difference on the probable degree of reproductive isolation is a method of inference that has long been applied successfully to isolated populations in sexual organisms. There is no reason not to extend its application to asexual types. (Mayr, 1963:28)

He therefore concluded that we may define sexual species in terms of interbreeding and asexual ones in terms of morphology. Now what Mayr says about inferences from morphology to reproductive isolation is quite right. In fact, finding morphological differences are among the best operational methods of determining reproductive isolation and probability of interbreeding. Accordingly, by the definition, they are useful marks of species differences. But they will not identify different species, if members of the different species in question do not *intra*specifically interbreed. Asexual species do not do this; in these cases, morphological differences would merely be operational marks of membership in an asexual taxon, one that does not satisfy the biological species definition.

Of course, both interbreeding and morphological differences may be joint effects of some third thing, so that variations in that single determinant could be cited in a new definition of species good for both sexual and asexual taxonomy. Indeed, Mayr seemed to be moving in that direction, for he justified the appeal to morphology on the grounds that genetic difference is responsible both for it and for reproductive isolation. In this view, reproductive isolation is not a cause of diversity, nor are morphological differences; rather, they are marks or signs of it. The cause turns out to be genetic differences. But this is to surrender the biological species notion, not to improve it by a minor qualification that will accommodate asexual species. A species now turns out to be a set of organisms that share a genotype or a disjunction of them, one whose differences from other genotypes are great enough to make for detectable morphological differences and/or reproductive isolation.

In his latest thoughts on the nature of species, Mayr has imported still another factor to solve the problem of asexuality:

A species is a reproductive community of populations (reproductively isolated from others) that occupies a specific niche in nature. . . . [T]he major biological meaning of reproductive isolation is that it provides protection for a genotype adapted for the utilization of a specific niche. Reproductive isolation and niche specialization (competitive exclusion) are, thus, simply two sides of the same coin. It is only where the criterion of reproductive isolation breaks down, as in the case of asexual species, that one makes use of the criterion of niche occupation. (Mayr, 1982:273, 275)

Morphology is dropped here in favor of niche specialization, and both it and reproductive isolation are no longer treated as the basic theoretical content of the idea of species, but as derivative consequences of a genetic concept of species. It will later be clear that this definition is an attempt to bring together several competing definitions of species, in the hope of aggregating their strengths while sloughing off their weaknesses. But the sequence of reformulations should really be understood as a series of admissions of theoretical inadequacy. They are admissions that the biological species notion is in fact inadequate to characterize and provide the explanatory framework for the biological diversity with which we are faced. A stark way of putting it is that the biological species definition does not carve nature at the joints. It does not provide a uniform, general, easily applicable way of identifying the natural kinds into which living organisms fall. Employing it will require arbitrary decisions that cannot be justified by biological theories already in hand, decisions in taxonomic classification that may differ from naturalist to naturalist and will not cite the explanatory basis of speciation.

These conclusions can be reinforced by considering a crucial operational problem facing this theory: that of applying it to the discriminations among species required for purposes of paleontology. These problems are operational because there seem to be no direct or even indirect operational connections between interbreeding and isolation on the one hand and the fossil record on the other. The problem is operational and not theoretical because the fact that the fossil record leaves no trace of interbreeding relationships is no reason to deny that it is such relationships that determine the species taxa that paleontology hopes to reconstruct. On the other hand, the isolation of the theoretical definition of 'species' from any operational test by which paleontologists distinguish species means at least that the definition cannot directly account for their decisions about species differences. There is, however, a more serious difficulty for the biological species notion reflected in the fossil record. Speciation may in principle proceed by anagenesis, that is, by the evolution on one line of descent from one species into another. In this case, each organism in the line of descent is reproductively connected to its immediate ancestors, even though organisms at the extremes of the line of descent are members of different species. Speciation may also proceed cladogenically, by the division of a line of descent into two lines that eventually become reproductively isolated from one another. In this case, the result of evolution is two species, either the original one and its offshoot or two new species. By contrast, anagenesis results in only one species at the end of a period of speciation. The differences between anagenesis and cladogenesis are schematically illustrated in Figure 7.1.

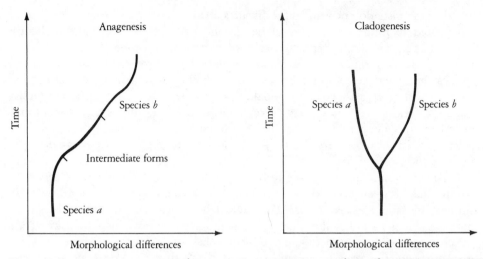

Figure 7.1. Anagenesis versus cladogenesis. In anagenesis, members of 2 species are connected in one line of descent, during which earlier species become extinct. In cladogenesis, there are 2 distinct lines of descent that constitute different co-existing species.

But gradual interspecies anagenesis seems incompatible with the biological species notion. If the potential for interbreeding between contiguous individuals is transitive, then populations at the end of a connected lineage are not isolated from populations at its beginning, and all must be conspecific. Of course, one could demand more for interbreeding than just the abstract potential provided by the transitivity of interbreeding. But how much more to require is the theoretical problem we have already noted. Our criterion for potential interbreeding cannot be actual interbreeding, so it must be some theoretically systematized potential or capacity. Two alternatives to providing such a theory are either simply to deny the occurrence of anagenesis and assert that only cladogenesis occurs or to deny that anagenesis is ever gradual.

The former tactic of denying anagenesis altogether seems not only arbitrary, but is in fact the establishment of an alternative species notion. For it requires, in addition to or instead of reproductive isolation, that every species is delimited by speciation events that split the lineages it belongs to or may issue in. Accordingly, it is the denial of gradual anagenesis that is more usually made. Indeed, the rejection of gradual anagenesis figures in quite another connection, in arguments favoring the so-called punctuated-equilibrium approach to evolution. According to this approach evolution proceeds by rapid bursts of change and geologically long periods of stasis and equilibrium. In such cases, reproductive isolation may not be transitive, and decisions about species membership can be made in a nonarbitrary way, at points of rapid evolutionary change flanked by reproductively isolated lines of descent, all within the same lineage. This view is convenient for paleontologists because their data seem to suggest stasis as the rule and evolution as the exception. The fossil record is often said to reveal no missing links and few intermediate forms, but rather uniformity over many layers, and then gaps, after which new morphologies can be found through several layers.

Paleontologists do not want their disputes about whether evolution is gradual or punctuated to hinge on matters of definition. Yet if they accept the biological species notion, and hold that anagenesis occurs often or occasionally, they are committed to denying that it can be gradual. The biological species notion forces us to make a choice between empirical questions of whether there is anagenesis or not and whether evolution as a whole is gradual or punctuated. Now any species notion has some theoretical content, and therefore adopting one does force choices about empirical matters upon us. But the ones the biological species notion confronts the paleontologist with are very drastic – so drastic as to suggest that it is not really relevant to his enquiries.

Gradualists and punctuated-equilibrium theorists dispute the rate of evolution within and between species. In expressing this dispute, they at least implicitly employ the same species notion. They agree that anagenesis and cladogenesis are both possible modes of evolution; indeed, they agree that both modes are actual. As the chief proponent of punctuated equilibrium has said, "I know nothing in evolutionary theory (nor can I envisage [sic] anything) that would render anagenetic transformation inconceivable a priori. Gradual phyletic transformation can and does occur. I claim merely that its frequency is low and that punctuated equilibrium is the predominant mode and tempo of evolutionary change" (Gould, 1981:84). It must follow, therefore, that punctuated-equilibrium theorists are not committed to the biological-species definition. And it must also follow that if their paleontological opponents employ the same implicit or explicit species concept, then this definition does not reflect the practices or the needs of this branch of evolutionary enquiry.

The natural kinds of paleontology, on which there must be minimal agreement to express and adjudicate the dispute about gradualism and its alternative, must surely be different from those generated through the application of reproductive isolation as a criterion of species. This notion cannot be operationalized for paleontology even indirectly, because doing so would force paleontologists to make theoretical choices that their data do not force upon them and that they do not want to make in any case. This is why in paleontology the species notion is employed to reflect detectable morphological discontinuities, differences in the hard parts whose marks are left in the fossils. Division points in paleontology between species are the geological gaps that separate morphologically distinct lineages. Of course, the assumption that such gaps are coincident with points of speciation is question begging so far as theoretical questions of gradualism versus punctuated equilibrium are concerned. But they serve paleontologists' purpose of organizing and classifying their data in a mutually agreed upon, replicable way. Because both parties to the dispute about gradualism versus punctuated equilibrium can understand each other's claims in terms of this description of the fossil record, this highly operational definition of species is adequate for many of their purposes. For describing patterns of morphological change, the division of the fossil record into morphologically defined species is an operationally unobjectionable approach. For it is distinct from the issue of whether the record of speciation thus described favors a gradual or nongradual evolutionary sequence. But the suitability of this morphological approach to species undermines the generality and applicability of the biological species notion. For reproductive isolation and interbreeding do not vary with morphological differences.

Adopting the biological species notion is a commitment to reproductive isolation as the cause of speciation. As we have seen, this claim is a hypothesis that cannot be reconciled with all biological findings both about sexual and asexual species; equally, it is a hypothesis that rules out certain possibilities biologists consider to be open questions, not decidable by the choice, even a theoretically self-conscious choice, of a species definition. Despite its unquestioned usefulness in many contexts, the definition turns out to be too strong and too weak, unable to answer some questions, presumptuous in providing unwanted answers to others. For all its virtues as the account of a natural kind, as a means of demarcating the basic facts about biological diversity, it appears to be seriously inadequate. It does not carve nature at the joints.

7.4. Evolutionary and Ecological Species

The second major alternative definition of species is known as the evolutionary species definition. Simpson has offered the following version of this notion:

An evolutionary species is a lineage (an ancestral-descendant sequence of populations) evolving separately with its own unitary evolutionary role and tendencies. (Simpson, 1961:153)

Simpson's preference for this definition is clearly theoretical. He holds that it reflects a more fundamental speciating force than reproductive isolation. The interbreeding of reproductively isolated populations fosters the "unitary evolutionary role" of a species, but it is not the sole determinant of this unity. And, as Simpson notes, it is entirely absent from the forces that promote unity of evolutionary role among asexual species. This fact reflects another feature by virtue of which the evolutionary species notion is more fundamental. It appears to accommodate sexual and asexual species under the same definition because it focuses on properties common to both. Additionally, the definition suffers from no strains in its application to paleontology. It is no more operational with respect to this subject than the biological notion, but it also seems quite neutral with respect to the subject's controversies.

But consider what constitutes a separate lineage: what the difference is between two distinct lines of descent in the same species and two lines of descent that are separated in distinct species. Of course separation of lineages cannot be necessary for speciation, without thereby ruling out anagenesis a priori, in the way that reproductive isolation rules out gradual anagenesis. There are some exponents of the evolutionary species notion who hold that only cladogenesis occurs and do deny the possibility of anagenesis. These taxonomists are known as cladists, naturally enough. Unless their claim is to be accepted as true by stipulation, not open to factual assessment, no definition of species can make genealogical separation of lineage necessary for speciation. This means either that the definition is inadequate or that it must be treated as asserting a more complex relation between species and separation of lineages than it appears to. It will have to be differences in evolutionary role that discriminate anagenically evolved distinct species and genealogical separated lineages in the same species. But this aspect of the definition raises problems that show the evolutionary species notion is inadequate and must be replaced.

These problems become particularly acute when we attempt to apply the evolutionary species notion to asexual species. Here the problem is not, as with biological species, that the whole phenomenon is left out of account, but that the definition leaves completely indeterminate what counts as speciation in evolution of asexual organisms. For all lines of descent among asexual organisms, intra- and interspecifically, are evolving in utterly separated ways. To appeal to morphological differences, as the biological species notion does, is either to change the standard of speciation from an evolutionary to a plainly morphological criterion or to admit that an asexual species is not a species at all, but only so-called by courtesy or confusion. Again, morphological comparisons may be a reliable sign, a good operational criterion, for taxonomy, but to establish them as constitutive of species differences is to surrender the evolutionary species definition. It is in effect to adopt the view that numerical taxonomists have failed to make a success of.

In the end, we must come to the same conclusion about the evolutionary species notion as about the biological one. Doubtless, it functions well in many taxonomic contexts and even dovetails nicely with the biological species notion in its systematic decisions. It had better, for it purports to be a deeper account of the matter that can explain the success of the latter notion. But it does not do the things that its status as an explanatory notion should enable it to do. It has neither the generality nor the uniformity required, and it cannot provide the basis for a uniform account of diversity. Apparently, what is needed is an even deeper, more theoretical account of the phenomena.

It is worth noting that the succession of accounts of the species notion thus far canvassed has moved in the direction of greater theoretical content and greater operational distance from the facts with which naturalists and taxonomists grapple. The sequence has reflected a willingness to trade off direct operational applicability in favor of greater generality. In doing so, it has moved in a direction common to classification in physical science. From the alchemist's classifications, through Lavoisier's, to Mendeleev's and atomic theory, the fundamental chemical categories became less and less operational and more and more theoretical. The rationale of this tradeoff was ultimately the justified expectation that the more generally theoretical account would lead to more and more powerful and diverse operational means of identifying the chemical elements. It enabled us to move from discriminating substances by taste and smell to discriminating them by spectrographs and nuclear-magnetic resonance. The same motivation seems to lie behind the succession of species definitions. They move from observable facts of morphology to a particular disposition, reproductive isolation, to a more general set of properties, evolutionary role, which explains this isolation and treats it as a special case. The last of the accounts of species we shall canvass is in fact a fuller and more theoretical elaboration of this notion of evolutionary role.

This last notion has been called the ecological species notion. According to van Valen, its proponent,

a species is a lineage (or a closely related set of lineages) which occupies an adaptive zone minimally different from that of any other lineage in its range and which evolves separately from all lineages outside its range. . . . An adaptive zone is some part of the resource space together with whatever predation and parasitism occur on the groups considered. It is part of the environment, as distinct from the way of life of a taxon that may occupy it, and exists independently of any inhabitants it may have. (Van Valen, 1976:233–4)

One respect in which this definition goes theoretically deeper than its predecessors is that it reaches beyond the classified organisms themselves, or their interrelations. It anchors their classification to states of an independently existing environment that determines their taxonomic status. It is worth recalling that Mayr's latest version of the biological species notion points in this direction by tying isolated interbreeding to niche specialization as "two sides of the same coin." Simpson has attempted to expound the notion of evolutionary role in this way as well.

However, the notion of an adaptive zone or niche faces several serious problems. First, if the notion is understood in any literal geographical sense, it renders unintelligible the taxonomic discriminations made between organisms living in the same region. Any two organisms in the same spatiotemporal region need not be conspecific. So obviously a niche, or an adaptive zone, is not a four-dimensional quantity of space-time. On the other hand, we have no idea of how many dimensions a niche can vary along, and so we have no way to identify niches or adaptive zones independent of their occupants. (This is what makes fitness impossible to define; see Chapter 6.) Construed as an argument from (temporary theoretical) ignorance, this point is not very telling. But insofar as a species notion should be operationally connectible to actual taxonomic discriminations, and should explain them, the objection is quite serious, for it shows that the proposal is at best a schema for particular species designations. In fact, the ecologists' actual strategy for identifying niches and enumerating their dimensions is to proceed from taxonomic decisions taken on other grounds – biological, evolutionary, even morphological – and inferentially construct a topography of niches. Of course, if such constructions can be generated in a clearly articulated, theoretically motivated, replicable way, then we can argue backward from the identification of niches to the systematization of biological diversity. We can thus explain the initial provisional taxonomy in terms of a theory of niches. However, the niches that ecologists identify have important properties that species should not have. For instance, the niche of a species can change into another one or disappear over time with consequences for the evolution or extinction of the *same* species. But taking the definition seriously means locating anagenic speciation at the points of niche change. Moreover, it is equally problematical to extend the notion of an adaptive zone from its synchronic – time-slice – application to coexisting species (as the definition employs it) to diachronic comparisons over time that are crucial for identifying the succession of species evolving in a single niche. Given a present distribution of niche spaces, we cannot tell either what niche spaces existed at prior times or whether they were filled or not. In fact, the character of niches suggests that not only can we not identify one "independently of any inhabitants it may have," but that they do not exist independently of their inhabitants. To see this, consider a hypothetical case constructed by Phillip Kitcher:

As van Valen's expression of the ecological notion reflects, part of the niche of any species is its predators and other species with which it interacts. In particular, a niche will be determined by the density of other organisms in the vicinity. Now consider two organisms, like the two species of butterfly, distinguished in their respective strategies of predator avoidance, in particular the avoidance of two species of birds, each of which normally predates only one of the butterfly species. However, suppose that these specialists will switch to the other butterfly species if their preferred prey becomes rare. In this case, the niche of each of the four species is a function of the relative population densities of the other three, and we will need to make a prior discrimination among organisms into species before we set out the niche that defines any one of them. This discrimination may not only involve presuppositions about other species, but even a prior discrimination of the species to be defined, insofar as its density bears directly on the niches of the other species and so indirectly on the identity of its own niche. This biological instance of a philosopher's problem of self-reference reflects the total interdependence of the notion of a niche and its species. These considerations do not by any means vitiate the notion of a niche or an adaptive zone, but they show that it cannot be viewed as a preexisting collection of nonevolutionary or even nonbiological, purely physical variables that would enable us to break out of a circle of taxonomic terms. No doubt there are niches, and the niche of any species is a region of some hyperspace supervenient ultimately on nonbiological factors. But this fact is of no consolation in the face of this notion's inability to ground the actual judgments of taxonomy.

Biologists, of course, do sometimes discriminate species by appeal to distinctions between niches and do explain differences in local levels of diversity in this way. Thus, a tropical rain forest contains more species than an arid desert because it is more variegated and thus provides more niches for species to inhabit. On the other hand, naturalists often identify distinct species occupying the same niche and struggling between then for its exclusive occupancy. This is particularly striking in the case of sibling species, reproductively isolated, though morphologically and evolutionarily indistinguishable. Niche competition between sibling species is indescribable in terms of the ecological notion. For it will multiply niches to match the numbers of distinct species, or assimilate biological, evolutionary, or morphological species to match the number of niches identified.

What really goes on in the diverse areas of systematics is that each of these different and incompatible characterizations of species is employed, not simultaneously of course, but in the particular contexts where they are appropriate, where they meet the needs of naturalists. These needs are not always the same. The paleontologist pursues taxonomic enquiries with different aims than the ethologist or the plant taxonomist. Of course, sometimes two or more criteria of species membership do seem applicable, and in these cases agreement is taken to reinforce a taxonomic judgment. Disagreement, on the other hand, is a spur to further enquiry, which will decide the particular taxonomic question – without of course resolving the question of which definition is ultimately correct.

It seems clear that the notion of a species is not amenable to anything like the physicists' agreement on the notion of an element. In this respect, the post-Darwinian opponents of essentialism have been vindicated. There appears to be no single schema within which any particular species characterization must fit, no universally applicable specification of the kinds of properties that determine species membership. There is no set of necessary and sufficient causal conditions for being a *Didus ineptus* or *Rattus rattus*. The basis of species discriminations is in practice disjunctive – either reproductive isolation or difference in evolutionary role, or niche, or morphological differences for that matter, make for species differences. Thus what makes organisms members of the same or different particular species will be disjunction of relationships or shared properties, no one of which is privileged as essential for species membership. Accordingly, species membership will be quite compatible with the wide variation characteristic of organisms in the same taxon. And there will be no reason to single out one property or statistical construction from the shared properties as the species' essence.

Many biologists, even participants in the debate about *the* best definition of species and *the* correct system of biological classification, appear to recognize that the debate may rest on a doubtful assumption that there are unique best and correct definitions and classifications for purposes of systematics. Thus Hennig writes:

> Each organism may be conceived as a member of the totality of all organisms in a great variety of ways, depending on whether this totality is investigated as a living community, as a community of descent, as the bearer of physiological characters of life, as a chronologically differentiated unit, or in still other ways. (Hennig, 1966:5)

This plea for tolerance is probably a useful attitude for the practical taxonomist and naturalist to adopt. Its agnosticism about the *truth* of highly theoretical matters forces systematists to focus on the immediate job at hand, classification. It allows them to employ whatever tools they think most useful in pursuing this task. These tools need not be viewed as the best theories of the matter. And it is only after a reasonably complete taxonomy has been provided and agreed to by informed classifiers that the theoretical question of how to explain the taxonomy becomes relevant. If, in addition, taxonomic enquiries are undertaken for several different reasons, with different aims, to answer different questions, then no decision may ever be needed about which definition of species and which classification is correct. They may all be correct, even mutually incompatible ones; that is, they may each be correct relative to the aims they meet and the questions to which they are addressed, though none is correct *simpliciter*.

7.5. Species Are Not Natural Kinds

If there is no single definition of species that can link taxonomic classification with a uniform theoretical explanation, then the challenge to Darwin, and to the theory of natural selection recorded in Section 7.1, can never be met. According to that challenge, Darwin never provided a theory of speciation, despite his aim to do so as betokened in the title of his *chef d'oeuvre*. The diagnosis given for this failure was his "uncertainty about the nature of species." But, as we have seen, this uncertainty persists, and more than a century of accelerating research has not brought us to the conclusion that there is a correct understanding of its nature. The upshot is either that the theory of natural selection must be admitted to have failed its most crucial self-imposed explanatory task or that the demand it explain the origins of species is misconceived.

Now although the second of these alternatives sounds as implausible as the first, it turns out to be the right view of the matter. Or so I shall argue. In brief, the argument is this. Our examination of various accounts of the species notion reveals that there is no single adequate account of what a species is. Therefore, there is no homogeneous class of items properly called species. There are of course very many species, a million or more of them in the earth's evolutionary history. But what we have seen is that these species do not constitute natural kinds, like, for example, the elements. There is no property common and peculiar to each of their members, nor is there a type of property that each species shares with all others, by virtue of which calling them all "species" reflects some common trait. By contrast, calling all the entries in the periodic table "elements" is based on their all sharing a type of property; that is, they all have atomic structure. What distinguishes one element from another is the difference between their atomic structure, the numbers and arrangements of electrons, protons, and neutrons out of which their atoms are uniformly composed. Now it is because all the elements have a common type of structure, and differ in the particular structures of this type that they manifest, that atomic theory can justify the taxonomy of the periodic table. It explains the chemical and physical differences between quantities of the various elements in terms of the differences in their structures: It explains transmutation and decay, the existence of isotopes, the character of chemical bonds and the stoichiometry of chemical reactions, the physical structure of quantities of the elements at various temperatures and pressures, their magnetic and electrical properties, etc.

If the elements did not have a common structure, then they would constitute a heterogeneous collection of items. Their behavior in these various respects could not be explained by a *single* theory but would be explainable only by a disjunction of theories of various kinds. If no manageable body of theories seemed capable of explaining all the diverse phenomena associated with the taxonomy of the elements, we might adopt a more radical conclusion: We might hold that the element names do not divide nature into its basic physical kinds; instead, they reflect local, unsystematizable, perhaps pretheoretical intuitions, like the division of nature by Greek science into earth, air, fire, and water. Or perhaps we might conclude that the chemical elements represent a methodologically credible but nevertheless factually unwarrantable taxonomy, like that of phlogiston chemistry, with its clax, dephlogisticated air, phlogiston, etc. In short, we can demand of a theory that it systematically explain a body of phenomena *described in certain terms* only if those terms divide up the phenomena in a way that reflects the distinct causal forces that generate it.

What we have seen in connection with the species notion is that it does not meet this condition; it does not divide up the phenomena of biological diversity into natural kinds. There is no single type of event properly called speciation, nor a single type of collection of organisms properly called species. There is no manageable set of regularities governing all the events and classes we describe in these terms. If there were, either we should have hit on an adequate definition of species by now, or we should keep looking, despite a century or more of failure. If we adopt the former regularities, we should not expect the theory of natural selection to provide an explanation of *the* origin of species. For there is no single origin of those "fortresses of stability" that the taxonomist reports. But if we give up the notion of species as a natural kind, we absolve the theory of evolution of the task its founder imposed upon it. Or rather, we give up the task described in the terms that traditionally have been used. The theory must still explain the course of evolution and must explain the facts of diversity. But it either must do so in terms of a new notion that replaces 'species,' in much the way 'oxygen' replaced 'phlogiston.' Or it must explain events of speciation and the characteristics of species not as general phenomena, but on a case-by-case basis, appealing to different ancillary and auxiliary theories and findings for each of the cases it explains.

The idea that our current taxonomic practices do not reflect natural kinds, do not divide nature at the joints, should not be an entirely foreign one. For the rationale behind the establishment of these practices by Linnaeus and his successors was exactly this sort. They considered previous taxonomies inadequate for the same reasons. What is more, we know that classifications embedded in ordinary language do not discriminate real biological kinds and are therefore obstacles to the discovery of generalizations about them.

Consider the term 'fish': If we take it to mean aquatic creatures, the class of organisms described by the term will be so heterogeneous that no worthwhile morphological generalizations about fish will be possible at all. Their differences in locomotion, respiration, anatomy, etc., will be so great that any attempt to say something generally true about fish will be frustrated by the need to add exceptions, qualifications, and exclusions for practically every sample we catch. Thus, if we substitute for the ordinary term 'fish' the class 'Osteichthyes,' restricted roughly to bony, aquatic, cold-blooded vertebrates with permanent gills and scales, thus excluding whales and crabs, jellyfish and tadpoles, we find ourselves better able to regiment aquatic diversity and to make useful general statements about it. If the theory of natural selection turns out to be unable to explain the character and the origin of species, or even higher taxa, then one response may be to search for a new classification, just as the inability to frame general statements about fish, ordinarily understood, led to the replacement of this notion by another.

One way of describing the fate of the nomenclature of taxonomy is to say that it fails because it is not essentialist. The proponents of contemporary species definitions are all agreed that species have no essence, no causally determinative, distinctive properties on the basis of which we can discriminate them. But if they are right, species have no properties on the basis of which we can explain them either, for discriminating features and explanatory properties are one and the same; they are the essential properties of any thing. It is only the appeal of operationalism, which demands that discriminative properties be accessible and independent of theories, that leads to the bifurcation of descriptive properties from explanatory ones. However, once the grip of operationalism is relaxed by the realization that species have no operationally discriminable properties, antiessentialism begins to look like a reasonable attitude. Moreover, its diagnosis of the failure of essentialist taxonomy takes on considerable force. With the removal of operationalism, the ground for distinguishing discrimination from explanation collapses. And if discrimination is nonessentialist, so too must explanation be nonessentialist. But to admit this is to say that no uniform explanation of the phenomena of diversity is possible at all.

The conclusion that species are not natural kinds is not an entirely satisfactory stopping place. For having concluded what species *are not,* the question of what they *are* is left open. Are we to conclude that they are gerrymandered, jury-rigged patchworks, artificial and accidental constructions, or superseded scientific concepts like 'phlogiston'? These conclusions are hard to credit, for species names, no matter whether they describe natural kinds or not, will continue to figure in the work of taxonomists, and justifiably so. For they do represent a satisfactory answer to a practical need. But we must reconcile this indisputable fact with the conclusion that, for all their practical merits, these terms cannot be expected to figure in general laws and theories or approximations to them.

7.6. Species As Individuals

In recent years, a radical proposal has been broached that will accommodate the important role that these notions have played and will continue to play in the description of the biosphere, despite their theoretical isolation. This is the view first advanced by Michael Ghiselin, and later championed by David Hull, that particular species are not kinds of organisms, whose members are instances of them. Rather, each species is an individual, spatiotemporally restricted (albeit scattered), particular object, whose members are its parts and its components, not its instances. Accordingly, species names are names of particular things. They will not require individual, general definitions that capture their essence, anymore than we can give the properties common and peculiar to all Mona Lisas, or Alexander the Greats, or Eiffel Towers. Because there is only one of each of these things, there is no basis for nor any point to a definition of it. If particular species are individuals, like these things, then the general category of species will be the name of a class of individual particular things that may differ among them so completely as to defy any definition for the general class term "species." Let us consider the arguments for this view of species and its ramifications for the theory of natural selection.

The first argument for treating species as individuals is of course the failure of a long tradition of attempts to treat them as classes, a tradition whose latest stages we have just canvassed. Now this catalogue of failure is no conclusive argument against any possible account of species that will combine generality, truth, and theoretical support. Similarly, the arguments in favor of the notion that species are spatiotemporally scattered, particular individuals do not conclusively demonstrate this conclusion either. But because it is hard to see what this could do for either view, the demand that any conclusion on this matter should be conclusive is probably excessive. We should keep this in mind in considering some of the objections lodged against the species-qua-individuals thesis.

Proponents of this thesis sometimes view themselves as taking contemporary antiessentialism to its logical conclusion. They deny not only that particular species taxa have essences, but also that the term 'species' denotes a property or class of items that have an essence or a distinctive causal basis. Like other antiessentialists, they trace mistakes about species back to Aristotle. Aristotelian logic and subsequent logic have always treated species names as paradigm class names, and they have offered statements of the form "all swans are white" as paradigms of general statements that link kinds. Such statements have of course long been recognized to be either very low-level generalizations, or perhaps definitions, disguised or undisguised, of their subject term. In either case, the unchallenged presumption, reflected in the universal employment of species concepts to describe kinds of things, or classes with instances, is the basis of the demand that biology produce laws about these kinds, including the evolutionary laws. It is also the basis of the search for definitions or generally applicable descriptions of them. In the philosopher's jargon, species names, like *Cygnus olor,* have been treated as predicates whose meaning is given in "open sentences": expressions like "x is an instance of *Cygnus olor* if and only if x has properties $P_1, P_2, \ldots P_n$," where the number of different items that satisfy x can range from zero to infinity. This view of the matter has persisted even though taxonomists have never been able to provide sets of necessary and sufficient conditions for being members of most species known to them. This failure of taxonomy has also been one of the most persistent points of criticism of biology as a nomological discipline, one capable of uncovering laws or approximations to them. For the failure to provide these essentialist definitions in terms of necessary and sufficient conditions reflects the fact that there are remarkably few generalizations of exceptionless sorts about particular species or about all species of a given kind, for example, arctic or aquatic species. "All swans are white" may serve the philosopher as a generalization for purposes of logic, but plainly has exceptions that deprive it of truth and/or nomological status.

Arguments that biology possesses no distinctive laws of its own trade on the inevitable exceptions and restrictions with which biologists must hedge their claims about general features of members of any species. As we shall see, such arguments reflect mistakes about the subject matter of biological laws: The general findings of biology and biological theory do not bear on regularities about particular species, like *Canis familiaris* or *Didus ineptus*; they bear on laws with which all species are in conformity. The subject matter of mechanics is the behavior of any body, but not any particular one. Similarly, the theory of natural selection tells us that, as a result of hereditary variation among the members of every species and of selection over this variation, every species *evolves*. The individual organism is of course the *immediate* (but perhaps not sole) unit of selection; it is the item on which environmental pressures operate and whose survival and reproduction determine the character of subsequent members of the species. The crucial feature of the theory of natural selection is that the unit of evolution is the *species;* it is they that evolve. Their evolution consists in changes in the relative proportions with which their members in successive generations manifest the varying hereditary characteristics, or phenotypes, determined by changes in the units of heredity and the forces of selection. This explains both why biologists can provide no necessary and sufficient conditions for various particular species and why there seem to be no exceptionless generalizations about particular species. Because species evolve, there is no trait that jointly meets the requirement of being hereditary and the requirement of being either necessary or sufficient for species membership through the course of its evolution: There are no essential properties of species. Similarly, the apparently general statement that "all polar bears are white" is not a lawlike consequence of evolutionary theory. If true, however, its truth can be explained by citing generalizations about the existence of adaptive variations in any species exposed to certain types of environmental conditions that survives in those conditions over the long run. Because, according to the theory of natural selection, species evolve, they should not be treated as classes whose members satisfy some fixed set of conditions – not even a vague cluster of them – but they should be viewed as lineages, lines of descent, strings of imperfect copies of predecessors, among whom there may not even be the manifestation of a set of central and distinctive, let alone necessary and sufficient, common properties. But a kind or sort for which *no* general properties whatever *could* be definitional is no *kind* at all. A kind that remains unchanged while any and *all* of the properties of its instances change over time is equally hard to comprehend. But neither of these difficulties arises for a *particular* object, which may change any of its biological properties over time, which may evolve, and about which the question of defining properties does not arise.

More importantly, biological theory dictates that an individuation and a unity be accorded to species that are unintelligible except on the assumption that species are particulars. Thus, the disappearance of all atoms of atomic number 79 does not entail the disappearance of a space in the periodic table of the elements. It only implies its temporary emptiness; the disappearance of all members of a species entails its extinction, the utter disappearance of the species. The appearance of new organisms qualitatively indistinguishable from the extinguished species' members does *not,* in biological theory, constitute the *reappearance* of the *same* species. It represents an entirely new one, just because it did not arise in any line of descent from the old one (which, having no issue, became extinct).

Moreover, just as individual organisms may undergo vast changes in properties, genotypic and phenotypic (may divide into two, fuse with another into one, or continue to exist while generating new individuals), so, too, evolutionary theory requires that species be capable of all these things, capable of anagenesis and cladogenesis. Thus, as binary fission makes two particular organisms out of one, geographic or other environmental isolation can make members of the same lineage reproductively isolated after a long enough period of separation so that, whether their appearance is similar or not, they must be classed as *new* and different species. This is of course how, in the view of exponents of the biological species notion, speciation proceeds. Where genetic change makes both branches reproductively isolated from the original lineage, there is cladistic speciation in which the parent species is said to become extinct. When substantial continuity and the possibility of reproduction between members of the original lineage and one of the subsequent lines remains, the phenomenon is akin to that of an individual generating a new individual while continuing to exist as a unit. This is a different cladistic sort of speciation. Similarly, introgression may obtain between two species, sometimes creating a third species interconnecting the lines of descent constituting each species. Now, although examples of one or more kinds of things may do any or all of these things, the kinds themselves cannot do any of them.

As Plato clearly recognized, kinds are immutable; it is only things, the instances of kinds, that change. Individuals are spatiotemporally bound; they are discrete entities with a location and a history, even though it is sometimes difficult to plot their boundaries and their beginnings or endings. So too with species. Temporal boundaries among species are marked by extinction, or the production of sterile offspring, or by the onset of reproduction among an isolated small group of founder organisms. Or again, they are marked by the process of polyploidy, in which meiosis, the multiplication of genetic material, occurs without cell division, doubling and thereby changing the chromosomal material that determines reproductive possibilities. Spatial boundaries are simply given by the distribution of members of the species. Spatial continuity is reflected in reproduction and in other sorts of species-specific behavior. Naturally, species do not have exactly the unity and spatiotemporal continuity of the usual example of a particular object, like a table or chair, but they certainly have enough to be so classified when we consider the vagaries of individuation for such particulars as nations, cultures, or even organisms, like the slime mold or sponge, with their peculiar biological potentials. Most important, species have the uniqueness that is necessary for being individual items and sufficient for not being general classes or kinds of items. Qualitative similarity up to any degree of completeness is neither necessary nor sufficient to determine whether two organisms are members of the same species (as it would be were species kinds or general classes); what is necessary for such determination is that the two organisms are links in the same spatiotemporally restricted chain or network of genetic inheritance. Without such a connection between the two organisms there cannot be an evolutionary path between them, and because species are the units in which such paths are laid out, the organisms cannot be members of the same species.

The conceptual status of species names as names of spatiotemporally extended particulars is reflected and reinforced in the character of biological laws, empirical generalizations, and statements of accidental universality or finite scope. Thus, apparently general statements like "beavers build dams" cannot count as biology's candidate for nomological respectability. For they are strictly speaking false and cannot themselves be linked to evolutionary theory as explainable empirical generalizations derived from it. The supposition that species are kinds gives such statements as "ducklings imprint within forty-eight hours of birth" the appearance of universality in form, whereas the physical possibility of exceptions as well as the restriction of their domains seems to deprive them of nomological force. That is why they are sometimes treated as accidental generalizations. It is why more than one provincialist philosopher of science has denied that biology has any distinctive laws and has asserted that it simply applies nomological generalizations from the "harder" sciences.

But, on the assumption that species are individuals, such statements are not the *laws* of biology. They are not even its *rough empirical generalizations,* that is, general statements with exceptions but enough nomological force to permit their explanation by the real laws of biological theory. Among the empirical generalizations of biology are the following statements, which mention no species whatever but are general claims about all species that are roughly correct:

Unspecialized species tend to avoid extinction longer than specialized ones.

Body size tends to increase during the evolution of a species.

Contemporary species living in the same environment tend to change in analogous ways.

The members of warm-blooded species are larger in colder regions than in warmer regions, though their extremities are proportionately smaller.

These are examples of the *lowest*-level generalizations of biology; they manifest exceptions, like other empirical generalizations. But these exceptions can be explained by appeal to the *same* more general claims that help explain the exception-ridden generalizations themselves. And these higher-level claims are the laws of biology. The empirical generalizations, unlike statements about swans or ducks or beavers, can be expected to obtain, ceteris paribus, on other life-supporting planets – and indeed on planets that manifest "life" that is so different from our own as to be unrecognizable to the anthropomorphic eye.

If statements like these are the lowest-level generalizations of biology, then the natural kinds of biology cannot be the particular species that the taxonomist constructs, for these laws mention no particular species. Rather, they mention the selectively significant properties that particular species exemplify. If particular species like *Ursus ursus* or *Canis lupus* are not natural kinds, because they do not figure in even the lowest-level generalizations of biology, what alternative is there for them besides being treated as names of particulars? We have recognized that we cannot treat these terms as we might treat 'phlogiston,' as ones that denote nothing. Species names are not kind terms that have been superseded in the course of scientific advance. They have a useful role: They enable us usefully to individuate and classify the vast number of individual organisms that populate our planet. They enable us to show conveniently the hereditary relations between various organisms, their evolutionary distance from one another, and the kinds of environmental factors that make for differences between them. How can we retain the advantages that the use of species terms provides, consistent with the recognition that they do not reflect natural kinds? Certainly not by according them the status of nonnatural kinds, like the nonnatural kind 'table.' Rather, we may accomplish this by treating species names as names of things instead of kinds, natural things, which are named in an organized way that reveals much systematic, biologically fruitful information about them, while recognizing their particularity and individuality.

Thus there are three arguments for treating species as individuals. First, it makes good sense of the notion that species evolve, whereas treating them as fixed kinds makes a conceptual mystery of this matter. If species were kinds, talk of extinction would be misplaced; on the other hand, the integral connection of this notion to the concept of species will not countenance the appearance of organisms indistinguishable from extinct species as a case of the reappearance of the same species. This, too, reflects taxonomic practice hard to accept if species are kinds. Secondly, species in fact behave in the way individuals do: They split and merge, they have fuzzy borders, and the point at which one species leaves off and another begins is often arbitrary, as it is with particular mountains, say, Everest and Lotze. Finally, treating species as individuals makes sense of the apparent inadequacies of purported generalizations about particular species. In offering them, biologists are actually expressing facts about particular, spatiotemporally scattered objects. None of these arguments is decisive, and each has been challenged.

Thus it has been pointed out that just because "members" of a species have no common and peculiar distinctive properties, their species may still formally be a set or class. In set theory, a set or class may be defined by its extension, that is, its members, and it may be entirely arbitrary: There is, for example, a perfectly well-formed set containing Darwin's nose, the number six, and the Andromeda galaxy. We may give this set a name and even agree that it has no "essence": There are no properties that characterize all its members. Similarly, species may be sensibly defined as sets with members, instead of individuals with constituents, even though they share no properties. All this is quite correct, and it would refute the claim that species *logically* must be individuals or that the heterogeneity of their members *entails* that they are individuals. But the thesis of individuality is not this strong, nor does the argument from the absence of an essence among organisms take this direct form. The form of the argument is an induction to the best explanation of why we can find no properties common and peculiar to organisms in the same species. What is more, opponents of the species-qua-individuals thesis do not hold that species are arbitrary classes, any more than proponents deny the logical propriety of describing them as classes at all. The issue is whether species are biologically nonarbitrary classes or not. If it could be shown that biologically nonarbitrary classes need not have any properties in common, then this would be a more serious objection to the conclusion that species are individuals.

A related objection holds that if organisms are parts or components of species instead of their members, then their parts, the organs of organisms, are also parts or components of their species as well. But, the argument goes, to say a dog's tail is related to the species *Canis familiaris* in the same logical way that the whole dog is related to it is absurd. To this complaint the reply is that the conclusion is indeed counterintuitive in some cases but is at worst an artifact of the reconceptualization that can do no real harm. On the other hand, there is nothing wrong with this conclusion when it comes to organisms like the hydra or the slime mold whose parts are related to their species in the same logical relation that the wholes manifest. At worst, this conclusion represents terminological inconvenience.

Of course, opponents of the view that species are individuals may also have to appeal to inconvenient but possible circumlocution. For example, they must hold that treating species as classes is no insurmountable obstacle to saying that species evolve. Anything we say about the evolution of a spatiotemporally restricted, particular object can be translated with a bit of formal ingenuity into equivalent if somewhat more complicated conjunctions and disjunctions of statements about a set of organisms. The question of whether these sets correspond exactly to species is not at issue. Because of the liberal conditions on set construction in mathematics, there is no doubt that such sets are in principle constructible for any species. These conditions will enable us to express their "evolution" in a cumbersome way. Again, what is required is formal ingenuity. But insofar as the need for inconvenient circumlocution is a sign of theoretical disconfirmation, the possibility noted by opponents of the species-qua-individuals idea is not a strong argument in their favor. We certainly could stipulate that species are kinds or sets like element names, but then we would need new terms to describe those lineages that evolve and those that become extinct.

The problem of how we describe extinction represents a potentially more serious problem for the conceptual revision advocated by species-qua-individuals proponents. For despite their brusque assurance that a new organism qualitatively identical to members of an extinct species would not be counted as a member of it, there seems room to disagree about the matter. And if a contrary conclusion could be unequivocally sustained, the claim that treating species as individuals is entirely compatible with taxonomic practice would be vitiated. The dispute may sound academic in view of the conventional wisdom that evolution does not repeat itself or go backward. Enshrined as Dollo's law, this generalization has greater evidential support than any other conclusion about the course of terrestrial evolution. The trouble is that, without going backward, evolutionary processes may give rise to morphologically or genetically identical lineages that are temporally separated. If one such lineage becomes extinct before the other appears the temptation to group them together in the same species may be very strong.

Some of these possibilities represent revolutionary advancements in genetic manipulation. Thus it may be possible to produce by recombinant DNA technology not just chimerical organisms but replicating, synthetically produced new species. Suppose such a line is constructed from naturally occurring microorganisms. Then, after several generations, it is allowed to become extinct. If the techniques of recombination are precise enough, it will not be difficult to reconstruct the same type of synthetic microorganism again. Here the sameness extends from the primary sequence of its genetic material through all its other traits, for all its other traits are dictated by this sequence. Are we to say that these new organisms are not members of the same species as the original ones? They will qualify for species membership under every criterion we have examined hitherto except the one that makes species particular, spatiotemporally restricted lineages. Are we to say that these synthetic microorganisms have no place in taxonomy and are not species at all? What if their characters are indistinguishable from those of naturally occurring microorganisms that have a place in the taxonomic scheme? This problem is introduced by the quite unexpected revolution in molecular biology. But this is a revolution that can be expected to produce problems of equivalent sorts for all the other conceptions of species. It can circumvent restrictions on reproductive isolation and rearrange distinct species to have the same evolutionary role or to occupy the same niche with equal efficiency. Thus recombinant DNA technology is a problem for all species notions. It is not a special difficulty for the thesis that species are individuals.

But what humans can do in the laboratory, nature can do beyond it. For example, occasionally two distinct species may hybridize to produce a third distinct species, which itself reproduces by parthenogenesis. This appears to be the case for lizards of the genus *Cnemidophorus* (see Kitcher, forthcoming). For all we know, extant parthenogenetic hybrids of this genus do not represent the first lineage to have been produced by hybridization between distinct species in this genus. If a previous genetically identical parthenogenetic line had arisen through hybridization and become extinct, the present organisms would represent a taxonomic problem for the claim that species are spatiotemporally particular. It could hardly be admitted that the new line is part of the same individual as the old. For treating species as individuals could not countenance such total spatiotemporal separation of a part of the individual. On the other hand, insisting that this second indistinguishable lineage is a distinct species seems unmotivated by any consideration save that of preserving the thesis. This counterexample is more than merely logically possible. There are independent, credible, biological reasons to suppose that such parthenogenetic hybrids might not be viable over the long term (just because of their restricted variability). Moreover, the sort of interbreeding that gives rise to them may recur with some frequency at the boundary or hybrid zones of distinct lizard species. We have no positive reason to believe that this particular type of speciation event has occurred more than once, but there is no general obstacle to supposing it has. This is what makes the argument particularly strong.

What weakens it is that at base it rests on the intuition that taxonomists would in fact agree that the proper course in the matter was to label the second lineage with the species name of the first. This universal agreement cannot be assumed. And there are reasons to believe that it would not be forthcoming. One reason is that in agreeing on this course taxonomists would be adopting a commitment to essentialism, at least for this species. Pointing this fact out to them, and showing its general discordance with the rest of their practice, would at least give them pause. Second, if the events of speciation occurred long enough apart under different environments, proponents of the evolutionary or ecological definition of species might have good reason to reject assimilation, as might exponents of the biological notion in virtue of the reproductive isolation of these unisexual organisms. In short, it is simply not clear what should be or would be said about this case. It may present a problem for all alternative accounts. The problem it presents for the notion that species are individuals seems perhaps less serious than for other views. After all, it reflects what is at most a terminological discomfort, a small degree of dissonance between the notion that species are individuals and some ordinary intuitions. This dissonance may be accepted in exchange for clarification on a large number of systematic matters. It must, however, be admitted that if there were taxonomic uniformity on this matter and it was contrary to individualists' conclusions, they would face a serious problem in this connection.

A final set of arguments attempts to undermine the individualist's claim that this thesis makes sense to the nature of evolutionary laws and of the absence of credible nomological generalizations about particular species. One such argument makes virtue out of vice, holding that sciences need not all emulate physics in searching for laws; that, in many areas of respectable science, advance does not consist in the uncovering of laws; and that the apparently general statements of biology about particular species are not disguised statements about individuals. Rather, they are accidental generalizations, the result of causal forces operating on local, initial conditions to generate rough, exception-ridden regularities not entrenchable in general theory.

The trouble with these sorts of claims is that they prove too much and do not solve the problems that the proposal they attack attempts to deal with. It is true that many scientific enquiries focus on case studies and make no explicit appeal to laws. On the other hand, insofar as their particular case studies contain causal assertions, they are implicitly committed to laws – perhaps derived from another science, as the provincialist would have it. But biology is not characterized by the absence of laws; it has generalizations of the strength, universality, and scope of Newton's laws: the principles of the theory of natural selection, for instance. The problem we face is to reconcile these laws with an account of the nature of species, both to legitimate this notion and reveal how the theory is applied in its intended domain. It is no solution to this problem to say that the individualist attempt to reconcile the species notion with the theory and the absence of laws at the level of species is a mistake, because there neither are nor need to be such laws. To say that the fact that beavers build dams or, more realistically, the fact that *Homo sapiens* have forty-six chromosomes is an accidental generalization and not a law, is to describe a symptom of the problem the individualist attempts to solve. Why is it that the best efforts of biologists to produce laws about what they take to be the observationally basic, natural kinds – the species – into which diversity is divided never seem to be any better than accidental generalizations? To admit that this is the best they can do is to admit that species are *not natural kinds*. This leaves only the possibility that they are artificial kinds, scientific mistakes like phlogiston, or that they are not kinds at all, though they still perform a useful task.

The best rejoinder to this sort of objection is to explore further the ramifications of the thesis that species are individuals for the nature of biological generalizations at all levels of theoretical systematization. This will enable us finally to bring the axioms of Chapter 5 into contact with the findings of the naturalist, paleontologist, and ecologist.

7.7. The Theoretical Hierarchy of Biology

If a statement about the color of swans is not even an empirical generalization and the claim that body size increases through evolution is no more than such a generalization, what are examples of biological laws? Some have offered the Mendelian principles of segregation and assortment and their refinements as the most fundamental laws distinctive of biology. But we have seen in Section 5.3 that the onset of the operation of these "laws" is a consequence of evolutionary forces – operating on the initial conditions of terrestrial life. These "laws" arise at the point when cytological evolution results in the appearance of meiosis, selecting this mechanism of variation and heredity as optimal in the circumstances. The point is not merely that these "laws" do not apply in asexual reproduction. Rather, it is that they do not follow from the theory of natural selection alone, but only from it and a description of the local terrestrial boundary conditions that obtained on the earth at a given time in its history. These "laws" are no less restricted in their generality than the far more exception-ridden statements about sets of species canvassed in Section 7.6. Mendel's laws, like these, are empirical generalizations about a large number of spatio-temporally restricted particulars, all those species whose members mostly reproduce sexually through the normal course of meiosis (as opposed to those exceptions that manifest nondisjunction or meiotic drive, for example).

The way to identify the laws of biology is to begin with principles of undoubted generality, unrestricted in expression or in bearing to the actual course of evolution on our planet. Unless we can find such laws, the provincialist will have been vindicated after all.

Such laws of course will include the axioms of the theory of natural selection, the general statements that can be deduced from them, and the general statements deduced from the combination of these axioms with other nonevolutionary laws. None of these laws will make reference, implicit or explicit, to the local conditions under which evolution operates on earth. Because our axiomatization is susceptible of alternative interpretations, which show it to govern genes, organisms, or populations, we may interpret it as a set of claims about the taxon of species, subject to the understanding that because a particular species does not pick out a natural kind, the axiomatization may not be able under this interpretation to explain everything we want to say about the evolution of terrestrial diversity. It will of course have to account for everything we want to say under one or another mutually compatible interpretation. Interpreted as a theory of the evolution of species, the theory's axioms take the following form:

1. Species are lineages of descent among organisms.
2. There is an upper limit to the number of organisms in any generation of a species.
3. Each organism has a certain amount of fitness with respect to its particular environment.
4. If D is a physically or behaviorally homogeneous subclass of a species, and D is superior enough in fitness to the rest of the members of the species for sufficiently many generations, then the proportion of D in the species will increase.

5. In every generation of a species (not on the verge of extinction) there is a subclass, D, that is superior to the rest of the members of the species for long enough to ensure that D will increase relative to the species and will retain sufficient superiority to continue to increase, unless it comes to constitute all the living members of the whole species at some time.

Unlike the merely empirical generalizations mentioned in Section 7.6, these principles are supposed to be exceptionless. These biological laws mention no particular species, and they treat species as particular evolving lineages, not as types, classes, or kinds of organisms. We cannot deduce anything about the evolution of any particular species from these laws, nor can we predict anything about the future of a given species from these laws *alone*. As we have noted, provincialists have criticized evolutionary theory on this score, condemning it as an empty and unfalsifiable account bereft of explanatory power. Autonomists have moved in the opposite direction on the tracks of the same argument, claiming that because evolutionary theory makes no predictions, and because it is clearly a theory of great explanatory power, the demand that biological theory must have predictive content, like physical theory, must be wrong. Both sorts of complaints reflect a misunderstanding of evolutionary theory and of the status of claims about particular species. They mistakenly suppose that statements about the properties of species or their members are general statements and that therefore they should follow from a theory about the evolution of species in general. Because any conditional statement that the theory alone could enable us to infer about particular species is either true because (close to) tautologous or so vague as to be unassessable, critics draw two different sorts of conclusions, depending on which axe they hope to grind: Either strictures on scientific explanation imported from physics are wrong because this clearly explanatory theory makes no specific predictions; or because such methodological strictures are correct, evolutionary theory is a scientifically disreputable enterprise. In fact, the relationship between the theory and particular statements about special species sustains neither of these two views. The theory cannot be expected to issue in the sort of singular statements that are open to test, without the addition of statements of initial or boundary conditions. And yet species-specific statements are of just the former sort. On the other hand, its derivative laws will enable us to make predictions up to the levels of accuracy of the statements of initial conditions supplied.

Such a derived law is the following interpretation of the equilibrium theorem of Chapter 5:

If D_1 and D_2 are distinct species within an environment of cohabitating organisms, D, and if D_1 is superior to D_2 as long as it constitutes less than e, a certain portion of D, whereas D_2 becomes superior to D_1 when D_1 comes to constitute more than that proportion of D, then the proportion of D_1 to D will either stabilize at e or oscillate around e.

Now if D_1 and D_2 are two species related as predator and prey, then our derived law will explain why their populations oscillate around fixed values. The determination of these values is the province of empirical generalizations and mathematical models in ecology. Given two particular species, say caribou and wolves, suppose it can be independently shown that the wolf population depends for survival on predation of the caribou, and that the survival of the caribou depends on avoiding the predation of wolves. Then the theorem enables us to derive the particular statements that the average fitness of wolves will decrease as their ratio to caribou increases, and thus their survival rate decreases, so that there must be a balance between the number of these species. In other words, if we can establish these initial conditions in the interrelations between wolves and caribou, the theory will enable us to predict that each of them will eventually exist at or around a fixed level of population, regardless of what population distribution the species began with.

Of course biologists, like other natural scientists, are not content with such a generic prediction, with the prediction of the existence of an equilibrium value of population for predator and prey species; they would like a prediction of the actual value where that value is unknown and an explanation of that value where it is known. The method of acquiring such predictions and explanations in biological theory has consisted in the construction of a series of mathematical models of increasing realism and sophistication, with a better and better fit both to actual data and to the factors that the theory tells us determine fitness and populations. The earliest of these models was due to Lotka (1925) and Volterra (1926) and took the following form:

$$\frac{dH}{dt} = (a_1 - b_1 P)H$$

$$\frac{dP}{dt} = (-a_2 + b_2 H)P$$

where H is the size of the predated or host species, P is the size of the predator or parasite species, a_1 and a_2 are the net growth rate per individual of the host or predated species in the absence of predation, and b_1, b_2 are parameters measuring the predation rate per individual of the predator and prey species respectively. Our interpreted version of the equilibrium theories is in fact a law that tells us there are values for H and P above and that their derivatives with respect to time are zero. This derived law together with the values of the parameters will enable us to generate a prediction from evolutionary theory about the population level of given species. The early model of course fails to take many evolutionary forces into account and so gives imperfect estimates of these population levels. Subsequent models attempted to correct this lack of realism in several directions. For example, we may add variables to the model to account for the presence and degree of available cover for prey, or for predation as a function of the food needs of predators instead simply of the density of prey. These more realistic models have been tested against such natural predator–prey systems as house sparrows and sparrow hawks, muskrat and mink, snowshoe hare and lynx, mule deer and cougar, white-tailed deer and wolf, moose and wolf, and bighorn sheep and wolf.

Even models more sophisticated than the Lotka-Volterra equations still make highly unrealistic assumptions. Thus, consider the following list of assumptions actually employed in and characteristic of models constructed to account for the relationships among different species:

1. The species under investigation occupies a spatially homogeneous environment and conditions are temporally constant.
2. The system is closed, so that interacting species are not reinforced by immigrations or depleted by emigration.

3. Each species responds to changes in its own and others' sizes instantly, without delay.
4. Variations in the age structure of the members of species do not occur or can be disregarded.
5. The interaction coefficient between each pair of species is unaffected by changes in the species composition of the remainder of the community.
6. The genetic properties and hence the competitive abilities of a species are independent of its size.

We know that every one of the six assumptions listed above is false. Each of the causal forces whose efficacy they deny actually does help determine the level of population of a species. So the models constructed with these assumptions can only provide estimates of this value even when selective forces are constant enough to ensure stability over a long period. But we also know, independently of the models and assumptions, that there is such a value. That is, we can be as certain about the existence of such a value as we are of the general laws of the theory of natural selection from which it is deducible.

When we interpret a model, like the Lotka-Volterra equations, as a claim about particular species, we are not deducing a restricted generalization from the derived or fundamental laws of the theory of natural selection. Rather, we are making a singular statement about a particular spatiotemporally restricted fact. This restricted statement is open to confirmation or disconfirmation; it describes facts about a biological state of affairs to greater or lesser degrees of accuracy. As in mathematical modeling generally, there will be a tradeoff in such descriptions between realism and computational tractability. But the crucial fact is that by the time we have interpreted a mathematical model as a claim about particular species, we are no longer formulating biological laws of any sort.

It should now be clear why statements of the sort cited in Section 7.6 are rough, empirical generalizations, to be explained as such by appeal to evolutionary theory. A statement like

contemporaneous species living in the same environment tend to change in analogous ways

or more precise generalizations like

the members of warm-blooded species are larger in colder regions than in warmer regions, though their extremities are proportionately smaller

are expressions of general tendencies, to which there are well-known exceptions. They and their exceptions are explained by bringing the axioms of the theory, interpreted as claims about species, together with a wide and heterogeneous variety of auxiliary hypotheses and theories from many different divisions of biological and physical science. Together with statements about boundary conditions in which groups of species may find themselves, these laws can explain general tendencies and their exceptions. In the case of the second generalization given above, some of these auxiliary hypotheses will be from physics. They will describe the differences in the rate of heat loss for objects with different surface-to-volume ratios. Those with higher surface-to-volume ratios come into thermal equilibrium with their surroundings faster than those with lower ratios. Because in cold regions fitness will be supervenient on the rate of heat loss, greater fitness will accrue to organisms whose rate of heat loss is lower. Therefore members of species that survive in cold regions will have relatively lower surface-to-volume ratios than conspecific members elsewhere. If these ratios are heritable, then we can expect the tendency described in the law. But we can also expect its exceptions. For whenever decreasing surface-to-volume ratios within a lineage over time are incompatible with some other adaptive response to the environment — like having extremities, lungs, or long legs — that permits escape from arctic predators, there will be tradeoffs. The relative contributions of each adaptation to the fitness of the *subclands*, the diverging populations in a lineage, will determine which in the long run predominates. In some cases, these tradeoffs will lead to exceptions to the generalization about the size of extremities. What is more, the theory of natural selection permits and leads us to expect short-run departures from such regularities as well. This is why they are tendency statements.

7.8. The Statistical Character of Evolutionary Theory

We cannot expect to upgrade such empirical regularities to exceptionless general statements, in part because they are about small collections of spatiotemporally restricted particulars, but also because of the statistical character of the theory of natural selection (more specifically because of the statistical character of Axiom 4). That the theory is a statistical one has long been claimed, but there have been disagreements about the sources of its inevitably probabilistic character. Its statistical nature has sometimes been traced to the stochastic features of Mendel's law of independent assortment; sometimes to the fact that we do not know all the causal forces that govern the actual direction of evolutionary change; and sometimes to the so-called founder principle, according to which speciation proceeds from the isolation of small, randomly determined collections of conspecifics bearing unrepresentative hereditary traits that come fortuitously to be isolated from the remainder of their species.

Now although all three of these facts do make for considerable uncertainty in our knowledge of the course of evolution, they do not reflect any probabilistic mechanisms within the operation of natural selection itself. For instance, confidence that we have accounted for all relevant forces in making an evolutionary prediction is almost impossible to come by because of the sheer numbers of such forces. In this respect, however, evolutionary predictions are no more inherently statistical than Newtonian predictions about systems of large numbers of bodies behaving in accordance with strict, nonstatistical, deterministic laws. For in this latter case we may also be in ignorance of all the forces because we cannot account for all the bodies in the system. Thus our predictions may be in error or may be given with varying degrees of statistically expressed confidence.

The indeterminism introduced into evolution by Mendelian randomness is also only an epistemic limitation, a restriction on the degree of refinement of our knowledge about the course of evolution. Leaving aside the fact that whatever randomness it introduces is not properly credited to the theory of evolution, Mendelian theory does not suggest that the segregation of genes is causally undetermined; it does not hold or imply that meiosis is an indeterministic phenomenon like radioactive decay.

Rather, it is predicated on the notion that the causal determinants of segregation, like those involved in the rolling of fair dice, are inaccessible to practical genetic enquiry, that they are large in number and uniform in their operation on the very heterogeneous molecular boundary conditions in which segregation occurs. Accordingly, we may be confident that genetic segregation is, like roulette, statistically random but also causally deterministic.

The same must be said about the statistical influences of widespread speciation through small, randomly chosen founder populations. The randomness is here again not of the ineliminable sort found in quantum mechanics. In this case, to say that the collection of founder organisms is a random one is not to say that its membership is not causally determined but to say that it is determined by nonevolutionary forces rather than evolutionary ones. It is to say that the initial collection is not the result of natural selection but of nonselective forces. The number and variety of these forces, our relative ignorance of them, and our inability to plot their effects together with those of selection introduces a statistical element into the application of the theory. Moreover, to the extent that the founder principle describes only one among several different vehicles of speciation, it is not an integral part of the theory of natural selection itself. Like a model for predator—prey relations, it is a model in which boundary conditions together with derived laws of the theory generate the effects described.

The real source of the statistical character of the theory of natural selection itself, as opposed to epistemic limitations on its application and probabilistic aspects of auxiliary hypotheses employed to apply it, is in Axiom 4. This axiom's claim, that fitness differences are large enough and the life span of species long enough for increases in average fitness always to appear in the long run, is of the same form as the statistical version of the second law of thermodynamics. According to this latter principle, thermodynamic systems must in the long run approach an equilibrium level of organization that maximizes entropy. Over finite times, given local boundary conditions, an isolated mechanical system, say the molecules of gas in a container or the bodies in motion in a region of the universe, may interact in such a way as to move the system's level of entropy away from this equilibrium level. But given enough interacting bodies and enough time, the system will always eventually move in the direction prescribed by the law. Accordingly, we may attach much higher probabilities to the prediction that for any system not in equilibrium its organization in any time period will reflect a greater level of entropy than in any previous period. We may be confident in this prediction even though its statistical nature makes it compatible with the highly improbable opposite outcome. As we increase the amount of time between two periods and the number of interacting bodies, the strength of our probabilistic prediction will increase. The basis of this probabilistic thermodynamic principle is of course entirely deterministic. It involves the supposition that the bodies behave in accordance with strictly deterministic principles of Newtonian mechanics, but that the mechanical values of position and momentum realized by any one of the bodies at a given time is independent of the mechanical values for the others at the same time and is different from them.

The same kind of generic claims that the second law of thermodynamics makes about a sufficient amount of time and a sufficiently large difference in causal values figures in Axiom 4. And their role is the same as the parallel expressions in thermodynamics: They reflect the possibility that evolution need not and does not move in a straight line toward equilibrium levels of population size for various species and their subpopulations. It asserts only that, over the long run, evolution must move in this direction and that the length of the long run is a function of these differences. Additionally, the axiom, like the second law of thermodynamics, rests on deterministic considerations: For fitness differences supervene nonevolutionary properties in the way that thermodynamic properties like temperature or entropy supervene nonthermodynamic properties like position and momentum. The statistical nature of evolutionary predictions and the character of its empirical generalizations as tendency statements both rest on suppositions of the mutual independence of the supervening nonevolutionary variables – and on the fact that the initial distribution of their variables among organisms that evolve is independent of the forces of selection that operate upon them.

Viewed in this light, the axiom's appeals to "sufficiently many generations" and to "sufficiently large" differences in fitness are not admissions of vagueness or uncertainty about how much is enough. They are reflections of the interrelationship between evolutionary variables and nonevolutionary ones on which they are supervenient. Failure to recognize the statistical character of this claim can lead to mistaken charges that the theory is false or unfalsifiable. These are charges that parallel mistakes about thermodynamics. The statistical version of the entropy law supersedes a deterministic version requiring that entropy always increase, even in the short run. The fact that there are closed thermodynamic systems that actually or potentially disconfirm this version of the law is one of the motivations for the statistical version. Similarly, if we treat Axiom 4 as a statement of what always occurs, and occurs in the short run, it is likely to be disconfirmed by facts uncovered in the field, or more rarely in the laboratory. If we refuse to treat such findings as disconfirming the theory, it may be accused of unfalsifiability. The mistake of supposing that such findings suggest either falsity or unfalsifiability rests on treating a statistical theory as a deterministic one.

It was Charles Sanders Peirce who first recognized that the apparent indeterministic or static character of the theory of natural selection was identical to that of statistical thermodynamics:

> Mr. Darwin proposed to apply the statistical method to biology. The same thing has been done in a widely different branch of science, the theory of gases. Though unable to say what the movements of any particular molecule of gas would be on a certain hypothesis regarding the constitution of the class of bodies, Clausius and Maxwell were yet able, by the application of the doctrine of probabilities, to predict that in the long run such and such a proportion of the molecules, would, under given circumstances acquire such and such velocities; that there would take place, every second, such and such a number of collisions, etc.; and from these propositions they were able to deduce certain properties of gases, especially in regard to their heat relations. In like manner, Darwin, while unable to say what the operations of variation and natural selection in every individual case will be, demonstrates that in the long run they will adapt animals to their circumstances. (Peirce, 1877:15)

This disanalogy here of course is that, although Clausius and Maxwell had a single underlying theory already in hand, Newtonian mechanics, to underwrite the statistical claims of thermodynamics, neither Darwin nor his successors have acquired a similar underlying theory for natural selection. We can now see clearly why not.

There is no single underlying theory; rather, there is a vast congeries of them, as many as are involved in the relations and properties on which fitness is supervenient. It is indeed one of the special strengths of the axiomatization of Chapter 5, and the associated treatment of fitness, that it provides an explanation for both the strong analogy between the statistical character of this theory and thermodynamics and for the important disanalogies between their foundations in more fundamental theories.

Pursuing the analogy for the nature of species, the role they play with respect to the theory of natural selection is equivalent to the role a particular container of gas, in a particular internal combustion engine, for example, plays in thermodynamics. Just as we can expect no mention of what exactly happens in your car or mine from thermodynamics *alone,* we can expect no mention of particular species in the theory of natural selection either.

7.9. Universal Theories and Case Studies

The upshot of our long discussion of the theory of evolution, of fitness, and of the nature of species is a very radical one. For in the end it suggests that, strictly speaking, there are in biology at most two bodies of statements that meet reasonable general criteria for being scientific theories. These will be the theory of natural selection and such general principles of molecular biology as are free from any implicit or explicit limitation to any particular species or indeed any higher taxon of organisms restricted to this planet. There are at most two theories because it may well be that there are no such distinctive statements of molecular biology and that its universal, unrestricted, general statements are all properly identified as within the province of organic chemistry. This, however, is merely a boundary dispute without biological ramifications. What does have ramifications is the fact that if species are spatiotemporally restricted, scattered, particular objects, then of course so are all the higher taxa in which species consist. Thus, the entire biomass of this planet is such a particular object, whose component parts at various levels of organization interact and evolve in accordance with principles that, if true, are true everywhere and always, on earth and everywhere else: on planets, if any, circling Alpha Centauri.

But this means that all those special branches of biology, ecology, physiology, anatomy, behavioral biology, embryology, developmental biology, and the study of genetics are not to be expected to produce general laws that manifest the required universality, generality, and exceptionlessness. For each of them is devoted to the study of mechanisms restricted to one or more organism or population or species or higher taxa and therefore is devoted to the study of a particular object and not to a kind of phenomenon to be found elsewhere in the universe. Each of the disciplines of biology, besides the most general molecular biology and evolutionary theory, is to be viewed as a *case-study*-oriented research program. The *model systems* on which each of these disciplines work are not to be expected to generate exceptionless regularities and laws in, for example, the way that working out the model system of the Bohr hydrogen atom or acid−base titration models in physical chemistry generates laws of nature. For these latter objects represent natural kinds, whose instances can and do occur throughout the galaxy. Particular species do not recur, and they cannot be expected to be much like particular lines of descent either elsewhere on the planet or elsewhere in the universe.

The pursuit of case studies, which the rest of biology constitutes, proceeds by the employment of the general laws of evolutionary theory and molecular biology. It appeals as well to whatever other regularities from the physical sciences it requires. It applies them all to particular facts about spatiotemporally restricted particulars, produced through the operation of natural selection over the course of several million years on an initial distribution of chemical matter, whose behavior is governed by laws of organic and physical chemistry. In this sense, there are but two distinctive biological theories.

This view of the character of theoretical biology seems prescribed by the notion that terrestrial evolution is the operation of the fundamental laws, the axioms of natural selection, on the biochemical starting material. This operation has, over the course of eons, generated the diversity we recognize and classify today. If everything on the planet is a reflection of universal laws operating on initial conditions, then the fundamental laws that account for the biosphere are just the ones that determine the consequences of the initial conditions. Therefore the study of the restricted phenomena that appear later in the sequence of events, starting from these initial conditions, must either appeal to these fundamental laws, or to their derived theorems, for the description and explanation of these subsequent phenomena. Because the identification of initial conditions cannot figure in general laws, because they vary from situation to situation, the description of their later consequences cannot figure in such laws either. We can express findings about particular model systems, about particular species, that have some application beyond their original domain. But we cannot expect them to be first approximations in a series of improvements that will result in universal laws of physiology, or of ecology, embryology, agronomy, etc. For each of these disciplines is dedicated to the study of particular groups of species. Even were they devoted to studying the physiology of all species, or their comparative anatomy, development, etc., they would still constitute the study of a particular object and not a natural kind. Accordingly, we must expect to always find both exceptions to our most general claims and limits on the degree to which they can be improved simultaneously in the direction of specificity and truth.

The conclusion that, beyond evolutionary theory and biochemistry, the life sciences must be viewed as the study of cases is no defect in them, nor any reason to suppose they differ methodologically, epistemologically, or metaphysically from the physical sciences; nor is it any reason to urge or expect that the biologist turn from the study of cases to the search for general theories. It is, rather, the basis for settling many of the outstanding questions of the philosophy of biology.

In other scientific disciplines, the subject matter is divided along theoretical lines. In physics, there have long been recognized relatively independent subdisciplines like mechanics, thermodynamics, and electromagnetism. In this century, the lines between these distinct subjects have disappeared. The lines originally reflected the relative independence of the laws of Newtonian mechanics, phenomenological thermodynamics, and Maxwell's theory of electromagnetism. The lines between these disciplines have begun to disappear because their laws have come to seem less and less independent from one another and to all reflect quantum-theoretical and relativistic regularities. In chemistry, the divisions are based in part on the employment of distinct laws and in part on the particularities of the chemicals dealt with. Thus, polymer chemistry is distinguished by the kinds of bonds it deals with, unlike hydrocarbon chemistry, which is distinguished not by a set of laws but by its subject matter: molecular combinations that include carbon and hydrogen. By contrast, aside from the nomologically characterized subdisciplines of biochemistry and evolutionary biology, the subdisciplines of biology are not characterized by distinctive laws that may or may not be reducible to those of evolution and/or biochemistry. They are determined and circumscribed by their subject matter. And furthermore, unlike a subdiscipline of chemistry, their subject matter is not a *kind* of substance or event we can expect to recur throughout the universe.

Even when the study of a general phenomenon like respiration reaches the molecular level of detail we know as the Krebs cycle, the biosynthetic pathways plotted are restricted in their domains of realization. The chemical reactions are of course universal, and the molecular mechanisms whereby the enzymes catalyze steps in glycolysis and oxidative phosphorylation are perfectly universal. But the complete details of any particular mechanism – its particular steps, chemical conditions, reaction rates, the primary structure of its catalyzing enzymes – will differ, sometimes only slightly, sometimes very greatly, from species to species. There will be important similarities, large enough in number and close enough in structure to make empirical generalizations worthwhile, even though we know they have exceptions that cannot be reduced or eliminated in a systematic way. This is, of course, characteristic of case studies and their use. They proceed by the application of general theories independently established and by the discovery of the local details of the system under study. Blocking out and then filling in the details of how the system under study functions results in a description that can usefully be employed to facilitate the study of a distinct but similar case. For example, studies of photosynthesis in plant chloroplasts and of the ATP synthesis in animal mitochondria have shed light on one another's mechanisms. But the reason such studies have never generated propositions that are on their way to being established as laws is that they focus on the study of processes that are spatiotemporally restricted in their descriptions. As such, they are likely to reveal that, within two organisms of the very same species, the detailed molecular mechanisms are slightly different, even when the more coarse-grained descriptions in physiological or functional terms are the same; that sometimes the same physiological process is generated by two different molecular mechanisms in conspecifics or across species; and that even coarse-grained physiological processes are not uniform across or within species. This, of course, is just what we would expect in the investigation of a heterogeneous collection of individuals brought together under a relatively arbitrary general category – or brought together under a name that is quite neutral on whether there are commonalities among the components of the thing it picks out.

If there are no laws about particular species or about groups of species, then why pursue the case studies of them and the individual components of them? Why not eschew the search for laws about species and search only for such evolutionary regularities as can be generated by the operation of Axioms 1 through 5 on whatever combinations and arrangements of the molecular stuff of life organic and physical chemistry permit? This is the recommendation that we will have to endorse, if biology must have the generality characteristic of physical science. But although this is, logically speaking, a coherent proposal, even if practically possible, it would serve only philosophical needs. Our practical biological need is for useful, implementible knowledge about this biosphere that can have a payoff in the short-term control of the environment, in the development of technologies for health, in the satisfaction of nutritional needs. This need will only be met by the case studies to which biology now devotes its attention. In so doing, it does indeed forgo its opportunity, real or perhaps only imaginary, to uncover generalizations intermediate between evolution and molecular biology, but it does so for a good, practical reason.

Moreover, if there are nomological regularities to be uncovered between the domains of these two theories, we have already seen that they are probably beyond our practical powers to implement — let alone discover. Even the restricted generalizations of Mendelian genetics, which we can now see to be themselves the reflection of the features of a small set of particular individuals, the sexually reproducing species, cannot be generated from the regularities of molecular genetics or biochemistry. Imagine how much more difficult it would be to find a set of regularities that would constitute a complete theoretical account of heredity in general. Such an account would have to be minimally compatible with all the particular vehicles of it throughout the universe. It would have to explain them all, given a specification of the initial conditions under which each has appeared. The chances of producing such a theory by reflection on the limited sample of methods of hereditary transmission accessible to us is vanishingly small. The chance that there is such a theory and that it is manageable enough to have explanatory or predictive utility for creatures of our intelligence and computational powers is even smaller. Yet this is the sort of theory we should have to look for if we are to take seriously a recommendation to search for more biological laws.

It is practical needs that have divided biology into disciplines along the lines of case studies, instead of along the lines of a hierarchy of laws, at many levels between the molecular and the evolutionary ones. The initial description and division of biology's subject matter into the taxonomy embedded in ordinary language, medicine, agriculture, animal husbandry, etc., did not divide nature at "its" nomological joints. It divided nature in the light of untutored observation and practical manipulation. The subsequent divisions were based on these original ones because our practical needs remained the same and continued to override our theoretical desires.

But there is another, more fundamental source for the disjunctiveness and specificity of our knowledge of the biosphere, one that is unavoidable, given that life is the product of natural selection operating repeatedly on successive levels of organization. Begin with the assumption that natural selection first operated on the constituents of the primal soup, a quantity of chemical compounds that came together through the operation of purely chemical and physical forces on the distribution of matter in the vicinity of the sun. Or we may assume, if we like, that natural selection's first opportunity to operate appeared with the arrival of DNA on a starship from elsewhere. In any case, provided only that there is a mechanism of reproduction and heredity for these molecular aggregates (one determined by purely physical factors and not at this first stage by evolutionary ones), natural selection will choose among them the fittest. In this simple setting, fitness is of course not just supervenient but directly reducible to or definable in terms of a small number of molecular characteristics and environmental variables. Surviving configurations will aggregate into more complex molecules, perhaps a handful of different arrangements. Selection will cull the fittest among these. But already at this level there may be ties for the title of fittest, and two or more arrangements may be permitted to persist by selective forces as equally optimal or satisfactory. And so on ad infinitum. At each level of organization there may be different arrangements, all equally fit. These differences may be small, and at the next level of aggregation they may result in arrangements only slightly and perhaps not even significantly different, so that their behaviors are roughly or even finely similar. At any given level of organization, what survives will be a function of natural selection operating on a disjunction of arrangements of equally adapted subsystems at the immediate lower level or at levels below this one. Thus in the course of phylogeny from the primal soup we can expect slight or considerable differences at one level to be selected and rendered indistinguishable to selection at a higher level. The result will be that when we come to uncover the underlying mechanism of a type of biological system identified, say, by its function at one level of organization, we shall discover a variety of slightly different subsystems operating within otherwise indistinguishable systems. Thus a general description of the mechanism of operation for all examples of the type of system under examination will have to be disjunctive if it is to be complete. If there are too many disjuncts for manageability, we may never attain completeness; or, having attained it we may be unable to employ it or even express it usefully. But an incomplete description, one with exceptions that cannot be progressively eliminated, cannot be a law. It can of course be extremely useful, and it certainly constitutes important scientific knowledge.

However, the crucial point is that the disjunctive character of biology is a reflection of the operation of general evolutionary laws. They are the source of the practical difficulty of providing other distinctive biological laws. It is to them that we must trace the persistence of teleological attributions in functional biology; it is to them that we must trace the ineliminable biological elements in even the most completely chemical explanations of biological processes; it is to the theory of natural selection operating on terrestrial boundary conditions that we must trace the supervenience of Mendelian genetics on molecular genetics and of fitness on its diversity of physical realizations. Finally, it is to them that we must trace our inability to provide an account of the notion of species that will link taxonomy with biological theory and therefore with the rest of science. So, although there are no other theories in biology besides the theory of natural selection and molecular biology, these two suffice not only to explain biology's peculiarities as a natural science. They also serve to substantiate its much more considerable commonality with physical science.

For the nature and status of evolutionary theory provides all we could want in a counterexample to the provincialist's denial that there are distinctively biological claims that meet the standards of physical science for being laws. And their role in explaining the apparently irreducible empirical generalizations of this science undercuts the provincialist's claim that what cannot in practice be reduced to physical science cannot be accommodated to it and must be repudiated as error or even pseudoscience. At the same time, the way natural selection forces biology to be a science of case studies substantiates the provincialist's oft-claimed treatment of biology as applied science. Provincialists have bracketed biology with engineering as disciplines without their own autonomous laws and theories but with a distinctive subject matter, one they treat by applying laws imported from the physical sciences. It is of course correct that biology does help itself to findings and theories from physical science. But biologists are practically interested in occurrences on this planet since life appeared. And because of the limitations imposed by their spatiotemporally restricted taxonomy, biology's natural unit of study is delimited in the way that engineers' concerns are directed to particular projects at varying stages of progress. Although this much of the provincialist's indictment of biology may be accurate, it turns out to constitute no scientific felony, nor even a misdemeanor. Biology is after all a perfectly "hard" science, whose differences in degree from physics are consequences both of the facts that it is called upon to systematize and of the practical constraints on useful systematization of these facts that physics imposes upon itself as well.

These conclusions no more vindicate autonomism than they substantiate provincialism. For the facts that distinguish biology from physics do not make for a metaphysical, or even an epistemological, difference between these subjects. Rather, they reflect limitations in our human ability to marshal and deploy nomological regularities that are logically as open to empirical discovery by biologists as are the laws of physics by physicists. They also reflect a divergence of interests between biologists and physicists. The latter are traditionally taken to be interested in uncovering the fundamental regularities that govern the universe everywhere and always. Having uncovered many of these regularities, they and others have applied them to rearranging and understanding physical processes on this planet. By contrast, though we must firmly distinguish the course of evolution from the general theory that describes its mechanism, biologists have been traditionally more interested in processes on this planet, in the course of terrestrial evolution. For them a generalization's interest depends on how much detail it can reveal about the intricacies of actual biological processes operating here and now, or here long ago. To answer questions about these processes, biologists need and produce the same sort of scientific claims as physics. And, just as in other sciences, these claims must actively cohere with those of other disciplines.

But biologists can and must sacrifice the assurance that the regularities they uncover are true everywhere and always to assure themselves knowledge of the greatest detail about the immediate facts before them. Once the status of their claims – as findings about particular facts instead of autonomous general theory irreducible to more fundamental physical laws – becomes clear, the autonomists' conviction that these findings establish the independence of biology from physical science is put into perspective. Many of the claims of biology are independent of the laws of physics only because they are like particular claims in physics, say, that a lunar eclipse has occurred, or that a particular pulsar is rotating at 10^6 revolutions per second. They are statements of individual fact and so do not follow directly from any generalization whatever. When we see that biological findings are in principle explainable as the result of the operation of general laws, evolutionary, biochemical, and physical, on given boundary conditions, the rationale for autonomism disappears as completely as that of provincialism. The last refuge of both autonomists and provincialists is the isolation of evolutionary theory from the rest of science. But if, as we have seen, its individual principles have reasonably direct physical explanation, detailed up to the limits of supervenience, the last stronghold of these two mistaken views is overrun. Yet because the physical explanation of evolutionary laws must, given the nature of the facts, be schematic in the extreme, both sides are likely to remain unconvinced. Both may hold that the reduction rests on controversial metaphysical presumptions like the finitude of nature and causal determinism, presumptions too weak to confer the actual physical reduction provincialists require and too strong to allow the epistemological independence autonomists claim. However, those who hold biology to be theoretically dependent on more fundamental physical facts – but practically independent of more fundamental physical theories – should find this account of the matter perfectly sufficient. For it demonstrates that biology is a natural science that does not after all differ in any significant *kind* of way from physics or chemistry.

Introduction to the Literature

Darwin's notion of species, and the subsequent history of the subject, are treated by one of the principals in the debate over its definition in Mayr, *Growth of Biological Thought*, pp. 265–97. The relevant portions of *On the Origin of Species* are quoted and analyzed.

The locus classicus of phenetic or numerical taxonomies is R. R. Sokal and P. H. A. Sneath, *Principles of Numerical Taxonomy* (San Francisco, Freeman, 1963). Its operationalist rationale is examined in David Hull, "The Operationalist Imperative: Sense and Non-sense in Operationalism," *Systematic Zoology*, 17(1968):432–59. Hull also examines alternative taxonomic theories in "Contemporary Systematic Philosophies," *Annual Review of Ecology and Systematics*, 1(1970):19–53. The most fervent critic of essentialism in biology is Mayr; *Growth of Biological Knowledge* contains many discussions of and arguments against it. Elliot Sober, "Evolution, Population Thinking and Essentialism," *Philosophy of Science*, 47(1981):350–83, is an excellent exposition and defense of Mayr's position.

My account of various species notions and their defects is largely derived from P. Kitcher, *Species* (Cambridge, Mass., MIT Press, 1985). My conclusions, however, differ from his. The biological species notion was first explicitly advanced in E. Mayr, *Systematics and the Origin of Species* (New York, Columbia University Press, 1942). G. G. Simpson, *The Principles of Animal Taxonomy* (New York, Columbia University Press, 1961), advocates the evolutionary species concept, and L. van Valen, "Ecological Species, Multispecies, and Oaks," *Taxon*, 25(1976):233–9, introduced the ecological notion of species.

The explicit suggestion that species are individuals was first made by Michael Ghiselin in *The Triumph of the Darwinian Method* (Berkeley, University of California Press, 1969). See also Ghiselin, "A Radical Solution to the Species Problem," *Systematic Zoology*, 23(1974):536–44. Hull elaborates and develops arguments for this view as well in "Are Species Really Individuals?" *Systematic Zoology*, 25 (1976): 174–91, and "A Matter of Individuality," *Philosophy of Science*, 45 (1978): 335–60. Exponents of punctuated equilibrium have embraced it for their own reasons. S. J. Gould, "Darwinism and the Expansion of Evolutionary Theory," *Science*, 216(1982):380–7, endorses the view. The fullest critical treatments of this view are to be found in Kitcher, *Species*, and in M. Ruse, "Species Are Not Individuals," manuscript.

CHAPTER 8

New Problems of Functionalism

It should be clear that neither autonomists nor provincialists have been vindicated by our results. So far as relatively abstract matters of metaphysics and epistemology are concerned, provincialist premises have been held to be more nearly correct. Accordingly, what there is in biology that cannot be accommodated in principle to physical science should be jettisoned. But it turns out that there is nothing of importance in biology that once properly understood cannot be integrated with physical science.

But this conclusion is perfectly consistent with a philosophically modest autonomism. Biologists and philosophers endorsing this latter view must hold that so far as the less abstract matters of methodology and the quite practical issues of research tactics are concerned, biological knowledge and its methods of acquisition are not soon to be absorbed by those of physics and chemistry. The practical interest of biologists in pursuing their case studies shows this claim, or some version of it, to be true, even while the metaphysical monism of the provincialists must also be accepted. But adopting the latter requires serious qualifications on autonomists' conclusions. Where there is no metaphysical difference there cannot be an epistemological one. So if apparent epistemological differences are found between two sciences that are not distinguished metaphysically, these differences are either spurious or philosophically inconsequential.

But to be philosophically inconsequential is not to be practically or technically inconsequential. And although the differences autonomists cite are sometimes nonexistent, and never philosophically consequential, they are in many cases of the greatest practical moment. In fact, to say they are practical instead of theoretical or philosophical is itself misleading. For we have seen that the impossibility of justifying some biological claims by grounding them in nonbiological ones reflects limitations on human intellectual capacities. Our brains are simply too small, too slow, and too fallible to store and manipulate the data that would enable us to proceed in biology from the bottom up. What is more, we do not need to do this to justify many distinctively biological claims. On both these counts autonomists come out appearing more nearly correct, provided their philosophically controversial claims are watered down into methodologically anodyne ones. So interpreted, autonomists turn out to be right for the wrong reasons. They repudiate the strong metaphysical basis their intended conclusions require; and, when suitably weakened, their conclusions are right and do not require vitalism or any other metaphysical view that provincialists could not accept.

226

On the other hand, provincialists' standards for scientific respectability turn out to be correct, but their criticism of biology for failing to meet them are gratuitous. For biological claims, even of the kind held by autonomists to be irreducibly biological, do meet these standards and need not be surrendered. Provincialists' errors are about biology, not philosophy. By misclassifying particular findings as laws, generic theories as tautologies, supervenient notions as irreducible ones, one may come to the conclusion that most of biology above the study of macromolecules needs to be replaced. For it cannot be brought into active coherence with the rest of science. Provincialists go wrong because of mistakes they make about biology, mistakes largely shared by autonomists. It is because they both make these mistakes that they see the alternatives as either the autonomy or the elimination of much contemporary biology. Once matters are seen in their correct light, the third alternative becomes inevitable: Biology as it stands is a perfectly respectable science, meeting all the same standards as physical science and theoretically coherent with (though tactically autonomous from) it.

These tactical differences are sufficient to ensure the continuing distinction of biological theorizing from physical theory, no matter how much we know about the physical mechanisms underlying the biological phenomena and how their organization eventuates in it. Biological thought will still be organized around the identification of functions, even when everything structural is known about the directively organized systems to which the functions are attributed. This is *not* because, as a matter of empirical fact, biological systems are irreducible, teleological ones. The persistence of this mode of attack on biological questions does not reflect such a presumption, especially not in the context of molecular biology and biochemistry, where it seems factually *false*. These descriptions will persist because the regularities uncovered at any given level of biological organization are invariably the result of a disjunction of regularities operating on a disjunction of component subsystems at the next lower level of organization. Because the number of regularities and the number of different types of subsystems are so large, we cannot conveniently describe, explain, or predict the behavior of the whole system at the subsystem level. We can only do so for the external determinants and especially the consequences of the whole system's operation. But to describe something in these terms is to adopt a functionalist perspective on the phenomena under investigation.

When we add the fact that such functional ascriptions are made in case studies of components of spatiotemporally restricted particulars, the practical irreducibility of such findings to general statements familiar from physics becomes a logical irreducibility. And this is the second reason such descriptions are destined to be fixtures of biology. For all the implicit generality that biological theory manifests, biologists' interests are parochial in a way physicists' interests almost never are.

It is not only paleontologists who have a special fascination with the course of evolution on this planet. This fascination is the occupational framework for almost all biologists, no matter with what systems they work. Only two kinds of biologists can escape a special concern with "history" — with the particular sequence of events that has occurred on the earth since its creation: those theoretical and mathematical biologists whose interests are in the purely formal elaboration of the theory of natural selection and those biochemists whose interests are indistinguishable from the research agenda of physical chemistry. Every other biologist is involved in attempting to understand a model system of greater or lesser generality: for example, the rat's liver, the chick limb bud, or the Hela cell line derived from one particular cancer patient and now used throughout the world in the molecular study of cellular regulation. But whether the research object is broad or narrow, its existence and character are a consequence of the operation of evolutionary forces on local conditions, obtaining back at the primal soup. Each object of study has therefore been the product of and the survivor of a particular, spatiotemporally restricted course of evolution. There is no reason to think each object exemplifies a natural kind of system, a type of object governed by regularities that are instanced wherever life, or even hydrocarbon life, has appeared throughout the galaxy. The features of a system that we hit upon to identify and distinguish it as an object of study are presumably shaped by this course of evolution. When the organization of these systems of study is broken down for purposes of compartmentalized study, for example, the mammalian circulatory system, the eucaryotic mitochondrion, or immunoglobin gene, the divisions are contingent on hypotheses about how parts of the organism were shaped by selective forces (including other parts and their developmental determinants) operating over the course of evolution to enhance fitness. As we have seen, fitness is consequential in its measurement and therefore would have the appearance of teleology, even without its ordinary semantical associations. Every biologist interested in structure must therefore be interested in history, for it turns out to be a guide to structure. This apparently innocuous claim has been a subject of great controversy in the very recent past, as we shall see in Section 8.2. But if it is right, it is another important tactical difference between biology and physical science, reflected in the employment of functional explanations and ascriptions.

It is probably obvious why there should be this difference in the type of research object between the life sciences and physical sciences. The practical payoff of the life sciences is more immediate in effect and more needed in the short run by individuals than payoffs from the physical sciences. What we can learn from biology about manipulating this particular environment, as opposed to any environment anywhere, answers our immediate needs to feed, clothe, and protect ourselves in a way that much general theory cannot do, just because it is so general. After a certain point in the understanding of our environment we cannot improve our practical abilities to control it without recourse to such general theoretical considerations. But we have not reached that point in the life sciences, and in biology it is probable that when we reach this point, the theory we need will be practically impossible to implement. For the moment, we have not exhausted the usefulness of case studies of restricted systems.

8.1. Functionalism in Molecular Biology

These considerations shed light on a curious inversion of modern biology. Until the postwar revolution in this field, the chief activity of the biologist was to search for structure that would account for previously identified function. The difficulties in finding such structure leads to the autonomist's attitude. But now biologists are increasingly searching for functions to attribute to structures identified independently. Why they should do this would be a mystery if functional ascriptions are viewed as merely surrogates for inaccessible structural ones. For in these cases we have the structural descriptions, and we still want the functional ones.

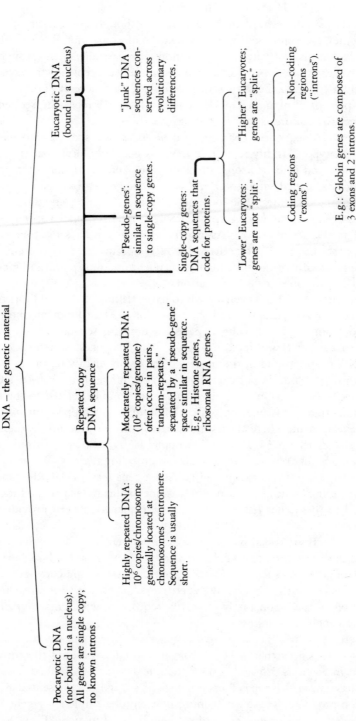

Figure 8.1. Structural differences among DNA sequences in genome. The nine distinct classes of DNA sequences distinguished here were structurally identified by molecular means, not by differences revealed in breeding experiments or protein assays.

In fact, the more that is learned about structure in the context of molecular biology, the more functional questions are raised. This striking fact deserves illustration and examination. We may do both by reviewing some of what has recently been learned about the structure of the molecular genome.

This information is summarized in Figure 8.1. Although both eucaryotic and procaryotic genomes are composed of DNA (except of course for RNA viruses), eucaryotic genetic material contains a large number of repetitions of the same base-pair sequences. For example, DNA from a single mouse has been shown to contain more than a million copies of the same sequence of base pairs several hundred units long. About 30 percent of mammalian DNA consists of such repeating sequences, whereas this structural feature is absent from procaryotes.

The repetitive DNA sequences characteristic of eucaryotic species are divisible into two further subtypes: Highly repetitive DNA is generally located at the chromosomal centromere and in bands some distance away from it. In some *Drosophila* species these bands, so-called satellite DNA, consist in a seven-base-pair sequence repeated a vast number of times. The location, primary sequence, and amounts of this highly repetitive DNA are known to be closely similar for related species. But what is it for? The absence of any RNAs associated with it and its location in the chromosome suggest that highly repetitive DNA is not involved in protein synthesis. Perhaps it is causally responsible for aspects of chromosome pairing in meiotic cell division and cell fusion during fertilization.

Moderately repetitive DNA has been shown to have striking structural properties. These genes do control biosynthesis, and the genes for the four kinds of ribosomal RNA are a good example. They occur in pairs, in tandem, separated by a spacer region of DNA that is not transcribed into RNA. Histone genes, which code for a protein (that DNA wraps around in the nucleus), are also tandemly repeated. In sea urchins, for example, the number of copies of this gene ranges from 300 to 1,000. The number of such genes within a species' genome seems to be related to the need for the rapid production of histones on which DNA may be wrapped during early embryological development. There are several different histone proteins, as there are several different ribosomal RNAs. Like the ribosomal RNA genes, the histone genes are clustered in groups that repeat, with spacers of noncoding regions. While in this case the need for large numbers of copies of the same gene seems clear, the tandem repetition and clustering, that is, the structure, is unexplained. That is, no function has been assigned to them, nor has an evolutionary explanation been provided for their occurrence.

Most protein products are coded by only one gene per genome. This is so even for proteins required in vast quantities over short periods of time. And the lion's share of the genome is composed of such single-copy genes. These single-copy genes are themselves interspersed with moderately repetitive sequences of 300 base pairs. Again, the function of most moderately repetitive sequences, which make up about 20 percent of the genome, is unknown.

But, even more striking, the single-copy genes in higher eucaryotes contain within themselves sequences that have no role in protein synthesis: the introns, already described in Sections 3.5 and 4.6. Although they are transcribed onto the RNA, the introns' transcription products are excised and go unrepresented in the protein-translation products. These intervening portions of a single-copy gene for a protein product seem to be irrelevant to the whole gene's function. They do not contain information needed for the construction of the body's macromolecules. They seem to have no effects on biosynthesis of proteins whatever. Accordingly, their function is a mystery.

As the structure of the genome is uncovered, more and more structural differentiation is revealed. These structural differences within the macromolecules that constitute the genome were not discovered by first identifying distinct functions and then searching for their underlying structural mechanisms. Functional information about other biological structures of course played a part in identifying these structures, but not information about *their* functions. Yet, once uncovered, molecular biologists made the identification of their function a highest priority. They have been treated in two different ways. Either attempts are made to find an immediate functional activity for them within the physiology of the organism. Or a wider evolutionary, adaptive, explanation of their presence is sought. The second strategy usually takes over when the first appears to be failing.

Thus, to illustrate the first tactic, the localization of highly repeated DNA to the centromere, and the absence of any RNA associated with it, leads to speculation that its function is to be found in meiosis and mitosis. But persistent failure to find such functions can lead to the conclusion that a quantity of DNA in the genome, even if repeated, and even if closely similar in sequence to a functioning gene, really has no regulatory role. It may be "junk" DNA, "selfish DNA," or a pseudogene, a parasitic "free rider" in the genome. However, this hypothesis requires an argument to show how features of the genome that meet no physiological need and confer no advantage can persist over long periods. For they incur costs in replication and risks of mutation or invasion by viral genomes, for example. The difficulty of establishing such conclusions gives the search for function greater force. Additionally, it widens the range of functions that are considered legitimate in explanation of structural properties of the genome. It calls into play the second of the two strategies to identify functions for known structures.

A good example of this approach is illustrated by the intron, the noncoding portion of the protein-coding gene in higher eucaryotes. Once discovered, introns posed grave puzzles to the molecular biologist: functional puzzles, not structural ones. Can they be, like the appendix, vestigial constituents, lost entirely in procaryotes, rare in lower eucaryotes, and still extant in higher eucaryotes, which are the youngest biological systems on the planet? The best recent speculation gives them a function, not on the time scale of the individual but on the evolutionary time scale. The existence of introns reveals that single-copy genes are divisible into segments that code for spatially distributed parts of proteins. If these spatially discrete parts of proteins are also distinct in their contributions to protein function and relatively stable in isolation from other parts of the complete protein, then each of ·the segments of a gene that codes for a part of the protein may be subject to the same selective forces the whole gene is subject to. If there are enough of these distinct, separated segments of DNA, they may be individually selected for, and may recombine, to produce new, more adaptive protein products, which will result in a feedback selectively favoring the integration of such independent sequences. Under these circumstances, introns, regions that separate such units, are to be expected, for the "shuffling" of coding sequences, 'exons,' will require spacers for separation and attachment after shuffling.

This is just the sort of hypothesis about evolution that has for so long appeared to be disreputable speculation, fueled by the impossibility of any evolutionary test of the matter. However, in this case, the methods of cloning and rapid sequencing have enabled molecular biologists to shed experimental light on the hypothesis that introns have an evolutionary function.

Among the earliest genes to reveal an intron were the globin genes. Cloning and sequencing these genes showed that they are composed of three exons, three coding regions, separated by two noncoding introns. This structure can be found across many species. This and its persistence as well suggest the evolutionary stability of the exons. Each of these exons codes for a primary sequence of amino acids. These three sequences are put together in the globin protein because the two introns are snipped out during RNA transcription. However, study of the primary sequence of the amino acids in the finished globin protein reveals not three but four stable amino-acid units in the hemoglobin molecule. Therefore, either introns are not needed to separate stable units that recombined to speed evolution, or else at least one intron has been excised from the globin gene in the course of evolution.

The second alternative may seem at first to be merely a theory-preserving ploy. But further structural research tends to substantiate it. In plants, there is a protein structurally very like globin, called leghemoglobin, which also functions as an oxygen carrier. DNA sequencing of the gene for leghemoglobin revealed a fourth intron in just the place required to separate stable units in hemoglobin. That introns can be excised precisely in nature is not controversial. Moreover, it had already been shown that two genes in the same genome coding for the same product may differ only in the presence of an intron in one and not in the other. The rat insulin genes differ in this way: One has two introns, the other none. It is equally clear that selective pressures can lead to a mixed strategy. Some introns may be preserved over time because of their ability to facilitate gene shuffling. Others will be eliminated because they threaten to break apart a particularly advantageous adaptation, like hemoglobin.

But how could the mammalian globin gene and the plant leghemoglobin gene be related? Clearly, they are not related in a line of evolutionary descent that handed down a globin gene from generation to generation from plant evolution all the way to mammalian species. For there are too many intervening species with no globin-like genes. What would have been required is transmission across contemporaneous species, instead of sown successive ones. Theories about such transmission are highly speculative at present. But one hypothesis has it that the leghemoglobin gene may have been transmitted from plants to animals by a plant virus infecting an insect. If this is the case, insect hemoglobin may be expected to have the three introns of leghemoglobin and not the two of mammalian hemoglobin. The sequencing of their globin genes is obviously the next order of business for the development of this functional hypothesis.

One of the results of coming to grips with the structure of the hemoglobin gene was the discovery of so-called pseudogenes. The mammalian genome contains five hemoglobin genes, each of which has a different function activated at a different state of development. But it also contains two sequences of base pairs that closely resemble the hemoglobin primary sequence but do not code for any gene product whatever. These are called pseudogenes. Molecular biologists claim with some confidence that such pseudogenes arise through mutations of just three different kinds. Their confidence that there are only three ways pseudogenes could have arisen is of course based on the fact that the three alternatives exhaust the major known ways in which the *function* of the hemoglobin gene can be destroyed. But the question remains, why do such pseudogenes persist? This is the functional question of what, given the costs of replication and the risks of lethal mutation, are the benefits associated with pseudogenes and to "whom" do they accrue?

Here again, the question is a demand for evolutionary function of a structurally identified system. And the current speculations attempt to satisfy this demand. First of all, the multiplication of functional genes is an evolutionary advantage in that the redundancy shields the gene family of repeated sequences from the effects of disabling point-mutations in any one member. This is the beginning of a functional explanation for the moderate repetition of some genes. The trouble with the explanation is that it is too good. For the same reasoning leads us to expect that many other genes for essential protein products should multiply within the genome for the same adaptive reason. But they do not. Therefore, treating gene multiplication as an adaptive strategy is not the whole answer to why it exists. This of course is just what the disjunctive character of evolutionary strategies would lead us to expect, however.

Although duplication provides evolutionary insurance to the family of genes, the redundancy it provides allows individual sequences to mutate and diverge in sequence much more widely than is otherwise compatible with survival. Pseudogenes may just be the product of these nonlethal mutations, hanging around with no current function. Still further structural knowledge strengthens this hypothesis of the evolutionary function of repetition and pseudogenetic multiplication. At least two noncoding pseudogenes have been discovered that are homologous to the α-hemoglobin gene. Both differ slightly in sequence from the active gene, and neither is located on the α-hemoglobin's chromosome. But one of these two pseudogenes is missing *exactly* those bases that constitute the introns on the normal α-hemoglobin gene. It has brought together the functioning parts of the hemoglobin gene. But in their searching of the genome for the location of these two pseudogenes, researchers have found evidence for many more such sequences generated from the globin-locus. This suggests that the production of moderately repeated genes, and their accompanying pseudogenes, is a common occurrence, common enough to be an important strategy of providing the variation required for evolution.

With so much structural information, the inevitable result is more functional questions, vastly more, than arose when only a little was known about the biochemical structure of the genome. We cannot answer many of these functional questions in much detail, and may never be able to do so. But the original structural information that raised them and the further structural information produced in attempts to answer them do enable the molecular biologist to provide a remarkably detailed account of the sequence of evolutionary steps taken by the hemoglobin genes from as far back as 90 million years ago to the present. We can rank current organisms in terms of their evolutionary divergence and closeness to hypothesized earlier ones. Then we can employ information about their hemoglobin-gene structures, the numbers of repeated hemoglobin genes and pseudogenes, to construct a natural chain of events that begins with a single globin gene and eventuates in the complex organization of the mammalian hemoglobin-gene system.

Contemporary molecular biologists know a tremendous amount about the hemoglobin molecule, as we recounted in Chapter 4. They also know a tremendous amount about the hemoglobin genes: their location; their nucleic acid structure; their similarities and differences; and now their evolutionary history, as well as that of other parts of the genome associated with them. This is not a reflection of the special importance of this protein or the genes that code for it. The two are at least as important as many other genes and gene products, but no more important. The depth of knowledge here is a consequence of the availability and suitability of the protein for study over the last hundred years; of the development of apparatus and research techniques; and probably also of the existence of some widespread and relatively severe diseases associated with hemoglobin malfunction, diseases whose symptoms provided clues and whose cures provided motives for all this research. But it is obvious that many questions remain on the hemoglobin-gene research agenda. Some of them are about this gene exclusively, and some are about any portion of the genetic material that is organized or that functions like it.

There are many functional and evolutionary questions about sequences of DNA that the understanding of the hemoglobin molecule does not touch on at all: Why are there highly repetitive sequences of DNA? Why do some of these sequences cluster, whereas others disperse? Why do these highly repeated sequences show cross-species similarities if they have no function? Why are they interspersed with only moderately repetitive DNA? How and why does the genome tolerate so much of this apparently functionless DNA? There are similarly a large number of questions about moderately repetitive DNA on which even good answers to questions about the globin genes will shed little light. Can there be one evolutionary explanation for all introns? If the past is any guide, the answer must be no. The same must be said for questions about pseudogenes. The more we know, the more functional questions we are led to ask, and yet the more our evidence undermines any general functional explanation for each of these features. The reason of course is that although biologists must pose functional questions, the answers to them can never be generally applicable, universal, exceptionless, the stuff of which laws and theories are made. And yet this is no defect, for, like other biologists, the molecular biologist is primarily interested in the particular facts about life on this planet. He is less interested in the nonfunctional theories that provide its general nomological explanation.

The reversal of the ancient functionalist problem is an important sign of biology's revolutionary maturity. But it is also a sign of the indisoluble bond between traditional biological concerns and its newest, most physically influenced department. Until the recent past, most of the history of biology was the pursuit of structures that would explain functions. Having pursued these structures to the natural stopping place of biochemistry and molecular biology, the discipline has neither gone further in hot pursuit across the border into physical chemistry, and beyond it to physics. Nor has it ceased at a dead end beyond which there are no further biological questions. Instead, it has turned around, and now it is actively engaged in the search for biological functions of chemically characterized structures. This turnabout generates the new problems of functionalism. The biological problems are obvious, and a few have been enumerated here. The philosophical and methodological problems generated by this *volte face* must now concern us.

The problem of justifying the pursuit of correct functional attributions to already identified structures is that we know that such functional identifications cannot eventuate in regularities combining generality, manageability, precision, and truth. This problem can be solved by applying the view of the nature of biology and its subject matter that has been defended in these pages. It is an important requirement for any philosophy of molecular biology that it allows for the recognized weakness of functional attribution and explanation, while rendering intelligible the continuing demand for it on the part of molecular biologists. Provincialism cannot do this, for it eschews functional questions as unfruitful because their answers are never laws of nature or first approximations to them. Autonomism cannot do this, for it treats functional attributions and explanations as universal and logically inelimiinable, instead of seeing them as the restricted and only tactically ineliminable sorts of statements that they are.

8.2. The Panglossian Paradigm

Most biologists' claims are implicitly about particular objects, earth-bound species or small and heterogeneous collections of them. Their questions and answers are dictated by considerations of human manageability at the expense of unwieldly and useless precision. They are primarily interested in the study of cases with a practical upshot in technological application or evolutionary history. These facts make the appeal to function both inevitable and perfectly respectable by any standard drawn from physical science. In this respect, the new philosophical problem of functionalism can be solved.

But there is an apparently more threatening problem, more threatening because more global and because advanced by some of the leading figures in biology today. Two such biologists, Stephen Gould and Richard Lewontin, have attacked the general adaptationalist strategy that has animated the biological study of the genetic material – and in fact the whole evolutionary approach to explaining phenomena and their origins. Their attack has been more influential than perhaps any other essay in the recent literature on the philosophy and methods of biology. It deserves detailed scrutiny. They call the adaptational approach the "Panglossian" paradigm, recalling the optimistic doctor of Voltaire's *Candide:* Dr. Pangloss, who found a function subserved for every distinct effect of an event or structure we could identify. Just as Dr. Pangloss found that the nose has its shape to enable us to wear glasses, adaptationalists maintain, according to Gould and Lewontin,

the near omnipotence of natural selection in forging organic design and fashioning the best among possible worlds. This programme regards natural selection as so powerful and the constraints upon it so few that direct production of adaptation through its operation becomes the primary cause of nearly all organic form, function, and behavior. Constraints upon the pervasive power of natural selection are recognized of course. . . . But they are dismissed as unimportant, or else, and more frustratingly, simply acknowledged and then not taken to heart and invoked. (Gould and Lewontin, 1979:584–5)

The defects of adaptationalism, in Gould and Lewontin's view, are that it makes false attributions of adaptive significance, and therefore obscures the course of evolution. Moreover, its ubiquity blinds biologists to alternative, better accounts of the origin and nature of biological phenomena. If they are correct, the adaptationalist strategy in molecular biology is the recapitulation of errors of method and theoretical presupposition common throughout biology.

The first step in the adaptationalist strategy to which Gould and Lewontin object is its division of organisms into traits, which are each to be separately explained as the result of different evolutionary processes optimizing for different ends. Of course the identification of such traits, and their individuation, represent empirical hypotheses: guesses that what our nomenclature distinguishes are in fact distinct and can respond to selection with some degree of independence. Often such guesses are wrong. Gould is well known for having identified one such implicit hypothesis: the belief that the chin is a single trait, instead of the interactive result of two separate growth fields, the mandibular and the alveolar (Gould, 1977c:381). In our terms, we may describe the division of organisms into traits as conjectures about whether they are the natural kinds out of which organisms are constructed. Every biological study of the components of a system must begin with such fallible conjectures. No one can object to such "atomization," in principle, for without it the life sciences could hardly proceed. But Gould and Lewontin seem to suggest that adaptationalists usually get it wrong, that their segmentation of organisms into parts each with its own distinct evolutionary path is almost always wrong. Often these erroneous divisions are revealed as such in the failure of attempts to give them evolutionary explanations. Gould and Lewontin seem to recognize this: "After the failure of part-by-part optimization, interaction is acknowledged via the dictum that an organism cannot optimize each part without imposing expenses on the others. The notion of 'tradeoff' is introduced, and organisms are interpreted as best compromises among competing demands. Thus, interaction among parts is retained completely with the adaptationalist programme. Any suboptimality of a part is explained as its contribution to the best possible design for the whole" (Gould and Lewontin, 1979:585). Gould and Lewontin admit that adaptationalists leave open the bare possibility that evolution may not proceed this way in the short run and in small populations. But they "maintain that alternatives to selection for best overall design have generally been relegated to unimportance by this mode of argument" (1979:585). This conclusion they attribute to a conceptual or methodological defect of adaptationalism: "The adaptionalist programme can be traced through common styles of argument":

(1) "If one adaptive argument fails, try another." Antlers are first viewed as adaptations for protection from predators; then, when this argument is surrendered, they are treated as adaptations for intraspecies dominance. (2) "If one adaptive argument fails, assume that another, as yet undiscovered one, exists." The failure to substantiate an adaptive claim is never viewed as a disconfirmation of the whole program but rather as a motive for a more extensive search for such an argument. (3) "In the absence of a good adaptive argument in the first place, attribute failure to imperfect understanding of where an organism lives and what it does." (4) "Emphasize immediate utility and exclude other attributes of form." In particular, adaptationalists are alleged to ignore nonadaptive hypotheses "both more interested and more fruitful than untestable speculations based on secondary utility in the best of possible worlds" (1979:586–7). Gould and Lewontin conclude:

We would not object strenuously to the adaptationalist programme if its invocation, in any particular case, could lead *in principle* to its rejection for want of evidence. We might still view it as restrictive and object to its status as an argument of first choice. But if it *could* be dismissed after failing some explicit test, then alternatives would get their chance. Unfortunately, a common procedure among evolutionists does not allow such definable rejection for two reasons. First, the rejection of one adaptive story usually leads to its replacement by another, rather than to a suspicion that a different kind of explanation might be required. Since the range of adaptive stories is as wide as our minds are fertile, new stories can always be postulated. . . . Secondly, the criteria for acceptance of a story are so loose that many pass without proper confirmation. Often evolutionists use *consistency* with natural selection as the sole criterion and consider their work done when they concoct a plausible story. (Gould and Lewontin, 1979:587–8; emphasis added)

By contrast with methodologically suspect adaptationalism, Gould and Lewontin cite passages from Darwin calculated to show that he was not himself wedded to this approach. Their point of scholarship is somewhat marred by the fact that the most extensive quote they cite from Darwin to show his rejection of natural selection as the only evolutionary force endorses Lamarckian use and disuse as an important evolutionary force.

They then turn to several alternatives to adaptationalism as explanations of "form, function, and behavior": (1) "No adaptation and no selection at all." The phenomenon of "genetic drift," in which a small number of unrepresentative genes isolated by nonselective, that is, random, forces may be the source of much speciation and even of the establishment of maladaptive traits; in addition, it will extirpate the effects of adaptive mutations as well. (2) "No adaptation and no selection on the part at issue." Where a trait is the result of selection on other features of the organisms' morphology, or is carried along as a phenotype despite its maladaptive character because of its linkage to an adaptive trait (by pleiotropy, for instance), "we come face to face with organisms as integrated wholes, fundamentally not decomposable into independent and separately optimized parts." (3) "The decoupling of selection and adaptation." Some adaptations, like the shapes of corals or sponges, are design solutions to conditions, like ocean currents, that are entirely or largely phenotypical. Equally, they show, it is at least theoretically possible for a mutation to become fixed that raises fertility while leaving average viability the same, thus decoupling selection from any adaptive consequence. (4) "Adaptation and selection but no selective basis for differences among adaptations." A population may respond to one selective pressure in two different ways, both of which are equally adaptive, so that the explanation for their distribution may itself be nonselective. (5) "Adaptation and selection, but the adaptation is a secondary utilization of parts present for [nonselective] reasons" (Gould and Lewontin, 1979:591).

In general, Gould and Lewontin prefer a developmental, holistic approach to evolution:

Developmental constraints, a subcategory of phyletic restrictions, may hold the most powerful reign of all possible evolutionary pathways. In complex organisms early stages of ontogeny are remarkably refractory to evolutionary change, presumably because the differentiation of organ systems and their integration into a functioning body is such a delicate process, so easily derailed by early errors with accumulating effects. . . . If development occurs in integrated packages, and cannot be pulled apart piece by piece in evolution, then the adaptationalist programme cannot explain the alterations of developmental programmes underlying nearly all changes of *Bauplan* [architectural design, construction scheme, or plan]. (Gould and Lewontin, 1979:594)

And further,

under the adaptationalist programme the great historical themes of developmental morphology and *Bauplan* were largely abandoned; for if selection can break any correlation and optimize parts separately, then an organism's integration counts for little. Too often, the adaptationalist programme gave us an evolutionary biology of parts and genes, but not of organisms. It assumed that all transitions could occur step by step and underrated the importance of integrated developmental blocks and pervasive constraints of history and architecture" (597)

Although Gould and Lewontin's critique of adaptationalism begins as a methodological attack, it ends by treating the approach as factually mistaken. For they advocate a factually more adequate approach to the explanation of biological phenomena. In and of itself there is nothing wrong with such a shift, so long as it is recognized that there is no real distinction between formal, methodological questions and material, factual, substantive ones. (See Chapter 1, for a discussion of this important matter.) The employment of an adaptationalist method is the expression of an implicit commitment to the truth of an adaptationalist theory. The rejection of this theory is an expression of commitment to equally factual claims. This means, however, that if Gould and Lewontin's methodological criticisms are just the suggestion that adaptationalism is pseudoscience because unfalsifiable, they cannot be taken seriously. For in their single-minded pursuit of the adaptationalist method, proponents of this approach are only doing what every rational scientist who embraces a theory must do: They are attempting to apply it and are treating its problems as puzzles to be solved, not anomalies that refute the approach.

The methodological imperative of the theory of natural selection – to search for adaptations everywhere, to assume them when not immediately revealed, to attribute failure to find such adaptations to the inadequacies of the scientist instead of the theory – is not only characteristic of many theoretical endeavors. It is an essential feature that the theory of natural selection shares with the most powerful scientific edifices of physics. Like Newtonian or quantum mechanics, the theory of natural selection is an *extremal theory*. Like Newtonian mechanics, the theory of natural selection treats the objects in its domain as behaving in such a way as to maximize and/or minimize the values of certain variables. This strategy is especially apparent in Newtonian mechanics when that theory is expressed in so-called extremal principles, according to which the development of a system always minimizes or maximizes variables that reflect the physically possible configurations of the system. In the theory of natural selection, this strategy is exemplified in the assumption that the environment acts so as to maximize the rate of proportional increase of the fittest, hereditarily similar subset of a species. This strategy is crucial to the success of these theories because of the way it directs and shapes the research and applications that are motivated by the theories. Thus, we hold that a system always acts to maximize the value of a mechanical variable. If our measurements of the value of that variable in an experimental or observational setting diverge from the predictions of the theory and the initial conditions, we never infer that the system is failing to minimize the value of the variable in question. Rather, we assume that our specification of the constraints under which it is actually operating is incomplete. For example, in the case of mechanics, attempting to complete this incomplete specification resulted in the discovery of new planets and eventually in the discovery of new laws, like those of statistical thermodynamics. Similarly, in biology, the assumption of fitness maximizing led to the discovery of forces previously assumed to have no effect on genetic variations within a population. It has led to the identification of adaptations and even to discoveries that explain the persistence in a population of apparently maladaptive traits, like sickle-cell anemia, for instance. It is because these theories are "extremal" ones that differential calculus may be employed to express and interrelate their leading ideas; and it is because evolutionary theory is an avowedly extremal theory, asserting that the systems it describes maximize fitness, that its mathematical development by Haldane, Fisher, Wight, and their followers was couched in the language of differential calculus.

It is by virtue of this extremal character that these theories are all committed *to explain everything in their domains.* By virtue of the claim that systems in their domains always behave in a way that maximizes or minimizes some quantity, the theory ipso facto provides the explanation of all its subject's relevant behavior and cites the determinants of all its subject's relevant states. There is no scope for treating such a theory as only a partial account of the behavior of objects in its domain or as a description of some of the determinants of its subject's states; for any behavior that actually fails to maximize or minimize the value of the privileged variables simply refutes the theory *tout court.* Thus, lengthy persistence of a maladaptive hereditary trait (which was genetically independent of an adaptive one) over a *large* population randomly interbreeding with another would not show that some other forces acted on such a population, in addition to and besides selective forces; it would be taken to refute the theory of natural selection altogether. And this is because the theory asserts that everything that happens to its subjects results in maximizing the rate of increase of their fittest subspecies. This is why a "story" can always be told that explains any possible increase or decrease of a hereditarily linked subspecies by appeal to selective forces, *known or unknown.* In other words, the pervasive character of extremal theories is but the other side of the coin from their insulation against falsification.

Now all theories are strictly unfalsifiable, simply because testing them involves the employment of auxiliary hypotheses. But extremal theories are not only insulated against strict falsification, they are also insulated against the sort of falsification that usually leads to modification of theories instead of auxiliary hypotheses. In the case of a nonextremal theory, falsification may lead to revision of the description of test conditions. Or it may suggest revision of the theory by the addition, for example, of new antecedent conditions to its generalizations or new qualifications to its ceteris paribus clauses. But this is not possible in the case of extremal theories. The assumptions of theories like Newton's or Darwin's do not embody even implicit ceteris paribus clauses. The latter does not, for example, assume that selection maximizes fitness ceteris paribus. With these theories, the choice is always between rejecting the auxiliary hypotheses – the description of test conditions – or rejecting the theory altogether. For the only change that can be made to the theory is to deny that its subjects invariably maximize or minimize its chosen variable. This, of course, explains why high-level extremal theories like Newtonian mechanics are left untouched by apparent counterinstances. It explains why they are superseded, not by qualified versions to which antecedent conditions are added, but either by utterly new extremal theories, in which the values of very different variables are maximized or minimized, or by new nonextremal theories.

Extremal theories have become an important methodological strategy because of the success of the earliest of them, Newtonian mechanics. In addition, their insulation against falsification has enabled them to function at the core of research programs, turning what otherwise might be anomalies and counterinstances into puzzles – opportunities for extending the domain and deepening the precision of the extremal theory. But the success of extremal theories and their associated research programs requires that there be at least some guide to or agreement on independent specifications, characterizations, or ways of identifying the determinants of their subject's extremal behavior. These guides are themselves theoretical claims. But it is essential that they are distinct from the theory's claims about its subject's maximizing or minimizing behavior. An example from mechanics will make this point clear. Treating the Newtonian force law, $F = ma$, as a synthetic proposition governing the behavior of bodies requires that an independent way of specifying mass, m, be proved. Hooke's spring law provides such a specification. This independent specification preserves the contingent character of the force law. It enables us to employ it to explain the operation of springs, projectiles, fluids, and much else that mechanics accounts for. As we have noted, in Section 5.2, it is a well-known criticism of the extremal claims of the theory of evolution that they are empty tautologies because the theory provides no independent specification for the concept of fitness, the variable whose rate of change is maximized within evolving populations. Because higher levels of fitness are cited by the theory to explain greater reproduction rates among populations, these same rates of reproduction cannot provide the independent specification of different degrees of fitness. Accordingly, unless such an independent specification is available, the theory is guilty as charged of being empirically empty. The supervenience of fitness reflects both the difficulty and the possibility of providing such an independent specification.

Thus, Gould and Lewontin's criticism of the adaptationalist program reflects real and unavoidable features of the theory that animates it: The refutation of one adaptationalist "story" does result in the provision of another; the failure to find any is considered a scientist's failure, not a theoretical falsification; and the criteria for acceptance of adaptationalist stories are very loose, a result of the nature of the theory they reflect. But the theory they reflect is the theory of natural selection, and neither Gould nor Lewontin wants to surrender this biological achievement itself (though Gould at least has rejected a view he calls "Darwinian evolution," the claim that evolution proceeds by continual selection over small variations at a constant rate). Accordingly, their methodological criticisms reflect a misunderstanding of the theoretical exigencies of an extremal theory in particular and of the fact that no theory can be strictly falsified in general. There is nothing methodologically wrong with constructing a new adaptationalist explanation of a given "form, function, or behavior" when an older one has been rejected; indeed, the theory of natural selection requires it by virtue of its extremal character.

There is, however, an ancillary and dispensable hypothesis, often uncritically employed in the application of the adaptationalist program, that is open to criticism. And the criticism may encourage the "developmental" alternative Gould and Lewontin favor. This doubtful assumption is the hypothesis that the items whose persistence is to be explained as adaptations are those individuated and identified in ordinary language or in one or another biological theory of relatively restricted generality. Gould and Lewontin cannot object to the division of organisms into independent traits, each of which can separately respond to selective forces. Their objections must be to the particular divisions adopted by evolutionary biologists. To see this, we need note only that Gould and Lewontin advocate or at least permit adaptive theorizing about "integrated packages," within which components constrain the development and presumably the evolution of one another. As they note, "differentiation of organ systems, and their integration into a functioning body is . . . a delicate process. . . . [D]evelopment occurs in integrated packages, and cannot be pulled apart piece by piece in evolution" (1979:594). Thus within a package, any one component evolves under the constraint of other components (and of its own developmental precursors), with which it must remain not merely compatible but functionally cooperative. Similarly, whole packages are equally constrained by the others together with which they compose the whole organism. Accordingly, we cannot assign adaptations and explanations based upon them to components and to packages that do not reflect the constraints under which they operate and evolve. These constraints may be so great that they overwhelm even moderately strong selective forces to which individual components and packages would respond were they independent. As a result, there might be much less evolutionary adaptation of individual components, packages, and organisms even during long periods of considerable environmental and genetic variation. From this Gould and Lewontin infer that "the adaptational programme cannot explain the alterations of developmental programmes underlying nearly all changes in *Bauplan*" (1979:594). But this will only be true of adaptationalism if it is not allowed to apply its explanatory scheme to the developmental packages and is not allowed to include among the selective forces shaping any one component the constraints of the rest of the "package" in which it participates.

Gould and Lewontin's conclusion only follows if adaptationalists are committed to the particular taxonomy of selectively independent traits that they have inherited or uncritically chosen or which their opponents have foisted upon them. Such uncritical adaptationalists differ from Gould and Lewontin not on a point of biological methodology, still less on a matter of general theory, but on a factual question of how organisms are to be carved up into parts that face selective forces as units. Freed from a superficial typology of traits, which divides organisms up into as many parts as we have names for, adaptationalists can avail themselves of each of the five explanatory approaches Gould and Lewontin sketch as potential alternatives of their "program." Only if Gould and Lewontin hold the radically holist view that organisms cannot be broken into components at all, are "not decomposable into independent and separately optimized parts," can we find anything in their attack on Panglossianism to which adaptationalists cannot, nay should not, reconcile themselves. But though they seem to endorse such a radical holism for these purposes, in their actual biological – and particularly genetic and paleontological – practice, both Gould and Lewontin repudiate any such commitment.

But it may be replied that these concessions are nothing less than the surrender of adaptationalism, for they require surrender of the thesis that "selection can break any correlation and optimize parts separately." In fact, they do not; these admissions require greater sophistication about what a "part" is and about what distinguishes components of a single unit from correlated but independent units. To see this, consider how we identify units that respond to selective forces and how we distinguish correlations from components by an adaptationalist approach. If the adaptationalist explanation for what is individuated as an organ or a part or a bit of behavior fails, this is one sign that we have not rightly individuated the components of the organism. Explanatory failure suggests that what we have focused on is perhaps not a "developmental package" but an artificial gerrymandering of the organism. Or again, suppose we hypothesize that two features are only correlated for nonevolutionary causes but we can find no reasonably independent adaptationalist explanations for each of them. One alternative is to deny that they are merely correlated and to treat them as components of a "package" that evolves as a unit. In fact, the only way to reach Gould and Lewontin's aim, "a pluralistic view [that] could put organisms with all their recalcitrant, yet intelligible, complexity back into evolutionary theory" (1979:597) is by pursuing an adaptationalist program.

The adaptationalist program is just the one that must be endorsed if, as has been hitherto argued, the only general biological forces that have operated to arrange each of the levels of biological organization are those of natural selection operating on successively greater aggregations of macromolecules. Such operations generate just the "developmental packages" that Lewontin and Gould urge us to make the objects of biological explanation. If these are the only general forces that have resulted in this hierarchy, then the only general theories to be uncovered in biology are the theories of natural selection and of molecular biology (to the extent it may be distinguished from organic chemistry). If these theories exhaust the theoretical possibilities and if research programs must reflect underwriting theories, then the only viable research program in biology is the adaptationalist one. The only alternative is to hold that there are distinct regularities and theories discoverable in biology intermediate between chemistry and selection. It is of course possible that there are such autonomous generalizations, say in developmental morphology or phylogeny. They will have to be generalizations with enough content, precision, and confirmation to be deemed laws, with enough coherence to constitute a theory and to animate a research program that may really compete with adaptationalism. The only way to undermine adaptationalist claims on biology would be to provide such a theory.

A developmental biologist, Stuart Kauffman, has reported studies that do seem to reveal constraints or, as he expresses it, "biological universals" intermediate between those of biochemistry and the principles of natural selection. But these constraints are not biological. As Kauffman notes, "the ultimate source of order in the living world must derive from basic properties of structural and dynamic self-organizing properties of complex molecular systems" (Kauffman, 1983b:292), which provide and limit the possibilities explored by variation and selection. There is nothing particularly mysterious about the "self-organization" Kauffman appeals to. Its simplest manifestation is the aggregation and layering of lipids we can detect when oil is mixed with water and allowed to stand. This "self-organization" is a consequence of molecular interaction, but will obtain among lipids no matter what their particular molecular size or composition. Similarly, other properties of macromolecular assemblies may appear just through stochastic forces or thermodynamic process, or as expressions of topological features of surfaces, no matter what their composition. Instances of such "universals" may recur at independent nodes in evolutionary phylogeny. They may appear not as a result of convergent adaptational evolution (though they may lead to it), but because at these nodes the purely physical initial conditions of the appearance of these structures obtain. These properties will serve as the building blocks on which selection operates, and their occurrence also constitutes constraints that selection cannot circumvent, any more than it can circumvent gravity. To the extent that such self-organizing properties can be systematized and predicted from purely physical and chemical considerations, they do not constitute a source of the sort of autonomous generalizations about development that Gould and Lewontin look to as an alternative to the adaptationalist approach in the explanation of biological phenomena. But to the extent they are significant constraints and/or a substantial part of the material evolution must operate on, they will provide much of the explanatory resources in biology. If what Kauffman calls biological universals cannot be explained by physical principles or by evolutionary ones, then a theory that treats of them may provide the kind of alternative to adaptationalism that Gould and Lewontin look to. (For further discussion of constraints under which selection operates, see Kauffman, 1983a, and Mayo, 1983.)

8.3. Aptations, Exaptations, and Adaptations

Gould has himself offered a terminological suggestion that may help clarify the adaptationalist controversy. Adaptationalism must be compatible with the possibility, indeed the actuality, that many evolved structures have a present function, but were "not built by natural selection for their current roles" (Gould and Vrba, 1982:6). In such cases, the identification of current role is no part of the evolutionary explanation of the origin and persistence of the structure. What is more challenging to the adaptationalist, such structures may not have arisen through any adaptive mechanism.

Gould illustrates this problem by appeal to several arresting examples, among them the problem of assigning a function and constructing an adaptationist explanation for highly repetitive DNA. Gould's first example is the evolution of birds and the appearance of feathers. The earliest known birds were fully feathered and yet capable of only very limited flight. Accordingly, the notion that feathers have their origin in the adaptation they provide for flight seems implausible. Alternatively, it has been suggested that feathers first appeared as an adaptation for thermoregulation. Later, they found a new role, unconnected with thermoregulation, in the trapping of insects, by increasing limb-surface area. Still later, together with other evolutionary changes, they found a role in flight. By this point, further changes in the nature of bird feathers may have come directly under the influence of selection for flight. But, on this theory at any rate, their origins have nothing to do with adaptation for their present role.

After sketching similar accounts for the origins of bone and of an essential mammalian enzyme, Gould turns to the puzzle of identifying adaptive functions and of tracing evolutionary histories of portions of the repetitive genetic material examined in Secion 8.1: "We are now, not even thirty years [after the decoding of DNA] faced with genes in pieces, complex hierarchies of regulation, and above all, vast amounts of repetitive DNA. Highly repetitive or satellite DNA can exist in millions of copies; middle repetitive DNA, with its tens to hundred of copies, forms about one quarter of the genome in both *Drosophila* and *Homo*. What is all this repetitive DNA for (if anything)? How did it get there?" (Gould and Vrba, 1982:10) Two explanations that Gould treats as inadequate are, first, that repetitive DNA plays an immediate regulatory role in cell division, a role for which it was selected, and, second, that such DNA has had a role in freeing genes to vary and to reshuffle in ways not bound by narrow selective forces. Gould suggests a third alternative:

Perhaps repeated copies can originate for no adaptive reason that concerns the traditional Darwinian level of phenotypic advantage. . . . Some DNA elements are transposable; if they can duplicate and move, what is to stop their accumulation as long as they remain invisible to the phenotype (if they become so numerous that they begin to exert an energetic constraint then natural selection will eliminate them)? Such "selfish DNA" may be playing its own evolutionary game at a genic level, but it represents a true nonadaptation at the level of the phenotype. Thus, repeated DNA may often arise as a nonadaptation. Such a statement in no way argues against its vital importance for evolutionary futures [that is, for subsequent phenotypical adaptation of repeated DNA]. When used to great advantage in that future, these repeated copies are *exaptations*. (Gould and Vrba, 1982:10; emphasis added)

Exaptation is Gould's neologism, one introduced to enable us to distinguish structures whose current functional roles are no part of the explanation of their origin from those whose roles are. Gould's terminological recommendation is as follows. 'Adaptation' should be reserved for a character with a current use, role, or function in meeting an organism's needs, if the character's origin hinges on selection for that role or function. An 'exaptation' is a character with a current role or function; but its origin is either the result of selection for a different role or function, or else the result of some nonadaptational process altogether. Gould suggests that both adaptations and exaptations come under the heading of 'aptation'; this term describes biological items that are normally described as having functions, or at least as having biologically significant effects. Thus, feathers are exaptations because they have been coopted for flight, though their adaptive origin is in thermoregulation; "selfish DNA" may now be an exaptation for increasing the rate of evolutionary variation, or for facilitating genetic recombination, though it arose through no adaptive process whatever. An exaptation is, according to Gould, what has hitherto been called a preadaptation. The notion of preadaptation has been employed among biologists with some discomfort. For the term suggests an immanent directedness of evolution toward as yet unrealized adaptations. The label preadaptation is given to a trait retrospectively, after an 'aptation' has become an adaptation. Such an aptation is, by these definitions, an exaptation.

This terminological suggestion of Gould's seems to afford scope in evolutionary studies for the antiadaptationalist alternatives that he and Lewontin have advanced. It also fosters the hypothesis of punctuated equilibrium as the "mode and tempo" of evolution. For he notes that "features co-opted as exaptations have two possible previous statuses. They may have been adaptations for another function, or they may have been nonaptive structures. The first has long been recognized as important, the second underplayed. Yet the enormous pool of nonaptations must be the wellspring and reservoir of most evolutionary flexibility" (Gould and Vrba, 1982:12). The second alternative is that features or structures have arisen and persisted for nonadaptational causes, that their character is a function of the constraints provided by the organization in which they are found. These constraints fix traits and characters over geologically long periods, and they are only broken in a way that permits adaptive diversity on limited occasions. These are the occasions that reflect bursts of evolution punctuating the otherwise changeless sequence of terrestrial history. It is only on these occasions that adaptation may come into play, revealing exaptations otherwise hidden from selective forces.

As we have seen in the previous section, this conclusion can itself result from adaptationalist research, once the right units that face selection are uncovered. This much is evident in Gould's own description of the potential exaptational character of "selfish DNA." He notes that "repeated DNA can originate for no adaptive reason that concerns the traditional Darwinian level of phenotypic advantage." But of course adaptationalist hypotheses are not limited only to the explanation of what can be described in the terminology of "traditional Darwinian phenotypes." Freed from this unreasonable restriction, adaptationalists can help themselves to Gould's own claim that "selfish DNA" may be playing its own Darwinian game at the genic level. Gould's inference that "it represents a true nonaptation at the level of the phenotype" does not follow if we are permitted to adopt a relatively sophisticated notion of the phenotype and of the items whose evolution is governed by the theory of natural selection. Because one of the interpretations of that theory makes it a theory about the evolution of segments of the genetic material, it empowers biologists to pursue an adaptationalist approach to molecular evolution. The identification of features as exaptations is no reason to surrender this program. Indeed, it enables us to apply the theory even where, as in this case, adaptationalist hypotheses at higher levels of organization must be viewed with suspicion.

Furthermore, it becomes apparent that adaptationalism and its motivating theory are perfectly compatible with, indeed might lead us to expect, punctuated equilibrium in evolution. Gould and Lewontin hold that the components of organisms and their parts are the chief evolutionary constraints on one another's responses to selection. If this is the case, the fitness of any one part and any one "package" supervenes heavily on other parts and "packages" with which it is grouped. By channeling other selective forces, those of the environment, into relatively narrow streams, these constraints will result in long periods of apparent stasis; periods during which almost any variation will prove inviable because it breaks the developmental programs required for life. Only rarely will massive environmental changes or fortuitous variations result in new forms that are viable. When it does, these new forms will have to differ very markedly from their predecessors, just because they break up significant constraints successfully. This is just what punctuated equilibrium demands. But it appears to be entirely the product of a multileveled adaptationalist mechanism that does nothing more than justice to all the selective forces governing evolution. Of course, the consistency of a punctuated-equilibrium approach and an adaptational one must not be misunderstood. It does not constitute an endorsement of the former view. Rather it is another argument to be added to those of Chapters 5 and 7, arguments that show the dispute between gradualism and punctuated equilibrium to be one within the ambit of the theory of natural selection, despite the labels that one party to the dispute employs to describe its opponents.

With this reconciliation in mind, we may usefully employ Gould's distinction between adaptations and exaptations to make clear the sorts of functional questions molecular biologists want to ask about the varied structure of the eucaryotic genome. Some of their questions are about immediate function: For example, what physiological needs of their hosts and of themselves are served by quantities of highly repeated

DNA? The answer to this question will probably be different from the answer to the question of what selective force resulted in their origin and persistence. We can mark this difference by calling them exaptations and thus distinguishing between the questions that biologists raise about them.

8.4. Information and Action Among the Macromolecules

Biology is thoroughly teleological, though it is so in a way perfectly consistent with its active integration with physical science. But although biology is teleological, it is usually distinguished from psychology and other behavioral science in that it is not an "intentional" science: It does not assume or attribute conscious or cognitive states to biological systems. It does not account for biological behavior by appeal to such states, either through simpleminded anthropomorphism or through appeal to the intentions and designs of an infinite power. Human sciences treat teleological behavior, as biology does, but they explain it not just as purposive behavior but as action – as the consequences of desires, beliefs, hopes, fears, wishes, conscious and unconscious intentions, recognition, and the information of agents who carry out this behavior. Biology has for obvious reasons eschewed such explanatory strategies. Who, after all, would suggest that plants maximize their exposure to the sun because they believe it will optimize their production of starch, and because they want to do so? One of the most salient effects of Darwin's revolution is that it ended forever the biological appeal of the argument from design, which founded the teleology of nature on the desires, intentions, and conscious designs of God. It has long been deemed a goal of biology to account for the intentional, cognitive states that social science cites in its explanations of human action by showing that they consist in the same kinds of organic states that suffice to explain the rest of biological behavior. There is a long history, stretching back to the seventeenth-century philosopher Thomas Hobbes, of attempts to show that cognitive processes – consciousness, sensation, perception, and the actions they eventuated in – are as amenable to nonteleological reduction as the rest of biological phenomena have been held to be. This program has hitherto met with little practical success, and among biologists the explanation has been that its failures are due largely to the prematurity of the enterprise. We do not know enough about neuroscience, even at far lower levels of phylogeny, to tackle the complex question of human neurophysiology. Among philosophers, there have been other arguments against this program, arguments about the mind and body that date from the work of René Descartes in the seventeenth century.

But one consequence of the failure to make progress in this area is the conviction that we do not understand the conscious or cognitive states of sentient creatures. Therefore appeal to them in explanations is for the foreseeable future unenlightening at best. This makes especially perplexing the degree to which terms normally employed only to refer to the powers and activities of human beings, the most complex of sentient creatures, have found their way into the descriptive vocabulary of molecular biology. Appeal to concepts like action, wants, or beliefs is resolutely eschewed in the explanation of highly complex biological systems. For these systems are not complex enough for us to be willing to attribute anything like *minds* to them. But when it comes to proteins, enzymes, and nucleic acids, such terms are used with apparently reckless abandon. We say that the genetic material contains *information,* *expressed* in a *code,* which it conveys to other macromolecules and employs to produce distinctive results. We attribute to transfer RNAs the power to *recognize* and *discriminate* amino-acid molecules from one another. We describe nucleases as *making* and *correcting mistakes* and *errors,* as *editing* and *proofreading.* Similarly, the specificity of an enzyme is described in terms of its power to *recognize* substrates by their molecular shape.

Now it is easy and common to explain away such descriptions as purely metaphorical. No one supposes for a minute that DNA carries information in the way people have beliefs or that a macromolecule recognizes another the way we recognize a familiar face. Macromolecules do not have minds. But merely acknowledging this usage as metaphorical or analogical ignores certain features of it. The first and most important of these is its *extreme naturalness:* The behavior of these simple substances simply cries out for description in intentional terms. Rendering explicit why this is the case may shed light on the meanings of these terms in their original locus of employment, the human case. It may even be supposed that the naturalness of intentional descriptions of molecular interactions, together with a detailed nonintentional description of these interactions, suggests the most promising avenues of development in neuroscience. A second feature of this usage is that it is not only natural; it is inevitable and unavoidable. If it were decided to altogether forgo apparently intentional descriptions in discussions of the behavior of proteins, enzymes, DNA, etc., the amount of circumlocution, and other increases in terminological apparatus, would render difficult further rapid progress in many areas of molecular biology.

Finally, even if metaphorical, the attribution to genes of properties normally restricted to sentient agents with minds has been claimed to have an important positive impact on the development of biological theory. Thus the attribution to genes of 'selfishness' and 'altruism' is an important part of certain sociobiological and ethological theories of animal behavior. Sometimes these attributions are announced as metaphorical. Sometimes these terms are introduced as redefined in explicitly nonintentional terms, which blocks any metaphorical or literal suggestions of mentality at the level of the gene. But such explicit redefinitions of terms that already have a fixed meaning in an old domain or in ordinary language raise the question of why the terms so redefined are useful, suggestive, and fruitful in their new area of enquiry. This question is a demand that theorists justify their choice of these terms, and their redefinition of them, instead of simply coining neologisms with no fruitful or misleading suggestions about matters outside the immediate theoretical concern. The justification must be something like this. A term like 'altruistic' is chosen for redefinition because, when its intentional component has been removed by explicit redefinition, it can still be used to describe human behavior as well as nonintentional, nonhuman behavior. As such, the theory in which a redefined term like 'altruistic' figures may unify diverse, human and nonhuman phenomena under the same set of laws. Why say that all genes are selfish, or that some genes are altruistic, unless it is intended to suggest that they are like humans in theoretically interesting respects? Surely it would be less misleading to avoid confusing homonyms and to coin neologisms free from misleading suggestions, unless the redefinition is meant to capture at least part of the original meaning of these terms, enabling us to understand the behavior of genes by analogy with the behavior of humans and vice versa. But some philosophers and biologists have argued that there are purely conceptual obstacles to this strategy, obstacles so great that any attempt to describe the behavior of the genetic material in intentional language is literally nonsensical. That a biological practice might be natural, indeed almost inevitable, and also logically incoherent is, to say the least, a serious problem.

The notion of the selfish gene is associated with a book of that name by Richard Dawkins. In describing genes as selfish, Dawkins makes clear that his claim is not to be construed as a metaphorical one, but in terms of the following redefinitions of selfishness and altruism:

> An entity is said to be altruistic if it behaves in such a way as to increase another entity's welfare at the expense of its own. Selfish behavior has exactly the opposite effect. "Welfare" is defined as "chances of survival," even if the effect on actual life and death prospects is . . . small . . . It is important to realize that the above definitions of altruism and selfishness are *behavioral*, not subjective. I am not concerned here with the psychology of motives . . . that is not what this book is about. My definition is concerned only with whether the effect of an act is to lower or raise the survival prospects of the presumed altruist and the presumed beneficiary. (Dawkins, 1976:4–5)

Attributing selfishness to genes thus turns out to be part of the claim that selection operates at the level of the hereditary material. In the long run, genes selfish in this sense have been selected for, because they have proved to be fitter than altruistic ones. One of the advantages of this terminology is that it enables us to conveniently express how mechanisms for altruism among genes, or among any biological entities, might have arisen. Because natural selection will advantage selfish genes above unselfish ones, in the long run all genes must be selfish. Of course, some genes act in ways that result in their own destruction but at the same time in the preservation of their copies. The copies of a gene produced by replication are related to it by the parent-of relation expounded in Chapter 4. It is for this reason that genes evolve in accordance with the theory of natural selection. Any gene whose character favors the survival of its copies at higher rates than genes without this character will do so because it is fitter than these genes. This sort of increased fitness may be described in terms of altruism, as Dawkins defines it, for these genes are altruistic with respect to their copies. Thus, altruism, as redefined, is rendered consistent with evolutionary theory. This is theoretically interesting, because it is normally supposed that evolution will ensure the persistence only of selfish genes.

The use of terms like selfish and altruistic to describe the behavior of genes has drawn objections amounting practically to outrage from some philosophers and strong criticism from biologists. S. J. Gould writes:

> No matter how much power Dawkins wishes to assign to genes, there is one thing he cannot give them – direct visibility to natural selection. Selection simply cannot see genes and pick among them directly. It must use bodies as an intermediary . . . Bodies cannot be anatomized into parts, each constructed by a single gene . . . Parts are not translated genes, and selection doesn't even work directly on parts. It accepts or rejects entire organisms . . . The image of individual genes, plotting the course of their own survival, bears little relation to developmental genetics as we understand it. (Gould, 1977a:24)

Much of this criticism of course reflects Gould's attack on adaptationalism, and as we have seen he does not do justice either to the alternatives of that "programme" or to the interpretative powers of the theory of natural selection. Moreover, his criticism treats the selfish-gene hypothesis as a metaphor, for he describes it as an image. But Dawkins insists that his claim is not to be construed metaphorically. Nevertheless, Gould's assertions reflect an objection he shares with philosophers who have examined Dawkins's views: the objection that saddles Dawkins's selfish genes with motives and then ridicules his view as false, silly, or unintelligible. This line is taken more explicitly by Mary Midgley: "Genes cannot be selfish or unselfish, any more than atoms can be jealous, elephants abstract or biscuits teleological" (Midgley, 1979: 439). In other words, the very notion of a selfish gene is incoherent, unintelligible, and useless for any cognitively respectable purpose. The reason is that selfishness, and altruism for that matter, are *motives;* they are essentially mental states or dispositions, which can only be attributed to systems capable of having wants, of calculating consequences, and of undertaking actions. Because it is evident that only sentient creatures have such capacities, the very notion of a selfish gene, or an altruistic one, is absurd: "Nobody attributes selfish planning to a paramecium. What then can Dawkins mean by attributing it to a gene?" (Midgley, 1979:46)

Of course this criticism is in part quite unfair, for Dawkins gives an explicit definition of these notions that makes no mention of such mental states and dispositions. It is a definition he describes as "behavioral"; the terms are used to describe behavior, not to identify its motivational causes. Selfishness and altruism are of course ordinarily attributed to behavior, but only in a diagnostic way, as teleological descriptions suggesting that its causes are certain intentional states, certain desires and beliefs. But this is just the sense of these terms that Dawkins eschews. And given his redefinitions, there is nothing absurd or unintelligible about the notion of a selfish or altruistic gene. The notion is on a par with attributions of 'charm,' 'color,' or 'strangeness' to microphysical quarks. On the other hand, the adoption of the redefined terms selfishness and altruism in these contexts seems natural. The behavior they are used to designate is at least to a first approximation the same sort of overt *behavior* caused in the human cases by motives of selfishness and altruism. Does the dropping of the "essential" motivational component of these terms make them simple homonyms of the ordinary words? Obviously Dawkins does not think so, nor does Midgley, for in her view dropping them produces self-contradictory notions, not homonyms. This debate must ultimately turn on linguistic intuitions, and it is therefore without much theoretical or methodological interest here. What is of interest is the question of whether the redefined terms facilitate the expression of fruitful theoretical ideas and hypotheses difficult or impossible to otherwise express. Additionally, it is worth considering whether such redefined terms may be reapplied in their original domains of application with the same payoff. If the answers to the first questions are affirmative, then appropriating them will be justified. If the answer to the second is yes as well, then the appropriation cannot be deemed even seriously misleading.

Dawkins answers these questions in the following terms:

What does [selection] choose among? The favoured answer is "individuals." In a sense this is correct, but only if we put it very carefully; what matters is not differential *survival* of individuals, but differential inclusive genetic fitness. The fitness of an individual . . . means its success in getting copies of its genes represented in future generations. Fitness is a difficult quantity to calculate and a difficult concept to understand. . . . My suggestion is that we can lessen the risk of misunderstanding if we shift our attention from the organism as an agent, to the gene itself. Inclusive fitness is . . . that property of an individual which will appear to be maximized when what is really being maximized is gene survival. We may say, with the majority of modern specialists, that maternal care is favoured by natural selection because of its beneficial effects on the inclusive fitness of the mothers concerned. Or, we may say what is essentially the same thing in terms of the selfish gene: genes that make mothers care for their young are likely to survive in the bodies of infants cared for; genes that make mothers neglect their infants are likely to end up in dead infant bodies; therefore the gene pool becomes full of genes that induce maternal care; that is why we see maternal care in nature.

In effect what I have done is to reject 'the selfish group' as an explanation of individual altruism, to say 'the selfish individual' is a better, but more complex and easily misunderstood alternative, and to offer the 'selfish gene' as a simple, correct alternative. . . . Individual behavior, altruistic or selfish, is best interpreted as a manifestation of selfishness at the gene level. (Dawkins, 1981:559–60)

Leaving aside the tautologous definition of individual fitness as a correctable slip, we need to clarify to what it is exactly that selfishness is being attributed. It cannot be segments of the particular genetic material in a single organism, for it is clear that what is described as selfish in the passage is to be found first in mothers and then in their offspring. Can a particular integrated quantity of DNA that is transferred in the germ line from one organism to its offspring be the selfish gene? The gene in question is indistinguishable from the same sequence of base pairs in every somatic cell of the parent. It cannot have a special, distinct property that they lack. Moreover, because the particular segments of DNA transferred in the embryo in fertilization may well degrade long before the embryo's own full development, let alone its offspring's production, it must be clear that selfishness is a property of lines of descent among genes, one supervient on relations of this line of gene copies to the behavior of the families of organisms they constrain. The property that particular individual genes have is something we might call "nepotism": the property of behaving in such a way as to increase the welfare of offspring at the expense of one's own or at the expense of nonoffsprings' welfare, where welfare is defined as "chance of survival."

Dawkins's claim is that attributing selfishness to genes explains the attribution of altruism to the organisms that carry them. To understand this claim, it is crucial that the terms be used in their redefined meanings. The claim is of theoretical interest because it appears to solve a theoretical problem faced by the theory of natural selection. The problem is that organisms do not behave in ways that appear to maximize their fitness. This failure is distinct from the short-run divergences that the theory can countenance because of its statistical character. For many nonselfish acts appear to be persistent features of the behavior of very many species. And the real problem for the theory is not the apparently altruistic acts of mothers toward their offspring. The problem is nonselfish behavior toward nonoffspring, towards organisms more distantly related that have a far smaller proportion of genes in common. Here we have phenomena seriously discordant with the theory. And if the selfish-gene hypothesis can help us reconcile behavior that reduces individual fitness with no nepotistic payoff, then it will have to be reckoned an important theoretical advance, not merely an illuminating turn of phrase. The puzzle is that natural selection seems to be incompatible with the persistence of nonnepotistic altruism. Such behavior seems persistent over evolutionarily significant time periods. It appears to follow that the theory is falsified because it rules out a possibility that we have independent reason to believe obtains. There are two responses to this puzzle about possibilities. One is to deny that alleged cases of nonnepotistic altruism are real cases of it; that is, to attempt to show that such behavior is either a side effect of other perfectly comprehensible selfish behavior or that, despite appearances, the genetic linkage is stronger than it appears, so that the case is really one of nepotism. The second and more favored strategy is to show that nonnepotistic, nongenetic altruism is after all compatible with natural selection. The easiest way to show this is to describe circumstances under which natural selection would lead to altruism of this sort, to show that these circumstances are themselves not merely abstract possibilities. Once this is done, of course, the biologist must turn to establishing their existence. This latter task is a long and difficult one, but merely showing that natural selection and altruism are jointly possible is itself an important theoretical advance.

Dawkins sketches a selective scenario that would lead to the persistence of altruistic behavior among organisms bearing only selfish genes. Suppose that there are three types of genes that code for the following three types of behavior of organisms: altruism, selfishness, and a third form of behavior Dawkins calls "grudging," which involves behaving altruistically toward altruists and selfishly toward selfish organisms. Depending on the costs of altruism and the benefits of selfishness, the strategy of grudging may be the most adaptive. As such, if once established in the long run, it will expand in the population and eventually fix itself so preponderantly as to give the whole population the appearance of all being altruists. Thus the persistence of nonnepotistic altruism is rendered consistent with natural selection.

However, to make this result bear any real evolutionary weight, much more must really be shown. It must be shown not just that once established this state of affairs can persist in the presence of selective forces. It must also be shown that once it appears it stands a chance of successfully competing against other behaviors, especially against selfishness; that it can overwhelm them; and that, once established, it cannot be invaded and overwhelmed by any alternative behavior. In effect, we require some assurance that such strategies can appear in conformity to the usual evolutionary sequence in the history of an adaptation. Additionally, we should demand that organisms evincing these sorts of behaviors interact frequently enough for the behavioral routines to actually have a selective effect on organisms. And we need an indication that such behavioral routines can be realized in organisms and in genes. For, on the definition Dawkins offers, the same properties of selfishness and altruism are attributable to organisms and to genes.

Without this sort of detail, an abstract possibility like Dawkins's will be subject to the antiadaptationalist complaint that such stories are easy to cook up and therefore of little interest. Without these details, there will be little motive to attempt to turn an abstract possibility into an evolutionary research program. In fact, these demands have been met in research reported by Robert Axelrod and W. D. Hamilton (1981:1390–6). This work is worth review because it is a paradigm of adaptational thinking of the best sort. Even if it is ultimately superseded, or its conclusions undermined, it highlights the way pure theory should proceed in evolutionary biology. Axelrod and Hamilton examine the appearance and evolution of strategies of cooperation and conflict by employing a well-known problem from game theory:

the prisoner's dilemma. Players in this game have the choice of working with each other or not, of cooperating or defecting. Payoffs are related to the behaviors players undertake in such a way that (1) mutual cooperation results in a higher payoff than mutual selfishness, but (2) unreciprocated cooperation (altruism) results in a lower payoff than joint selfishness, and (3) selfishly free riding on the altruism of the other provides the biggest payoff. Accordingly, if one organism behaves selfishly, then maximizing fitness requires the other to do so, even though the fitness of each would be greater if both cooperated. On the other hand, if one behaves altruistically, then maximizing fitness requires the other to behave selfishly. In consequence, maximizing fitness must always lead to selfish behavior in a prisoner's-dilemma context. Highest average fitness would result from cooperation, but maximizing individual fitness in a single game encounter will never produce this outcome, so that cooperation cannot arise in the face of evolutionary forces. The selfish strategy of defection is an *evolutionarily stable strategy* in that no other strategy can do better than it when strategies compete. Should one arise by mutation, it will never displace defection. But in nature situations that reflect prisoner's-dilemma payoffs between the same two organisms will certainly arise repeatedly, given the lifespans of two or more organisms and the opportunities to compete or cooperate open to them. Axelrod and Hamilton's evolutionary account of altruistic behavior hinges on this fact.

It is important to note that the introduction of game theory, of the notions of strategy, competition, and cooperation, does not fall afoul of charges of incoherence like those lodged by Midgley. As Axelrod and Hamilton note:

An organism does not need a brain to employ a strategy. Bacteria, for example, have a basic capacity to play games in that (i) bacteria are highly reponsive to selected aspects of their environment, especially their chemical environment; (ii) this implies that they can respond differentially to what other organisms around them are doing; (iii) these conditional strategies of behavior can certainly be inherited; and (iv) the behavior of a bacterium can affect the fitness of other organisms around it, just as the behavior of other organisms can affect the fitness of a bacterium. . . . The model of the iterated [repeated] prisoner's dilemma is much less restrictive than it may at first appear. Not only can it apply to interactions between two bacteria or interactions between two primates, but it can also apply to the interactions between a colony of bacteria and, say, a primate serving as a host. (Axelrod and Hamilton, 1981:1390)

Axelrod and Hamilton consider the following questions: Given the supposition that organisms are repeatedly faced with prisoner's-dilemma-type situations, what strategies will be *robust*, that is, will thrive in the presence of a wide variety of other strategies employed by organisms in the repeated games? Are there conditions that will ensure that a robust strategy once established cannot be defeated by a new one? And, most crucially, how can any cooperative, nonselfish strategy arise in the presence of an already established general strategy of noncooperation? This last question is the most crucial, for if it cannot be answered Dawkins's theoretical possibility will forever remain just that.

To determine what strategies are robust, they ran fifteen different strategies for deciding a prisoner's-dilemma game against one another 200 times on a computer. The payoff matrix for a player is shown in Figure 8.2. Each player has the same payoff matrix. For each player defection is the dominant strategy, no matter what strategy the other player chooses. The highest average score in this and other computer tournaments was gained by a strategy very like Dawkins's "grudging"

Player A	Player B Cooperates	Player B Defects	
Cooperates	3	0	Numbers represent payoffs to A in terms of increased fitness and subsequent reproductive success.
Defects	5	1	

Figure 8.2. Payoff matrix for repeated prisoner's dilemma game.

rule. Axelrod and Hamilton call it "tit-for-tat": Cooperate on the first move, and thereafter do what the other player did on the immediately preceding move. In a series of competitions against a large number of strategies, including selfish defection, in which the worst strategies are weeded out turn by turn, which simulates the action of natural selection, tit-for-tat continued to do well and eventually became fixed. In other words, beyond a certain point the only players left in the simulation employed tit-for-tat and all players cooperated in every game in the iteration.

Axelrod and Hamilton establish by a mathematical proof the stability of tit-for-tat, that is, its resistance to invasion and preemption by a new strategy arising from variation, mutation, or immigration of a new organism. The conclusion of the proof is a conditional claim: Tit-for-tat is evolutionarily stable just in case the interactions between individuals have a high probability of continuing. More precisely, this probability must be greater than or equal to the quotients of differences between the fitness-level payoffs of the two behavioral choices available in the prisoner's-dilemma game. (The size of the probability is a statistical feature of the fitness differences in the same way Axiom 4 of the theory makes the rate of evolution a statistical feature of size of fitness differences between subclands.)

Like tit-for-tat, the strategy of always defecting, of always being selfish, is also an evolutionarily stable strategy, and it is so independent of the frequency of interaction among organisms. Therefore, it must be shown how the cooperative behavior characteristic of tit-for-tat can ever have arisen in the face of selfishness. If this cannot be shown, its robustness and stability count for little. To show it, Axelrod and Hamilton have recourse to the kin-altruism of nepotistic selfish genes. If frequent interactants are closely related, and so have a large number of genes in common, then the first appearance of tit-for-tat as a mutant strategy will result in cooperation between the cooperating parent and the tit-for-tat offspring. But this offspring will defect in all interactions with other defecting organisms (after the first encounter, of course). Thus, closely related organisms among whom the tit-for-tat strategy appears as a mutation from the always-defect (except against close kin) strategy will be fitter than those among whom it does not appear. Once tit-for-tat appears in the context of even limited kin-altruism, it can be expected to compete successfully against the strategy of selfishness:

Once the genes for cooperation exist, selection will promote strategies that base cooperative behavior on cues in the environment. Such factors as promiscuous fatherhood and events at ill-defined group margins will always lead to uncertain relatedness among potential interactants. The recognition of any improved correlates of relatedness and use of these clues to determine cooperative behavior will always permit advance in inclusive fitness. When a cooperative choice has been made, one clue to relatedness is simply the fact of reciprocation of the cooperation. Thus modifiers for more selfish behavior after a negative response from the other are advantageous whenever the degree of relatedness is low or in doubt. As such conditionality is acquired, cooperation can spread into circumstances of less and less relatedness. Finally, when the probability of two individuals meeting each other is sufficiently high, cooperation based on reciprocity can thrive and be evolutionarily stable in a population *with no relatedness at all*. (Axelrod and Hamilton, 1981:1934; emphasis added)

Axelrod and Hamilton suggest another mechanism that will lead to the invasion and overthrow of preponderant strategies of selfish defection by tit-for-tat. This requires that the initially small numbers of organisms employing the latter strategy interact with each other in a cluster far more often than with other selfish organisms. This will permit initial viability and eventual expansion.

The chronological story that emerges from this analysis is the following. All D [that is, always defect] is the primieval state and is evolutionarily stable. This means it can resist the invasion of any strategy that has virtually all of its interactions with all D. But cooperation based on reciprocity can gain a foothold through two different mechanisms. First there can be kinship between mutant strategies, giving the genes of the mutants some stake in each others' success, thereby altering the effective pay-off matrix of the interaction when viewed from the perspective of the gene rather than the individual. A second mechanism to overcome total defection is for mutual strategies to arrive in a cluster. . . . Then the tournament approach demonstrates that once a variety of strategies are present tit-for-tat is an extremely robust one . . . . And if the probability that interaction between two individuals will continue to be great enough, then tit-fot-tat is itself evolutionarily stable. Moreover, its stability is especially secure because it resists the intrusion of whole clusters of mutant strategies. Thus cooperation based on reciprocity can get started in a predominantly noncooperative world, can thrive in a varied environment, and can defend itself once fully established. (Axelrod and Hamilton, 1981:1934)

Axelrod and Hamilton go on to apply their theory to a variety of biological phenomena: the origins of cooperation among hermaphrodite sea bass; various sorts of interspecies mutualism, like the hermit crab and sea anemonies; a tree and its fungus; insects and bacteria; fig wasps and trees; cleaner fish and larger fish that could be expected to prey on them; cooperative territoriality among song birds; and, more speculatively, the etiology of cancer and nondisjunctive chromosomal irregularities.

Two things are especially important about this attribution to genes of strategies usually ascribed to humans. First of all, it is more than a merely plausible story. It involves a certain amount of replicable evidence. And it answers crucial questions, not only about how an apparently maladaptive character like altruism can persist under selective constraints, but also about how it could arise as an adaptation and displace another evolutionarily stable strategy. Secondly, it meets the demand we imposed on the use of terms redefined away from their usual meanings: It not only helps explain how cooperation could have evolved, but it also can be applied in the explanation of what is normally called altruistic behavior and cooperation among humans. In this application, it is not meant to displace the explanation of particular acts of cooperation in terms of motives, beliefs, desires, etc. It can explain the universality and inevitability of this sort of behavior, however, as an institutionalized feature of human life. It can explain why we are not all egotists or free riders, why we find ourselves using the strategy of tit-for-tat not on the basis of calculation, nor through the establishment of a formal social contract. As such, the biologist has every right to appropriate and redefine terms from ordinary language and to reapply them with some justice to the phenomena these terms, in their ordinary meanings, describe. This reapplication is innocent of the charges of either incoherence or misleading tendentiousness.

8.5. Metaphors and Molecules

Attributing selfishness to genes turns out to need be neither metaphorical nor anthropomorphic. Selfishness and altruism in biological contexts are explicitly defined as sorts of behavior with distinctive effects on individual fitnesses. As such, there is nothing biologically unusual about these notions. The only questions surround the warrant for appropriating and redefining these notions from ordinary language instead of some others. But the use of intentional terms to describe *the causes of behavior,* instead of *the behavior itself* or its effects, cannot be similarly justified. When a mutation is said to be removed from the DNA because a "proofreading" enzyme "recognized" it, the statement cannot be understood in terms of explicit redefinition of these notions. This appropriation of terms is usually justified as merely metaphorical. But the naturalness, and inevitability, of this usage make its attraction for the biochemist too compelling to be merely metaphorical.

Let us review some of the things molecular biologists find themselves saying about molecular interactions. Consider the enzymes that bind amino acids to their appropriate transfer RNAs, the aminoacyl-tRNA synthetases:

These enzymes are highly selective in their *recognition* of the amino acids and of the prospective tRNA acceptor. . . . tRNA molecules that accept different amino acids have different base sequences, and so they can be readily *recognized* by the synthetases. A much more demanding *task* for these enzymes is to *discriminate* between similar amino acids. For example the only difference between isoleucine and valine [two amino acids] is that isoleucine contains an additional methylene group . . . The concentration of valine *in vivo* is about five times that of isoleucine, and so valine would be *mistakenly* incorporated in place of isoleucine 1 in 40 times. However, the observed *error* frequency *in vivo* is only 1 in 3000, indicating that there must be a subsequent *editing* step to enhance fidelity. In fact the synthetase *corrects* its own *errors.* Valine that is *mistakenly* activated is not transferred to the tRNA specific for isoleucine. Instead this tRNA promotes the hydrolysis of valine-AMP, which thereby prevents its *erroneous* incorporation into proteins . . . How does the synthetase *avoid* hydrolyzing isoleucine-AMP, the *desired* intermediate? Most likely, the hydrolytic site is just large enough to accommodate valine-AMP, but too small to allow the entry of isoleucine-AMP . . . It is evident that the high fidelity of protein synthesis is critically dependent on the hydrolytic *proof-reading action* of many aminacyl-tRNA synthetases. (Stryer, 1981:644–5; emphasis added)

Passages like this can be multiplied from research reports, textbooks, and technical presentations ad libitum. There is in addition the semantic vocabulary of DNA transcription and translation, which makes use of terms like information, translation of a code, sense and missense readings and misreadings, start and stop signals, and the proofreading of DNA polymerases. Similar terms come naturally to molecular immunology, to the description of substrate-specificity among enzymes, to bacterial chemotaxis, to membrane transport, etc. This terminology is widely employed in molecular biology without the slightest precaution of definition, nor with any caveat against taking it literally. There are two reasons for this: The first is that the use of this terminology appears so transparently nonliteral that any announcement of its metaphorical application seems quite superfluous. The second is that molecular biologists are confident that, wherever the intentional terms of ordinary language are employed, the actual biochemical mechanism can be described and explained in terms that surrender enlightening metaphor for literal, though unwieldly and complicated, biochemistry.

But this explanation and assurance do not seem enough to dispose of the naturalness and inevitability of intentional descriptions of molecular interactions. To point out that macromolecules engage in extremely intricate functional, that is, teleological, relations, and that humans' recognizing, proofreading, making mistakes, or avoiding undesired outcomes are also extremely complex teleological activities, is only part of the explanation for the naturalness of these metaphors.

To be sure, some of this terminology, especially the language of information and codes, can be given a clearly nonintentional interpretation by appeal to the formal apparatus of the mathematical theory of information, so no problem of inappropriate or misleading metaphor arises. It is indeed the case that reflection on the probabilistic theory of an information-carrying channel does lie behind the theory of the genetic code. The purely formal, mathematical constraints that must be satisfied by a signal system carrying just twenty different messages (one for each amino acid) did help suggest a mechanism for the genetic code that employs four symbols, four bases to construct sixty-four possible message units that redundantly specify twenty amino acids. In this context, 'information,' 'signal,' 'message,' and 'code' do not have their usual meanings but are assigned redefinitions of a wholly nonmental kind by the axioms of a branch of applied mathematics, information theory. Here the problem of explaining the naturalness of metaphor does not arise, because we are faced with literal description.

However, although most of the terminology the molecular biologist employs is nonliteral, its naturalness and inevitability cannot be explained as merely the result of an extremely apt metaphor. For metaphors can always be dispensed with. It is not clear that biochemistry can really proceed without reliance on the intentional vocabulary it has found so unavoidable in the description of its domain.

In fact, the usage involves neither metaphor nor explicit redefinition, but rather implicit, covert, unnoticed change in meaning. This change has important ramifications both for the limits and the significance of contemporary molecular biology. To see that the use of intentional vocabulary involves undetected definitional shifts, we must first explore the nature of this vocabulary on its home base, in the description and explanation of human cognition and its consequences.

Cognitive or intentional states have a logical property that distinguishes them from all other states and especially from merely physical, or even goal-directed, states of nonintentional teleological systems, plants, thermostats, organs of the body, or simpler forms of biological life or macromolecules. It is a property that hinges on their representative character, on the fact that they "contain" or are "directed" to propositions about the way things are or could be. The property they have is this: When we change the descriptions of the states of affairs they "contain" in ways that seem innocuous, we turn true attributions of intentional states into false ones. For instance, consider two presumably true statements:

(D) Oedipus desired to marry Jocasta.
(B) Oedipus believed that Jocasta is the queen.

These two sentences report relations of desire and belief between Oedipus and the propositions repectively that Oedipus marries Jocasta and that Jocasta is the queen. These two statements are the contents of Oedipus' intentional states of desire and belief, respectively. But now Jocasta is the queen and is Oedipus' mother, so anything true of the queen or of Jocasta is true of Oedipus' mother, and we should be able to refer to Jocasta indifferently as the queen and as Oedipus' mother. But when we refer to her as Oedipus' mother in describing Oedipus' desires and beliefs, we produce two presumably false statements:

(D') Oedipus desired to marry his mother.
(B') Oedipus believed that his mother is the queen.

By making innocent substitutions of terms that refer to the very same objects, we have turned truths into falsehoods. But this is something we simply cannot do to any expressions that report *nonintentional* relations, such as those characteristic of physics, chemistry, or biology. I may refer to the earth and the moon in an infinite number of ways, and the statement that their gravitational attraction varies as the inverse of the square of the distance between them will be true no matter how I refer to them. Any statement about the physical or chemical properties and relations of the Hope Diamond will remain true no matter how I refer to it, as the largest diamond in the world, as the stone that brings bad luck to its owners, as the only gem mined in Katanga in March of 1914, etc. Similarly, the goal of photosynthesis is the production of polysaccharides, and this statement will remain true no matter how I refer to them: as starch; as the substance with the chemical formula $(CH_2O)_n$; as the chief constituent of papier-mâché; as an important input in the TCA cycle; or as the main ingredient of my favorite junk food. But such substitutions are not permissible in intentional contexts. For they may, and often do, change a true statement about an agent's beliefs or desires into a false one.

This special logical feature of intentional states has important consequences for the ways we identify and distinguish them. We can in fact only do so by reference to the propositions they contain, to the states of affairs they represent. To see this, consider again this statement:

(B') Oedipus believed that his mother is the queen.

The only temptation we have to suppose that (B') is true is that at the end of *Oedipus Rex* Oedipus does come to believe that Jocasta, the queen, is his mother. In consequence, he plucks his eyes out. But this belief at the end of the play is a different belief from that which Oedipus has in the middle of the action. His belief that Jocasta is the queen, (B), is a *different* belief from (B'). The difference in the beliefs is based on the difference in the contents. It is not based on the facts about Jocasta. These have not changed throughout the play. It is not the case that at some point in the play it was false that Oedipus' mother was the queen and that later it became true that she was. Accordingly, we cannot identify Oedipus' beliefs by reference to the facts about Jocasta. It was not on these facts that his beliefs depended, and it is not by appeal to them that we can identify his beliefs or trace the changes in them. Much the same can be said about his desires. Jocasta was the object of his desire, but were he to have been asked whether his mother was the object of his desire, he would have rejected the suggestion violently. Yet Jocasta was his mother. Accordingly, his desire cannot be identified or disclosed by any examination of the actual person at whom it was "aimed," but only by appeal to its content, a proposition about that person. This is even clear in the case of unattainable desires: Ponce de Leon desired to reach the fountain of youth. But we cannot find out what he desired by any inspection of the fountain of youth or his actual, physically describable relation to it, for there is no such thing as the fountain of youth. We can only identify his desire by its content.

If the identity of an intentional state is determined by the proposition it contains, and if beliefs can be false and desires unattained, then we cannot decide what beliefs or desires an agent has by finding out whether any particular proposition about the world independent of his intentions is true or false. Of course, we can and do discover peoples' intentional states by asking them. But this method is itself intentional: Their responses to our questions will only be counted as replies, as the utterance of meaningful speech, as actions, if we assume that they are sincere and understand the meaning of the language in which we put our questions; that is, if we assume they desire to answer truthfully and believe that the utterances they emit will attain this desire. In other words, our normal means of identifying the causes of actions are in terms of their effects, the actions themselves; and the identification of an event, a movement of the body, as an action presupposes the attribution to it of intentional causes. Now there is nothing improper about this intentional circle so far as our everyday, nonscientific purposes are concerned. But science seeks to sharpen its explanations and predictions of behavior beyond commonsense levels of accuracy. Therefore, if it employs intentional terms, it must break out of this intentional circle: We must find a way of identifying movements as actions without assuming that the movements were caused by intentional states; we must find a way of identifying intentional states without appeal to their effects in action. The reason is simple, and it can be illustrated by considering the explanation and prediction of how alcohol thermometers work.

To explain why the column of alcohol in a thermometer rises, I note that the substance it is in contact with has become hotter. This increase in temperature causes an expansion of the alcohol in the closed tube of the thermometer and thus a rise in its level. But to ascertain that the substance has become hotter, I require a thermometer. If I employ an alcohol thermometer to establish this initial fact, my explanation will be open to the criticism that it presumes what it sets out to explain. More seriously, without an alternative means of measuring temperature, say a mercury thermometer, a bimetalic bar, or a gas-diffusion thermometer, I will be unable to test, correct, and improve my measurements of heat by the use of an alcohol thermometer. Consequently, I will be unable to make improving predictions about the effects of changes in temperature, effects like changes in state, electrical resistance, etc. And, without alternative measures of temperature, I will be unable to relate my regularities about the relation between heat and linear expansion in a closed tube to the rest of thermodynamics. Indeed, the rest of thermodynamics would have been undiscoverable without an alternative means of measuring what an alcohol thermometer measures by linear expansion in a closed tube.

The situation with respect to the intentional notions is identical, and it has long been recognized. The recognition that intentional variables needed nonintentional anchors if our understanding of intentional or cognitive states is to improve beyond the level of commonsense "folk psychology" was long sensationalized as the "problem of other minds." The suggestion that if detached from behavior the intentional variables were empirically empty, theoretically barren, or methodologically suspect, that we had no good grounds to attribute such states to others, is the real force of the picturesque philosophical skepticism about the knowledge of other minds.

The need to anchor the intentional in the nonintentional spawned behaviorism. Unfortunately even the most sophisticated sort of behaviorism cannot provide the connections to the nonintentional that are required. No matter what the setting, no mere movement of the body can reveal a desire, unless we already know a person's relevant beliefs; no mere movements of the body can reveal a belief, unless we already know a person's relevant desires. So any definition of a desire will have to mention beliefs in the definiens, and any definition of a belief will have to mention desires in the definiens. Definitions free of such intentional residues will not correctly identify the paradigm beliefs, desires, and actions that folk psychology can identify. But such nonintentional definitions are just what behaviorism demands.

Behaviorism's failure should come as no surprise. It is foreordained in the fact that intentional statements have the logical property of forbidding otherwise innocent substitutions, whereas any statement about mere movement permits all such substitutions without provoking a change in truth or falsity. Therefore any nonintentional definition of the intentional was bound to leave something crucial out. The only alternative to behaviorism as a way of anchoring mental states to nonintentional ones is a neuroscientific reductionism: The idea is that we should be able to find classes or types of brain states that are identical to states of belief and desire. Although such states will be practically useless in the identification of intentional states, they will provide a theoretically possible way to check our ordinary attributions and perhaps even to improve them in artificial, clinical situations. In this respect, a neurophysiological approach to mental states would be like a highly complicated and inconvenient thermometer, which cannot replace our ordinary ones in any practically important context but provides a theoretical reassurance that we can in principle test and correct the alcohol thermometer's accuracy. Unfortunately, neuroscience has revealed no brain states that do march in lockstep together with mental states. At the level of detail that neuroscience currently operates on, there are no changes in the brain and/or parts of the brain that seem at all to vary systematically with changes in the mental states that experimental subjects *report;* of course, tomorrow someone may report the required parallelism. But it is highly unlikely, and even when it happens the gulf between nonintentional brain states that can be described in an indefinite number of ways and intentional mental states for which there are only privileged descriptions will remain to be explained. The recalcitrance of intentional terms to analysis or definition in terms of behavior or neurophysiology is not explained by their logical peculiarities. But it is reflected in them.

Descriptions that preserve reference cannot be freely substituted into sentences reporting intentional states, without the risk of changing them from truths into falsehoods. But this risk is never to be met anywhere in natural science, all the way from microphysics to evolutionary biology. This divergence is symptomatic of conceptual differences not to be taken lightly. Though it has the appearance of a merely grammatical or semantic difference for which a clear explanation should be readily forthcoming, no such satisfactory explanation has yet found any wide acceptance among philosophers or cognitive psychologists. There are of course many differences between cognitive states and purely physical states, and presumably this grammatical difference between their descriptions is not the most crucial and explanatorily most central one. But it is a difference that we can agree upon and state clearly. It does seem especially indicative of the difference between mental states and purely physical ones, providing a practical test that discriminates the classes of mental from nonmental kinds with greater precision and accuracy than any other test. Accordingly, it seems reasonable to conclude that the failure to allow free substitution of co-referring expressions in the description of a state is at least a necessary condition for the state's being a cognitive one. States that do not have this feature, no matter what words we use to describe them, cannot be cognitive states.

It is important to see that this argument about logical peculiarities in the meaning of intentional terms like belief, desire, and action infects a much wider circle of notions. Thus, consider the statement

(R) Oedipus recognized that Jocasta was the queen.

We can no more substitute coreferential terms into (R) without changing its truth value than we can do this to (D) and (B). This is partly because recognition is "built" up out of beliefs about classification. Actions too are intentional and brook no free substitution of coreferring terms in their description. To edit or to proofread is a kind of action; it is distinguished from merely changing the incorrect spelling of a word by accident, because it is the result of beliefs about the correct spelling and desires to correct a misspelling. An error or a mistake is an action undertaken as a result of a false belief. All these intentional terms have in common the fact that we cannot identify them singly except by their propositional content. We cannot reliably identify the cognitive process of recognizing a face by reference to a class of bodily movements it results in, or the neural events that constitute it, because there is no finitely specifiable class of such movements. Only by adding the constraints of other intentional states can we narrow the class of such movements down to a size that has real implications for what movements will follow the event of recognition. This generates the intentional circle, because the same problem daunts the identification of the other states.

But now return to biochemistry: Compare the constraints on identifying the recognition of a face by an agent with the recognition by a tryptophan-tRNA of a tryptophan molecule. Recognition in this latter case can be strictly characterized in terms of molecular interactions of one and only one kind; what is recognized, what the consequences of recognition are, and the physical process by which recognition is accomplished can be unambiguously described in biochemical terms that make *no* appeal to any intentional notion. *There is no intentional circle in employment of cognitive terms by molecular biology.* Recognition in this case is definable by appeal to a formula from organic chemistry.

In fact, every state of a macromolecule that can be described in cognitive terms has both a unique, manageably long, purely physical characterization and a unique, manageably describable disjunction of consequences. This is something intentionally described human cognitive states lack altogether. Because they lack it, we can only describe intentional states by reference to their contents or to other intentional states operating with them to determine action, that is, intentionally caused behavior. This claim that human intentional states are not reducible to physical, neural states, nor specifiable by reference to a fixed type of behavior they result in, is not to deny that each and every particular cognitive state is identical to some particular neural state or other. It is rather to admit that regularities employing the intentional description of mental states can never be theoretically linked up with neuroscience, or with nonintentional behavioral psychology. They must therefore remain isolated from the remainder of natural science. Here we have a clear difference in kind between human cognitive states and the states of macromolecules. For these latter even when described in cognitive terms can be linked to physical science: We can dispense with intentional descriptions and explain intentional regularities about molecular recognition, information, etc., in wholly nonintentional terms. This is why the application of such terms (with their ordinary definitions) to macromolecules is nonliteral. But it is also why the molecular, biological employment of these terms is nonmetaphorical as well; why it constitutes what is in fact a disguised *redefinition* of them, identical to the *explicit* behavioral redefinition of terms like 'selfishness' and 'altruism' in the work of Dawkins and others surveyed in the last section. It differs from Dawkins's redefinition in being unnoticed even by the biologists who have unknowingly effected the redefinitions.

Consider the consequences for the meaning of intentional terms if we employ them in contexts where they can always be eliminated through the substitution of complex physiochemical descriptions of the intentionally described phenomena. Whenever we say that a tRNA recognizes its appropriate amino acid, we can substitute a description of what happens in terms of biochemical specificity and chemical bonding. This is tantamount to saying that a purely behavioral characterization can be given for amino-acid "recognition." Similarly for all other such terms: A molecule of DNA polymerase 'proofreads' if and only if it has a specified chemical interaction with the base sequence in the DNA it is polymerizing, etc. But intentional, cognitive terms are incapable of behavioral definitions. If the terms of molecular biology are so definable, then they must be employed here not under metaphorical license but through redefinition, albeit redefinition unnoticed by those who employ these redefined terms in the description of macromolecular interactions. Suitably redefined, it is no surprise that these terms function naturally and inevitably in the description of molecular interactions. They have been implicitly tailored to this purpose. When pressed into biochemical service, intentional vocabulary retains its teleological, functional implications for behavior. This together with the intricacy, plasticity, and variability of the biochemical interactions is what makes for its appropriateness in this context.

This is in fact a theoretically and methodologically more preferable conclusion than the suggestion that intentional terms are used in these contexts in their original meanings and only metaphorically. Metaphorical description, for all its heuristic advantages, is always at best potentially misleading. For following out the metaphor far enough is necessary to exploit its heuristic advantages, and following it too far is sufficient to lead theoretical development into dead ends. The trouble with metaphors in science is that there is no reliable mark of when a metaphor has been carried too far.

Aside from this general problem of the use of metaphor, the metaphorical exploitation of intentional descriptions in the development of biological theory is fraught with special problems. One obvious feature of the intentional terms we employ to describe human cognition and its effects in action is that scant progress has been made in literally several thousand years in the formulation of a theoretical science of these states. Psychology and the other social sciences of the intentional are little further along than they ever were. Whatever power our commonsense "folk psychology" has to explain and predict human action, it has had since the unrecorded past. It remains an important philosophical problem to explain this crucial difference between the lack of improvement in the human and behavioral sciences that employ this intentional vocabulary and the progress in natural sciences that do not. One explanation for this want of improvable generalizations about the relations of intentions to actions turns on the logical property of prohibiting substitution of equivalent descriptions in such generalizations. Because of it, intentional terms cannot be replaced in laws by descriptions of underlying physical states and processes. Their intentionality thus precludes the formulation of any laws connecting intentionally and nonintentionally described states of human agents required for a theory of human behavior. The lack of such laws enabling us to entrench, explain, and improve potential regularities about human cognition, and its effects on behavior, in a wider body of theory suggests that there are no such regularities. There are no laws about human behavior under its intentional description. This explains the lack of progress in the human sciences.

If there are no laws about intentional human behavior, and no attainable theory couched in this language, then these intentional notions should not be exploited in the attempt to improve another theory; at least they should not be exploited in their intentional meaning. For there will be no regularities about intentional variables to guide our search for regularities governing molecular interactions; unlike other useful analogies and metaphors in science, we do not have in this case a reliable guide to theory development, because on its home ground the intentional theory is without merit. The fact that this theory is implicitly entrenched in ordinary "folk psychology" shared by molecular biologists and laymen, gives it a spurious aura of real explanatory content and encourages its application to the molecular context.

The appropriation of cognitive terms from ordinary language and their implicit redefinition in molecular biology thus parallels the appropriation of motivational terms like 'selfishness' and 'altruism' to describe macromolecules and their behavior. The difference is that although the latter appropriation involved explicit redefinition, the redefinition involved in the former appropriation has been hidden from the view of the users of this terminology themselves.

But if there really is a parallel here, we should demand that molecular biology's implicit behavioristic redefinition of cognitive terms have an eventual payoff for the understanding of humans and other organisms to which the terms apply under their original intentional definitions. After all, one reason to countenance the redefinition of terms like selfishness to describe genes is that it enables us to explain the possibility of altruistic behavior in humans. Does the attribution of "behavioralized" cognitive language to genes enable us to explain human behavior? Under one interpretation of the question, the answer must be no; under another one, the answer is that contemporary biochemical research seems to be moving in precisely this direction.

The answer to the question must be no, if the human phenomena that molecular biology, and its neurophysiological application, are supposed to illuminate is intentionality itself. If the aim of neuroscience is ultimately to provide biological, biochemical explanations for regularities governing intentional states and their causes and effects, then this is an aim it will never attain. The reason is that intentional states are not related by causal or structural regularities to physical states, either in behavior or physiology. They divide up human neural phenomena into classes that are not linked to specifiable behavior or physiology. This is why behaviorism fails as a method of access to intentionally describe states; and why physicalism fails as well. Molecular biology will never generate a science of the mental under the intentional terms in which it is now described. And if molecular biology will not do this, neither will any other division of biology, including sociobiology – for it too is a nonintentional discipline.

But the failure of behaviorism, foreordained by the intentionality of cognitive states, has spawned an entirely different approach to behavior, one rigorously nonintentional. Contemporary behavioral learning theory constitutes a research program that begins by explicitly redefining terms. Learning, discrimination, and recognition are introduced in purely behavioral terms, in terms of stimuli, responses, reinforcers, and operants. Behavioral psychology attempts to uncover regularities about behavior described by these redefined terms. As such, it purports to search for regularities that can be linked to the remainder of science. It is the explanation of these regularities about conditioning, operant and classical, that molecular biology is already fast approaching. To the extent that regularities of behavioral psychology, and especially its theory of learning, are applicable to human behavior, we can expect the cognitive terminology of molecular biology, under its nonintentional redefinitions, to be as fruitful as the redefined terms of Dawkins's *The Selfish Gene*.

A recent example of this sort of application is provided by the research of Kandel and Schwartz (1982:433–43). It has been known for some time that certain invertebrates can be conditioned to respond to stimuli in ways that demonstrate learning, that is, habituation, sensitization, classical and instrument or operant conditioning, of the type well confirmed in the study of mammals and of course humans. The nervous systems of such organisms, and especially the marine mollusk *Aplysia californica*, are much simpler than those of mammals and are particularly suited to biochemical study. The neurons in their nervous systems are both much larger in size and far fewer in number than those of mammals. Physiological research has revealed that the pathways of simple forms of sensitization learning always converge on synaptic connections between specific neurons in the abdominal ganglion of the *Aplysia* nervous system. Accordingly, Kandel and Schwartz claim that

a behavioral system in *Aplysia* can be used to examine the mechanism of several forms of learning at different levels of analysis: behavioral, cell-physiological, ultrastructural, and molecular. At the behavioral level we can rigorously characterize various forms of learning and obtain the time course for the short-and-long-term memory of each form. At the physiological, ultrastructural, and molecular levels it is possible to specify, for each form, the locus and mechanisms in individual identified neurons. *More important, the information from any one level of analysis can be related to information obtained at the others.* (1982:433)

Kandel and Schwartz's studies show that molecular changes at synapses in the abdominal ganglion determine classically conditioned learning and that these changes depend on a relatively well-known type of biochemical reaction mediated by cyclic AMP (a macromolecule we have already encountered with several other roles in Chapter 4). What is actually involved is a biochemical cascade involving a sequence of steps whose details have been worked out in enough detail to offer a biochemical theory of the differences between long- and short-term memory. Memory here is of course not defined intentionally, as "true belief about the past that may eventuate in certain actions," but is explicitly defined in terms of a response curve for a stimulus. Sensitization produces short-term memory when the conditioned stimulus is followed by a response for only a small number of trials after the unconditioned stimulus is removed. Long-term memory is the persistence of the response long after the unconditioned stimulus is removed.

In brief, Kandel and Schwartz began by first conditioning *Aplysia* snails for long-term and short-term memory. They then uncovered morphological and biochemical differences in the neurons of these animals, and these differences suggest that the additional conditioning producing long-term memory does so by stimulating the expression of a new regulatory enzyme. This new enzyme reacts to much lower concentrations of cyclic AMP at the synapse than is required for the normal firing of the neuron. In other words, the difference between long-term memory and short-term memory is on the point of being explained in purely biochemical terms. But what is being explained is not memory differences as we understand them in ordinary language, but differences under new explicit redefinitions.

Kandel and Schwartz conclude their review of research in this area with a discussion of "Reductionist Strategies for Studying Behavior":

Because its nervous system is advantageous for biophysical and biochemical analysis, *Aplysia* has provided molecular insights into a simple form of learning . . . the mechanisms . . . we have considered here may not be unique to *Aplysia* or even peculiar to sensitization and classical conditioning but operate in other animals and, with variations, in more complex forms of learning. . . .

Reductionist strategies have been successfully applied to the analysis of various biological mechanisms including muscle contraction, the genetic code, protein synthesis, secretion, active transport, membrane excitability and electrical and chemical transmission between nerve cells. Still, the reductionist approach to learning has not yet been fully accepted. Some psychologists and biologists still hesitate: they believe that only higher animals exhibit complex forms of learning and that the analysis of learning in any simpler animals may not be relevant to humans. Within the last year this view has been challenged in two studies that have compared the details of two forms of classical conditioning in invertebrates and animals. [One] found that *Aplysia* shows a conditioned association between a neutral stimulus and a central defensive motivational state that resembles the conditioning of fear in mammals. Even more striking [it has been] found that the land snail *Limax* can learn higher order forms of associative learning (including blocking, second order conditioning, and preconditioning) previously thought to be restricted to animal intelligence. These studies show that invertebrates, like mammals, can form central representation of the CS [conditioned stimulus] and learn about its predictive properties. (1982:441)

Molecular biology stands on the threshold of providing the underlying biochemical mechanisms for the generalizations about learning that have been uncovered in behavioral psychology. Its exploitation of a behaviorally redefined class of intentional terms will therefore turn out to have all the warrant that the redefinition of terms of ordinary language ever needs. Thus we may conclude that, although the revolution in molecular biology has raised several new versions of traditional problems in the philosophy of biology, it has not, despite appearances, raised the entirely new one of the need to justify or explain away psychologism and anthropomorphism at the level of the gene.

This account of the role of apparently intentional terms in biology has an important and controversial corollary. The social and behavioral sciences are all thoroughly intentional, and as we have seen the intentionality of their explanatory and descriptive concepts is ineliminable and irreducible. If biology is a thoroughly nonintentional science, in which intentional terms appear only under redefinitions – implicit or explicit – then biology can hope to shed no systematic light on, still less absorb, the science of human behavior. This does not mean that humans are anything more or less than complex biological systems, only that the view of these systems as irreducibly intentional ones must remain forever biologically autonomous or provincial.

Bibliography

The following list contains all works mentioned in the literature introductions that follow each chapter. It also includes books and papers referred to in the book but not mentioned in these introductions. The list does not pretend to be anything more than a compilation of sources for the quoted or described claims I have attributed to various writers, as well as bibliographical information about works recommended for further reading. Many of the works cited below contain extensive scholarly bibliographies. For example, Ernst Mayr's massive history, *The Growth of Biological Thought,* contains an equally lengthy bibliography of about fourteen hundred items bearing on the history of biology and its philosophy. Additionally, R. Burian and M. Grene (eds.), *Philosophy of Biology in the Philosophical Curriculum,* contains a usefully annotated bibliography of recent work, especially journal articles, in the philosophy of biology. Elliot Sober (ed.), *Conceptual Issues in Evolutionary Biology,* reprints many of the papers cited in the present work, along with other extremely influential documents on this subject.

Because of the revolutionary character of contemporary biology and because of the accelerating pace of work in the philosophy of biology, any bibliography becomes quickly dated. The only way to keep informed about these subjects is in the scholarly journals devoted to them. Among the journals in which the work of philosophers of biology regularly appears are the following: *Philosophy of Science, The British Journal for the Philosophy of Science, Systematic Zoology,* and *Studies in the History and Philosophy of Science.*

The following list gives the place, publisher, and date of the most recent or most accessible edition of the work cited.

Abelson, J., and Butz, E. (eds.). 1980. "Recombinant DNA." *Science,* 209:1317–1478.

Alberts, B., Bray, D., Lewis, J., Raff, M., Roberts, K., and Watson, J. D. 1983. *Molecular Biology of the Cell.* New York: Garland.

Axelrod, R., and Hamilton, W. D. 1981. "The Evolution of Cooperation." *Science,* 211:1390–6.

Ayala, F. 1968. "Biology as an Autonomous Science." *American Scientist,* 56 (1968): 207–21.

Ayer, A. J. 1959. *Logical Positivism.* New York: Macmillan.

1961. *Language, Truth, and Logic.* New York: Dover Books.

Barnes, B., and Shapin, S. (eds.). 1979. *The Natural Order*. Oxford: Blackwell.

Barrett, P. H. 1974. "Darwin's Early and Unpublished Notebooks." In H. E. Gruber (ed.), *Darwin on Man*. New York: Dutton.

Beadle, G. W., and Tatum, L. 1941. "Genetic Control of Biochemical Reactions in *Neospora*." *Proceedings of the National Academy of Science*, 27:499–506.

Beatty, J. 1983. "The Insights and Oversights of Molecular Genetics." In P. Asquith and T. Nickles (eds.), *PSA 1982*, vol. 1. East Lansing, Mich.: Philosophy of Science Association.

Bennett, J. F. 1976. *Linguistic Behavior*. Cambridge: Cambridge University Press.

Bloor, D. 1979. *Knowledge and Social Imagery*. London: Routledge and Kegan Paul.

Braithwaite, R. B. 1953. *Scientific Explanation*. Cambridge: Cambridge University Press.

Burian, R. 1984. "Adaptation." In M. Grene (ed.), *Dimensions of Darwinism*. Cambridge: Cambridge University Press.

Carlson, E. A. 1966. *The Gene: A Critical History*. Philadelphia: Saunders.

Crick, F. 1966. *Of Molecules and Men*. Seattle: University of Washington Press.

Cummins, R. 1975. "Functional Analysis." *Journal of Philosophy*, 72:741–65.

Darden, L., and Maull, N. 1977. "Interfield Theories." *Philosophy of Science*, 44:43–64.

Darwin, C. 1859. *On the Origin of Species*. London: Murray. Reprinted 1976; Baltimore: Penguin Books.

Dawkins, R. 1976. *The Selfish Gene*. Oxford: Oxford University Press.

1981. "In Defence of Selfish Genes." *Philosophy*, 56:556–73.

Dickerson, R. E., and Geis, I. 1969. *The Structure and Action of Proteins*. Menlo Park, Calif.: Benjamin/Cummings.

1983. *Hemoglobin: Structure, Function, Evolution, and Pathology*. Menlo Park, Calif.: Benjamin/Cummings.

Dobzhansky, T., Ayala, F. J., Stebbins, G. L., Valentine, J. W. 1977. *Evolution*. San Francisco: Freeman.

Driesch, H. 1914. *The History and Theory of Vitalism*. London: Macmillan.

Eldredge, N., and Gould, S. J. 1972. "Punctuated Evolution: An Alternative to Phyletic Gradualism." In T. J. M. Schopf (ed.), *Models in Paleobiology*. San Francisco: Freeman.

Feigl, H., and Brodbeck, M. (eds.). 1953. *Readings in the Philosophy of Science*. New York: Appleton, Century, Crofts.

Feyerabend, P. K. 1962. "Explanation, Reduction and Empiricism." In Feigl, H., and Maxwell, G. (eds.), *Minnesota Studies in the Philosophy of Science*, vol. 3. Minneapolis: University of Minnesota Press.

Freifelder, D. 1978. *Recombinant DNA*. San Francisco: Freeman.

Gabriel, M., and Fogel, S. 1955. *Great Experiments in Biology*. Englewood Cliffs, N.J.: Prentice-Hall.

Ghiselin, M. 1969. *The Triumph of the Darwinian Method*. Berkeley: University of California Press.

1974. "A Radical Solution to the Species Problem." *Systematic Zoology*, 23:536–44.

Goudge, T. 1961. *The Ascent of Life*. Toronto: University of Toronto Press.

Gould, S. J. 1977a. "Caring Groups and Selfish Genes." *Natural History*, 86:21–4.

1977b. *Ever Since Darwin*. New York: Norton.

1977c. *Ontogeny and Phylogeny.* Cambridge, Mass.: Harvard University Press.

1981. *The Mismeasure of Man.* New York: Norton.

1982a. "Darwinism and the Expansion of Evolutionary Theory." *Science,* 216: 380–7.

1982b. "The Meaning of Punctuated Equilibrium." In R. Milkman (ed.), *Perspectives on Evolution,* 83–104. Sunderland, Mass.: Sinauer.

Gould, S. J., and Lewontin, R. 1979. "The Spandrels of San Marco and the Panglossian Paradigm: A Critique of the Adaptationalist Program." *Proceedings of the Royal Society of London,* B, 205:581–98.

Gould, S. J., and Vrba, E. 1982. "Exaptation – A Missing Term in the Science of Form." *Paleobiology,* 8:4–15.

Hanson, N. R. 1958. *Patterns of Discovery.* Cambridge: Cambridge University Press.

Harvey, W. 1962. *On the Motion of the Heart and Blood in Animals.* Chicago: Regnery.

Hempel, C. G. 1952. *Fundamentals of Concept Formation.* Chicago: University of Chicago Press.

1965a. *Aspects of Scientific Explanation.* New York: Macmillan.

1965b. *The Philosophy of Natural Science.* Englewood Cliffs, N.J.: Prentice-Hall.

Hennig, W. 1966. *Phylogenetic Systematics.* Urbana, Ill.: University of Illinois Press.

Hull, D. 1968. "The Operationalist Imperative, Sense and Non-sense in Operationalism." *Systematic Zoology,* 17:432–59.

1970. "Contemporary Systematic Philosophies." *Annual Review of Ecology and Systematics,* 1:19–53.

1974. *The Philosophy of Biological Science.* Englewood Cliffs, N.J.: Prentice-Hall.

1976. "Are Species Individuals?" *Systematic Zoology,* 25:174–91.

1978. "A Matter of Individuality." *Philosophy of Science,* 45:33–60.

Huxley, J. 1942. *Evolution: The Modern Synthesis.* London: Allen and Unwin.

Judson, H. F. 1979. *The Eighth Day of Creation.* New York: Simon and Schuster.

Kandel, E., and Schwartz, J. 1982. "Molecular Biology of Learning." *Science,* 218:433–43.

Kauffman, S. 1983a. "Developmental Constraints: Internal Factors in Evolution." In B. Goodwin, N. Holden, and C. Wyle (eds.), *Development and Evolution.* Cambridge: Cambridge University Press.

1983b. "Filling Some Epistemological Gaps: New Patterns of Inference in Evolutionary Theory." In P. Asquith and T. Nickles (eds.), *PSA 1982,* vol. 2. East Lansing, Mich.: Philosophy of Science Association.

Kim, J. 1978. "Supervenience and Nomological Incommensurables." *American Philosophical Quarterly,* 15:149–56.

Kitcher, P. 1978. "Theories, Theorists, and Theoretical Change." *Philosophical Review,* 87:389–406.

1982. "Genes." *British Journal for the Philosophy of Science,* 33:337–59.

1984. "Species." *Philosophy of Science.* 51:308–33.

Forthcoming. *Species.* Cambridge, Mass.: MIT Press.

Kuhn, T. 1970. *The Structure of Scientific Revolutions.* 2d ed. Chicago: University of Chicago Press.

Laudan, L. 1981. "The Pseudo-Science of Science." *Philosophy of Social Sciences,* 11:173–98.

Lerner, I. M. 1958. *The Genetic Basis of Selection.* New York: Wiley.

Levins, R. 1968. *Evolution in Changing Environments.* Princeton, N.J.: Princeton University Press.

Levins, R., and Lewontin, R. 1980. "Dialectics and Reductionism." *Synthèse,* 43:47–78.

Lewontin, R. 1974. *The Genetic Basis of Evolutionary Change.* New York: Columbia University Press.

    1983. "The Ghost in the Elevator." *New York Review of Books,* 29:34–7.

Lewontin, R., and Levins, R. 1982. "The Problem of Lysenkoism." In H. Rose and S. Rose (eds.), *The Radicalization of Science.* London: Allison and Busby.

Loeb, J. 1912. *The Mechanistic Conception of Life.* Chicago: University of Chicago Press. Reissued 1964; Cambridge, Mass.: Belknap Press.

Lotka, A. J. 1925. *Principles of Physical Biology.* Baltimore: Williams and Wilkins.

Mainx, F. 1939. *Foundations of Biology.* Chicago: University of Chicago Press.

May, R. M. 1973. *Stability and Complexity in Model Ecosystems.* Princeton, N.J.: Princeton University Press.

Mayo, O. 1983. *Natural Selection and its Constraints.* New York: Harcourt Brace Jovanovich.

Mayr, E. 1942. *Systematics and the Origin of Species.* New York: Columbia University Press.

    1963. *Animal Species and Evolution.* Cambridge, Mass.: Harvard University Press.

    1970. *Populations, Species and Evolution.* Cambridge, Mass.: Harvard University Press.

    1976. *Evolution and the Diversity of Life.* Cambridge, Mass.: Harvard University Press.

    1982. *The Growth of Biological Thought.* Cambridge, Mass.: Harvard University Press.

Midgley, M. 1979. "Gene-Juggling." *Philosophy,* 54:439–58.

Milkman, R. (ed.). 1982. *Perspectives on Evolution.* Sunderland, Mass.: Sinauer.

Mills, S., and Beatty, J. 1979. "The Propensity Interpretation of Fitness." *Philosophy of Science,* 46:263–86.

Monod, J. 1971. *Chance and Necessity.* New York: Knopf.

Monod, J., Jacob, F., and Changuex, J. P. 1963. "Allosteric Proteins and Cellular Control Systems." *Journal of Molecular Biology,* 6:306–29.

Nagel, E. 1961. *The Structure of Science.* New York: Harcourt, Brace and World. Reissued 1979; Indianapolis: Hackett.

    1977. "Teleology Revisited." *Journal of Philosophy,* 74:261–301.

Peirce, C. S. 1877. "The Fixation of Belief." *Popular Science Monthly,* 12:1–15.

Peters, R. H., 1976. "Tautology in Evolution and Ecology." *American Naturalist,* 110:1–12.

Pielou, E. C. 1969. *Introduction to Mathematical Ecology.* New York: Wiley.

    1979. *Population and Community Ecology.* New York: Gordon and Beach.

Popper, K. 1963. *The Logic of Scientific Discovery.* New York: Harper & Row.

    1974. "Darwinism as a Metaphysical Research Programme." In P. A. Schillp (ed.), *The Philosophy of Karl Popper.* LaSalle, Ill.: Open Court.

Prout, T. 1969. "The Estimation of Fitness from Population Data." *Genetics,* 63:949–67.

    1971. "The Relation Between Fitness Components and Population Prediction in *Drosophila.*" *Genetics,* 68:127–49, 151–67.

Provine, W. 1971. *The Origins of Theoretical Population Biology.* Chicago: University of Chicago Press.

Quine, W. V. O. 1961. *From a Logical Point of View.* New York: Harper & Row. 1960. *Word and Object.* Cambridge, Mass.: MIT Press.

Rosenberg, A. 1978. "Supervenience of Biological Concepts." *Philosophy of Science,* 45:368–86.

1982. "Causation and Teleology in Contemporary Philosophy of Science." In *Contemporary Philosophy,* vol. 2:51–86. Hague: Nijhoff.

Ruse, M. 1973. *The Philosophy of Biology.* London: Hutchinson.

1979. *The Darwinian Revolution.* Chicago: University of Chicago Press.

1981. *Is Science Sexist?* Dordrecht, The Netherlands: Reidel.

Ms. "Species are Not Individuals."

Schaffner, K. 1967. "Approaches to Reduction." *Philosophy of Science,* 34:137–47.

1969. "The Watson-Crick Model and Reductionism." *British Journal for the Philosophy of Science,* 20:325–48.

Schopf, T. J. M. (ed.). 1972. *Models in Paleobiology.* San Francisco: Freeman.

Simpson, G. G. 1949. *The Meaning of Evolution.* New Haven, Conn.: Yale University Press.

1961. *The Principles of Animal Taxonomy.* New York: Columbia University Press.

Smart, J. J. C. 1963. *Philosophy and Scientific Realism.* London: Routledge and Kegan Paul.

Sober, E. 1981. "Evolution, Population Thinking and Essentialism." *Philosophy of Science,* 47:350–83.

Sokal, R. R., and Camin, J. H. 1965. "The Two Taxonomies: Areas of Agreement and Conflict." *Systematic Zoology.* 14:175–95.

Sokal, R. R., and Sneath, P. H. A. 1963. *Principles of Numerical Taxonomy.* San Francisco: Freeman.

Sommerhoff, G. 1950. *Analytical Biology.* New York: Oxford University Press.

Stebbins, G. L. 1982. "Modal Themes: A New Framework in Evolutionary Synthesis." In R. Milkman (ed.), *Perspectives in Evolution.* Sunderland, Mass.: Sinaur.

Steele, E. 1979. *Somatic Selection and Adaptive Evolution.* Toronto: Williams and Wallace.

Strickberger, M. 1968. *Genetics,* 1st ed. New York: Macmillan.

Stryer, L. 1981. *Biochemistry.* 2d ed. San Francisco: Freeman.

Suppe, F. 1974. *The Structure of Scientific Theories.* Urbana: University of Illinois Press.

Taylor, C. 1964. *The Explanation of Behavior.* London: Routledge and Kegan Paul.

Thompson, P. 1982. "The Structure of Evolutionary Theory: A Semantic Approach." *Studies in the History and Philosophy of Science,* 14:215–30.

Van Valen, L. 1976. "Ecological Species, Multispecies, and Oaks." *Taxon,* 25:233–9.

Volpe, E. P. 1967. *Understanding Evolution.* Dubuque, Iowa: W. C. Brown.

Volterra, V. 1926. "Variazione e Fluttuazini del Neuro D'individui In Specie Animali Conviventi." *Memoria Academica Nazionale Lincea,* series 6, no. 2:31–113.

Waddington, C. H. 1968. *Towards a Theoretical Biology.* Chicago: Aldine.

Watson, J., and Crick, F. H. C. 1953. "A Structure for Deoxyribose Nucleic Acid." *Nature,* 171:737–8.

Williams, M. B. 1970. "Deducing the Consequences of Evolution." *Journal of Theoretical Biology,* 29:343–85.

1973. "Falsifiable Predictions of Evolutionary Theory." *Philosophy of Science,* 40:518–37.

1982. "The Importance of Prediction Testing in Evolutionary Biology." *Erkenntnis,* 17:291–306.

Wilson, E. O. 1975. *Sociobiology: The New Synthesis.* Cambridge, Mass.: Harvard University Press.

Wimsatt, W. 1978. "Reduction and Reductionism." In P. Asquith and H. Kyberg (eds.), *Current Issues in the Philosophy of Science.* East Lansing, Mich.: Philosophy of Science Association.

Woodfield, A. 1976. *Teleology.* Cambridge: Cambridge University Press.

Woodger, J. H. 1937. *The Axiomatic Method in Biology.* London: Routledge and Kegan Paul.

1952. *Biology and Language.* Cambridge: Cambridge University Press.

Wright, L. 1976. *Teleological Explanation.* Berkeley: University of California Press.

Index

accidental generalizations, 108, 207, 211; *see also* boundary conditions

adaptation, 31, 33, 121, 126, 129, 151, 152, 216; and functionalism, 46, 47, 48, 50, 58; and levels of organization, 149–50; and molecular biology, 66, 67; as research program, 228, 236–43, 245–46

adaptationalists, 150, 178, 235–46, 248; theory, 238–40

adaptive zone, 198–99; *see also* niche

adenosine, 3',5'-monophosphate (cyclic AMP), 104–5, 264

adenosine triphosphate (ATP), 61, 104

adequacy, of axiomatization, 144–52

ad hoc corrections, 159, 160; devices, 191

agendas in philosophy of biology, 15

allosterism, 54, 79, 80, 81, 83; *see also* enzymes; macromolecules; proteins

altruism, 247–51, 252–54, 255, 261, 262; defined, 248

amino acids, 38, 74–79, 84, 247, 255, 256, 261; allosteric shifts in, 80; differences in hemoglobins, 82–83, 118; linear sequence, 74, 77, 79, 80, 102, 232

anagenesis, 194–96, 197

antiessentialists, 188–91

antireductionist, 88–93, 116, 122; *see also* autonomists

Aplysia californica, 262–64

aptations, 243–44; defined, 244

Aristotle, 3, 45, 63, 70, 183, 204

artificial selection, 93, 170–71; analogical role, 171

asexual species, 192–94, 197, 198

assortment, principle of, 31, 93, 96, 100, 106, 130, 134–35, 176, 216; stated, 93, 132

astronomy and cosmology, 27, 174

atomic theory, 16, 24, 84, 181, 186–89, 190, 198

autonomists, 126, 213, 223, 228; and antiessentialism, 190; and reduction, 90, 111, 119; and teleology, 45–46, 47, 49, 60, 62, 63, 64, 65

autonomy, 59, 63, 117, 164, 182, 224–25, 226, 227, 235, 242, 265; defined, 16; *see also* practical autonomy

autoradiography, 176

auxiliary hypotheses, 174–77, 178, 202, 239

Axelrod, R., 251–54

axioms, 131, 135, 219, 220, 221; one, 140, 167, 212; two, 140, 152, 167, 212; three, 141, 152, 167, 212; four, 143, 147, 152, 167, 212, 216–17; five, 144, 148, 152, 167, 212; advantages, 136; defined, 137; of theory of natural selection, 136–44, 151, 180, 212

Ayala, F., 36, 67, 152

Ayer, A. J., 11

backward causation, 43, 50, 52

balanced polymorphism, 82, 146, 147–48

Barnes, B., 36

bases, *see* DNA, structure

Bauplan, 237, 241

Beadle, G., 100

Beatty, John, 134, 161–63, 179

behavior, 52, 71, 74, 248, 249, 251, 255

behaviorism, 261, 262–65

Benesh, R., and Benesh, R., 78

Bennett, Jonathan, 68

Benzer, S., 93, 95, 97, 98, 112

biconditionals, 115, 165

biochemistry, 19, 23, 29, 72, 134; of directively organized systems, 39–41, 54, 58, 66; and genetics, 93, 98, 100, 104, 118, 227, 234; of hemoglobin, 73, 81, 83; and neuroscience, 260, 261, 264

"biological entity," 134–38, 148, 150, 152, 168, 178

biological laws, 23, 31, 32, 35, 69, 205–7, 212–13

biological species notion, 191–97; definition, 191, 194; in paleontology, 194–96

biosynthesis, 30; disjunctive, 104–5; and pathways, 102, 111, 134, 221
biotechnology, 84
Biston batularia, 142, 170
blood: circulation of, 46–47, 74; diseases of, 83; glucose maintenance, 53–57, 58, 60; transport function, 77–81, 83
Bloor, D., 36
Bohr, C., 78, 81; effect, 78
boundary conditions, 135, 150, 169, 177, 211, 212, 213, 215, 217, 224, 228
Braithwaite, R. B., 67
breeding experiments, 93, 94, 112; and identity of genes, 97, 98, 100, 102, 106–7
bridge principles, 131, 175
Burian, R., 179

Camin, J. H., 184
Canis lupus, 208
Canis familiaris, 171, 190, 205, 209
caribou, 148, 213–14
Carlson, E., 120
case studies, 211, 219–21, 227
causal chains, 42, 48, 51, 54, 55, 57, 84, 100, 104, 105, 192
causal descriptions and explanations, 33, 45, 48
causal laws, 48, 58, 63, 90, 126
causal mechanisms, 9, 10, 32, 41; from DNA to protein, 83, 101–3; *see also* heterogeneity
central dogma, 30
ceteris paribus clauses, 60, 105, 239
Changeux, J. P., 80
chemistry, 9, 13, 14, 22, 24, 26, 29, 31, 83, 121, 176; and molecular biology, 41–42, 74, 76, 234, 257, 261; and periodic table, 186, 188, 200, 202, 221, 227; and reductionism, 69, 71, 89, 90, 96, 243
chromosomes, 94–95, 102, 183, 206, 211; mapping, 94–95; structure, 230–33
circularity, of theory of natural selection, 127–29, 145, 154–55, 160, 173
cis–trans test, 98–99
cistron, 95–96, 97, 101, 112
cladogenesis, 194–96, 206
clan, 139
classes, 208
classification: as conventional, 186–87; *see also* systematics
Clausius, R. J. E., 218
clocks, 113–14
cloning, of genes, 85, 88, 232
Cnemidophorus, 210
cognitive processes and states, 246–47, 256, 260; *see also* "folk psychology," intentionality; recognition

competitive exclusion, 194; *see also* niche
complementary DNA (cDNA), 85–87
computer, 84, 109
conditioning, 263–64
confirming evidence, for evolutionary theory, 174–78; in taxonomy, 186–87
conformation, and molecular shape, 73, 75, 76, 81; *see also* primary sequence and structure
connectibility criterion in reduction, 91–93, 107, 110, 111, 117
conserved structure, 89, 168
constraints: on adaptation, 48, 50, 237, 244; physical, 241–43
content of mental states, 257–58
convenience, of definitions, 186–87
correction: in measurement, 156–59, 173, 258–59; in reduction, 96
Crick, F., 24, 36, 66, 69, 73, 88, 94, 112, 119
crossing-over, 111
Cummins, R., 68
Cygnus olor 183, 204
cytology, 112, 175–77

Darden, L., 106
Darwin, Charles, 1, 2, 46, 47, 132, 134, 150, 151, 153, 170, 172, 179, 182, 183, 190, 201, 218, 225, 237, 239
Darwinian subclan, 140–46, 179
Darwinian theory, 30, 31, 89, 118, 119, 123, 127, 130, 133, 150, 153, 177, 201, 218, 240; *see also* theory of evolution; theory of natural selection
Dawkins, R., 248–51, 252, 261, 263
deamination of cytosine, 39–40, 59, 61, 168
deduction of theorems, 93, 213
defection, 252–53
defined terms, 137–38
definitions, 5; of adaptation, fitness, selection, 127; behavioral, 248; essentialist, 188; of species, 181, 187, 192, 204–29; *see also* redefinition; *under individual species notions*
Descartes, R., 246
descent with modification, 145, 180
design criterion, 157–58, 159, 173
determinism, 7, 8, 62–64, 110, 111, 116, 217, 225
development, 30, 37, 51, 71, 175, 219, 237–38, 240, 241–43
dialectical materialism, 126
Dickerson, R., 119
Didus ineptus, 183, 188, 190, 200, 205
differences of degree vs. kind, 26, 28, 29, 65, 225
differential: perpetuation, 145; reproduction, 174

dimensions: of evolutionary environment, 156; of niche, 200

2,3-diphosphoglycerate (DPG), 79, 81

directively organized system, 53–57, 58, 59, 61, 65, 66, 67, 80, 118, 227; conditions on, 53, 55, 57

disjunction of chromosomes, 134

disjunctiveness: and fitness, 164; of gene definition, 95, 101; of genes, 102–4, 106; of nonteleological accounts, 61, 62, 260; of pathways, 101, 104, 106; as result of evolution, 149, 169, 182, 222–23, 227, 233; and species, 200; and supervenience, 118, 125, 164, 223

disposition, fitness as, 160–64

diversity, 129, 149, 150, 170; and species, 183, 186, 187, 190, 194, 198, 199

DNA, 33, 66, 77, 101, 102–103, 106, 222, 247, 255, 261; complementary, 84–86, 88; hybridization, 66, 176; structure, 38, 69; synthetic, 84; and thymine, 38–42, 49–50, 59, 61, 68

DNA-polymerase, 42, 61, 85, 96, 255, 261

Dobshansky, T., 152

Dollo's law, 209

Driesch, Hans, 2, 3, 4, 8, 9, 11, 36

Drosophila melanogaster, 11, 95, 97, 98, 100, 101, 115, 134, 183, 243

Duhem, P., 12

ecological species notion, 198–200

ecology, 18, 25, 27, 30, 146, 175, 179, 213, 219

Einstein, A., 91–92, 96

Eldredge, N., 153

electromagnetism, 8, 30, 70, 71, 140, 220

electrophoresis, 30, 86–87, 176

element, 34, 181, 187, 188, 202; compared to species, 189, 190, 200, 201–2

embryology, 2, 30, 31, 37, 38, 71, 130, 219

empirical content, 124, 128, 136, 154–55, 158

empirical generalizations, and regularities, 135; biological, 207, 212–16, 219; see also biological laws; laws

endonuclease, 66, 96, 102; in genetic engineering, 84–87

Engels, F., 29

engineering, 24–5, 157, 223, 224

English moth, 142, 170

entelechy, 3, 8, 9

entropy, 143, 217, 218

environment, 140, 152, 155–56, 157, 164–66

enzyme kinetics, 27, 30

enzymes, 54, 67, 73, 84, 255, 264; allosterism of, 54, 81; in glucose–glycogen conversion, 53–57; as phenotype, 100, 101

epinephrine, 54–56

epistemology, 9, 10, 11, 62, 65, 110, 111, 112, 178; and autonomy vs. provincialism, 13, 20, 21, 22, 23, 220, 224, 226; double standard, 21, 25, 90

equilibrium theorem, 145–46, 148, 213; Hardy-Weinberg, 132–33

Escherichia coli (E. coli), 85, 173, 183, 190

essentialism, 187–91, 201; defined, 188; impediment to evolutionary thinking, 189–90; and periodic table, 188–89

eucaryotic genome, 104, 135, 230–31

evolution: course of, 124–27; mechanisms of, 130–31; see also natural selection

evolutionarily stable strategy, 252–53

evolutionary forces, 31, 33, 89, 118, 119, 136, 140, 142, 149, 171, 212, 216, 228, 231, 241

evolutionary laws, 35, 122, 125–26, 211, 215, 223

evolutionary roles, 197–98

evolutionary species notion, 197–98, 210

evolutionary theory, 25, 34, 46, 47, 48, 50, 58, 169–79; axiomatized, 136–44; existence of, 122–26; as history, 124–26; as implicit definition, 128–29; see also axioms; theory of evolution; theory of natural selection

exaptation, 243–45; defined, 244

existence claims, 140, 144, 146, 170, 214; see also generic claims

expansion of the fitter, 143

expected number of offspring, 162–63

experimentation, 27, 67, 83, 98, 108, 232; see also laboratory experiments

explanation: trade-off with prediction, 66–67; weakness of evolutionary theory, 174; weakness of teleological, 67; see also prediction

extinction, 144, 206, 209–10

extremal theories, 238–40

fair die, 158

falsifiability, 4, 6–8, 239; see also circularity; unfalsifiability

feedback and feed-forward loops, 53, 79, 80–81, 104

Felus domesticus, 171

fetus, 74, 80–81

Feyerabend, P., 35, 119

finitude of nature, 62–64, 110, 111, 116, 177, 225

Fisher, R. A., 143, 239

fitness, 33, 35, 51, 118, 121, 125, 143, 152, 252–54; definition, 127, 129, 154, 161; differences, 144, 147, 150, 151, 158, 162, 169, 218; as disposition, 160–64; measured by effects, 156–60, 162; operational defini-

fitness (*cont.*)
    tion of, 141–42; predictive weakness, 175; as
        primitive term, 141, 152, 154, 159; super-
        fluousness of, 173–74; supervenience of, 163,
        164–66, 168, 174, 218, 219, 240
Fogel, S., 3
"folk psychology," 259, 262
force, 128, 141, 240; *see also* evolutionary forces
fossil record, 195–96
founder principle, 216–17
Friefelder, D., 119
function, 9, 73, 110; as effect, 40–41; indepen-
        dence from structure, 89, 117–18; in mo-
        lecular biology, 39–42
functional explanation and description: bio-
        chemical, 39–42, 223, 228–34; discharged,
        58, 59, 81, 168; and evolution, 168, 223
functional laws, 32; *see also* teleological laws
fundamental theorem, of natural selection, 143

Gabriel, M., 3
Galileo, 43, 69, 70, 91
game theory, 251–54
gas law, 57, 92, 96
Geis, I., 119
gene complex, 98, 100, 105
gene expression, 109
gene frequencies, 130, 132, 133
generality and precision, 18, 21, 29, 66, 105,
        125, 220, 235
generalizations, 148, 205, 212–16; *see also* em-
        pirical generalizations
gene regulation, 81, 102–3, 134, 168
generic claims, 48, 51, 65, 147, 151, 174,
        214, 217; *see also* existence claims
genes, 31; definition, 95; *see also* DNA; globin
        gene; localization; Mendelian genes; molecular
        genes
Genesis, 174
genetic drift, 143, 162, 237; *see also* statistical
        character of evolutionary theory
genetic engineering, 84, 88
genome, 132–33, 148; molecular, 101–4; as
        program, 51–52
genotypes, 132–33, 148; and species, 192–94
Ghiselin, M., 204, 225
globin gene, 66–67, 89, 232–34
Glucagon, 54–56
glucose and glycogen, 53–57, 60
goals, 9, 32; direction, 32, 34, 37, 60, 138
God, 1, 44, 246
Goudge, T., 36, 124, 153
Gould, S. J., 36, 153, 179, 196, 225, 235–45,
        248
gradualism, 145, 149, 150–51, 177, 178, 195–
        96, 240, 245
"grudging," 251, 252

Haldane, J. S., and J. B. S., 78, 80, 81, 239
Hamilton, W. D., 251–54
Hanson, N. R., 35
Hardy-Weinberg law, 132–33, 176
Harvey, W., 46–47, 74
heart, 46–47, 50–51, 60
hemoglobin gene, 168, 232, 234
hemoglobin molecule, 51, 74–83; conserved
        amino acids of, 77, 89, 117, 168; fetal, 80–
        81; function, 77–81, 117; in sickle-cell ane-
        mia, 82–83; structure, 75–77
Hempel, C. G., 35, 67, 119, 152, 179
Henning, W., 201
heredity, 149; mechanisms, and materials, 132–
        33; required by theory of natural selection,
        133, 136, 144, 147, 152, 169, 173, 176
heterogeneity, 64, 65, 105, 107, 125, 145,
        155, 221; of boundary conditions, 217; of
        fitness-determining features, 158, 164–65;
        gives way to uniformity, 169; *see also*
        disjunctiveness
heterozygote superiority, 82, 146, 147–48
highly repeated DNA, 230, 231, 234, 243–45
histone gene, 231
history, theory of evolution as, 124–26, 228
Hobbes, T., 246
Homo sapiens, 171, 190, 243; obstacles to study
        of evolution, 27, 170, 211, 243
Hooke's law, 240
host vs. parasite, 146
Hull, D., 35, 36, 67, 119, 155, 179, 204, 225
human action, 44, 246, 254, 260, 262
human mind: computational, informational lim-
        itations, 23, 106, 109, 110, 111, 116, 223,
        224, 226, 235
Huxley, J., 130, 132, 153
hydra, 209
hypothetico-deductive system, 131, 132, 135

ideal gas law, 91, 92, 94
identity conditions, 257–58, 260; *see also* breed-
        ing experiments
ideology, 19–20
immunology, 33, 134, 255
implicit definition, 128, 163
inclusive fitness, 249–50
independent assortment, law of, 93, 94; stated,
        93, 132; *see also* Mendel's laws
indeterminism, 63, 216–17
indirect confirmation, 177–78
individuals, species as, 204–8
information, 247, 255, 261; theory of, 256
insulin, 54–56, 73, 83, 87
intentionality, 33, 246, 249; analysis of, 256–
        58; and biochemistry, 256–65; circle, 258–
        59, 262
interbreeding, 192–97; and anagenesis, 195–97

intercalation, law of, 49
interpretations of axiom system, 138, 146–48, 150, 151, 167–69, 212
intervening sequence, *see* intron
intron, 66–67, 103–4, 230–31

Jacob, F., 80
Judson, H., 36, 77, 119
"junk" DNA, 67, 231

Kandel, E. R., 263–65
Kaufmann, S., 242–43
Kepler, J., 43, 70, 109
Keynes, J. M., 83
Kim, J., 120, 179
kinds, 34, 181, 204, 205, 206, 211, 219; nonnatural, 203; *see also* differences of degree vs. kind; natural kinds
kinetic theory of gases, 28, 70, 96, 108, 173
Kitcher, Philip, 97, 98, 120, 191, 210
Kreb's cycle, 221
Kuhn, T., 12, 35

laboratory experiments, 27, 108, 147, 172; and selection, 173–74; superfluousness of evolutionary theory, 174
Lamarkian theory, 129, 133–34, 175, 237
Laudan, L., 36
Lavoisier, A., 198
laws, 4, 7, 34, 67, 72, 82, 109, 111, 113, 136, 163, 219, 220–22, 262; *see also* biological laws
learning theory, 263–64
leghemoglobin, 232
Lerner, I. M., 126
levels of organization, 22, 48, 64, 94, 100, 117–19, 150, 164, 222–23, 227, 242
levels of selection, 145; *see also* levels of organization
Levins, R., 36, 179
Lewontin, R., 19, 36, 153, 235–42, 244, 245
Limax, 264
limb bud, chick, 37, 228
linear sequence, 74, 77, 79, 80, 89; *see also* primary sequence and structure
Linnaeus, 183, 203
liver cells, function of, 53–57, 58
localization, of genes, 94–95, 112, 228–35
Locke, John, 1
Loeb, J., 36
Logical Empiricism, *see* Logical Positivism
Logical Positivism, 2, 3, 4, 7, 9, 10, 11, 12, 14, 16, 17, 19, 21, 22, 110, 111
long run, 83, 174, 218
Lotka-Volterra model, 214

macromolecules, 27, 29, 32, 54, 67, 70, 72, 74, 81, 83, 231, 242, 261
macromutation, 175
Mainx, F., 36
maladaptation, 82, 175, 237, 239
malaria, 82; parasite, 82–83
Malthusian axiom, 140, 167
mammals, 170, 263
manageability: of explanation and prediction, 156; of measurement, 159; *see also* human mind
Marx, Karl, 29, 126
mass, 91–92, 128, 240
materialists, 9, 10, 29, 72
mathematical models in ecology, 213–15
Maull, N., 120
Maxwell, J. C., 6, 8, 218, 220
Mayo, O., 243
Mayr, E., 23, 24, 36, 46, 50, 51–52, 67, 88–89, 112, 115, 124–25, 191–97
meaningfulness, principle of, 3–4
measurement: of fitness, 157–60, 162; of temperature, 156
mechanism, 8, 9, 22, 28, 29, 30, 358; and mechanical causation, 44
meiosis, 93, 108, 206, 216, 231; as phenotype, 134–35, 175–76
melanin, 104–5
Mendeleev, D., 181, 186, 188, 189, 198
Mendelian genes, 31, 93, 94–95, 96–97, 100, 101
Mendelian theory, 30, 108; reduction of, 90–96, 152; and theory of evolution, 130–36, 147–48, 173, 222, 223; *see also* Mendel's laws; supervenience
Mendel's laws, 31, 96, 110, 111, 136, 148, 149, 216; assortment and segregation, 73, 93, 94, 132; consequences of evolution, 132, 134–35, 176; as local generalizations, 108, 132, 212; and meiosis, 108, 175–77, 212; stated, 93, 132
mental states, 260; *see also* cognitive processes and states; intentionality
messenger RNA (mRNA), 66, 84, 101, 102, 103
metaphor, 44, 52, 171, 247, 248, 249, 255–56, 261, 262
metaphysical commitments, 45, 63, 65, 72–73, 88, 110–12, 116, 224
metaphysics, 1, 5, 178, 220; and autonomy vs. provincialism, 11, 13, 20, 21, 22, 23, 24, 28, 29, 224, 226; and Logical Positivism, 6, 7, 8, 9, 10, 11; and reductionism, 74, 90, 111
methodology, 13, 20, 23, 72, 117, 220, 226, 234, 236–37, 238, 241, 261
microscope, 175–76

Midgley, M., 249, 252
Mills, S., 161–63, 179
mitochondria, 42, 51, 61, 104, 118, 228
mitosis, 93, 108, 175–76, 231
model systems, 109, 146, 219, 220, 228; see also case studies
moderately repetitive DNA, 230, 233, 234, 243
molecular biology, 152; and autonomy vs. provincialism, 16, 18, 19, 22, 27, 30, 32, 34, 36; and chemistry, 22, 41–42, 74, 76, 234; as sole biological theory, 219, 220–22, 242, 246, 263; and teleology, 37, 38, 41, 42, 49, 54, 59, 64
molecular genes, 102, 105, 106
molecular genetics, 62, 82, 88, 93, 165; and mendelian phenotype, 97
Monod, J., 36, 80, 81, 112, 119
Morgan, T. H., 97, 98
morphology, 30, 94, 130, 193, 194, 196, 198, 242
motives: for reduction, 69–73; vs. reasons, 18–20
mutation, 39, 40, 49, 59, 61, 67, 94, 95, 96, 97, 98–99, 102, 134, 231, 232–33, 237, 253–55
muton, 95, 96, 101, 112
myxomatosis, 170

Nagel, E., 35, 67, 119, 152
narrative view of theory of evolution, 124–26, 151
natural kinds, 34, 194, 196, 197, 201, 202–3, 211, 236
natural selection, 52, 89, 118, 126, 148, 164, 166, 242; and artificial selection, 170–71; compatibility with altruism, 250–51
Neospora, 100
nepotism, 250, 253
neuroscience, 246–47, 259, 261, 263
neutral variations, 178
new synthesis, see synthetic theory of natural selection
Newtonian mechanics, 14, 24, 28, 30, 31, 43, 70, 71, 91–92, 116, 122, 128–29, 130, 131, 177, 220, 238–40
Newton's laws, 6, 91, 109, 211, 240
niche, 148, 166, 194, 199–200
nondisjunction, 134–35, 176, 212, 254
nucleotides, see DNA; primary sequence and structure; sequence
numerical taxonomy, 183–86

Ohm's law, 6–8
operant, 263
operational definitions: of fitness, 141–42, 162; of species, 183–85, 191–92

operationalism, 159; philosophy of, 184–85, 203
optimal design, 157–58, 179
organic chemistry, 32, 40, 41, 52, 59, 71, 221
organicism, 21, 22, 24, 35
organization, 21, 24; see also directively organized system; levels of organization
orthogenesis, 175
osteichthyes, 203

paleontology, 11, 25, 30, 31, 51, 130, 151, 174, 177, 179, 194–96
pangenesis, 132
"Panglossian Paradigm," 243–46; see also adaptation; adaptationalists
panmixia, 132, 133, 192
"parent of" relation, 138, 150, 152, 178, 248
Pauling, L., 81
Peirce, C. S., 218
periodic table, 181, 186, 187, 206; see also element
Perutz, M., 75
phenetic taxonomy, 183–86
phenotype, 51, 95, 96, 97, 98, 99, 100, 101, 102, 106, 134, 150, 175, 205, 244
phlogiston, 202, 203, 211
photosynthesis, 257
physicalism, 45, 246; see also mechanism; provincialism; reductionism
physical theories and laws, 124, 130–31, 141, 148, 220; see also laws; Newtonian mechanics; quantum mechanics; theory
physics, 9, 23, 24, 25, 29, 30, 33, 41, 45, 52, 69, 110, 234, 243; nonteleological, 43–44; see also physics and biology
physics and biology: differences, 13, 14, 15, 16, 17, 22, 23, 26, 27, 28, 65, 122, 224, 227; similarities, 15, 224, 227; see also practical autonomy
physiology, 3, 30, 42, 53, 71, 74, 77, 81, 88, 90, 118, 134, 140, 221, 231, 245, 263
piecemeal explanations, 89, 108, 111, 117, 164, 173; and corrections, 158–59, 162
Pielou, E. C., 153
plasmids, 85–86
Plato, 206
Pleiotropy, 237
polyploidy, 206
polysaccharides, 257
Popper, K. R., 4, 12, 124, 153
population genetics, and theory of natural selection, 130–36, 147; see also Mendelian theory
Positivists, see Logical Positivism
postpositivists, 14, 15, 17, 18, 30, 35, 153
practical autonomy, 29, 62, 64, 110–11, 119, 226–27; and human needs, 221–23, 228

preadaptation, 244

precision, of ecological models, 214; of teleological description, 32, 61, 64

predation and prey, 61, 140, 146–47, 166, 213, 214, 217

prediction, 14, 16, 32, 64, 65, 69; and theory of natural selection, 126, 156, 173–74, 178, 213

primal soup, 169, 222–23, 228

primary sequence and structure, of macromolecules, 30, 61, 76–77, 80, 81, 83, 84, 85, 88, 89, 100, 102, 109, 118, 221, 232

primitive term, 137, 138, 140, 151; fitness as, 141, 164

prisoner's dilemma, 252–54

probability, 158, 161; epistemic, 216–17; in thermodynamics and evolution, 216–19

procaryotic genes, 229, 230

proofreading, 255, 261

propensity definition of fitness, 161–64, 179; stated, 161

proteins, 38, 75, 81, 85, 87, 88, 89, 102; synthesis, 103, 230; see also macromolecules

Prout, T., 153, 179

provincialism, 59, 265; defined, 16

provincialists, 45, 47, 48, 65, 72, 90, 107, 108, 126, 182, 207, 211, 212, 223–24, 226–27, 235; and essentialism, 190

Provine, W., 152

proximate vs. ultimate causation, 46, 50, 59

pseudogenes, 231, 232–33

psychology, 246, 262

punctuated equilibrium, 145, 149, 150–51, 153, 177, 178, 194–96, 244–45

purposive phenomena, 9, 32; see also function; goals; teleology

quantum mechanics, 7, 24, 28, 42, 43, 62, 63, 70, 71, 84, 94, 96, 109, 116, 129, 140, 217, 220, 238

quaternary structure, 77, 82

Quine, W. V. O., 12

radium, 189

Rattus rattus, 200

recessive traits, 30

recognition, 255, 256, 260, 261

recombinant DNA, 209–10

recombination, 94, 95, 96, 97, 98, 111

recon, 95–96, 101, 112

redefinition: in evolutionary biology, 247, 248; in molecular biology, 256, 260–61, 262, 264, 269

reduction, 35: definition of, 90–92; and genetics, 89–96; as a millennial program, 83, 88; opportunistic, 74, 81; and recombinant DNA,

84–88; revisions of definition, 111–12; schematic and piecemeal, 89, 109, 110

reductionism, 19, 23, 24; Mayr on, 88–89

reductionists, 88–91, 108, 110, 111, 122

redundancy, 65, 102, 233; see also disjunctiveness

reproductive isolation, 191–96, 210

reproductive rates: as estimates, 157, 158–59, 162, 165; and fitness, 155, 164, 165, 166–67

restriction enzymes, see endonuclease

reverse transcriptase, 84, 85

ribosomal RNA, 101

ribosomes, 85, 102, 118

RNA, 84, 102–3; and uracil, 38–42; see also messenger RNA (mRNA); ribosomal RNA; transfer RNA (tRNA)

Rosenberg, A., 67, 179

Ruse, M., 35, 47, 67, 120, 130–32, 147, 153, 154, 173, 179, 225

saltation, 175

Sanger, F., 83, 87

satellite DNA, 230, 243

Schaffner, K., 119, 120

Schlick, M., 11

Schopf, T., 153

Schwartz, R. H., 263–65

sea urchin, 2, 230

secondary structure, 76–77, 89, 100, 168; see also conformation

second law of thermodynamics, 16, 133, 141; statistical version, 143, 217–18

segregation: principle of, 31, 96, 106, 130, 134–35, 176; stated, 93, 132, 217

selection, 31, 133, 141; see also adaptation; natural selection

selective forces, see evolutionary forces

selfish DNA, 231, 244–45, 247–51

selfish genes, 249–51

selfishness, 247–51; defined, 248, 255, 261, 262

sequence: high-speed determination, 87, 232; molecular, 73

set theory, 208–9

sexual species, 109, 135, 176–77, 192, 212

Shapin, S., 36

short run, 174, 218

shuffling of introns, 231–32

sickle-cell anemia, 52, 82–83, 239

Simpson, G. G., 127, 197, 199, 225

single-copy genes, 230

skin color, 104–5

slime mold, 206, 209

Smart, J. J. C., 24–25, 36, 119

Sneath, P. N., 185, 186, 225

Sober, E., 190, 225, 266

social sciences, 262, 265

sociobiology, 174, 247, 263

Sokal, R. R., 184, 185, 186, 225

Sommerhoff, G., 67

special theory of relativity, 6, 43, 91–92, 116, 129

speciation, 149; causes of, 196–97, 206; explanations of, 201, 202; not a natural kind, 202

species, 33–34, 35, 137, 139, 170; as individuals, 204–8; not natural kinds, 202–3; selection of, 146, 148; *see also under individual species notions*

splicing of genes, 86

sponge, 206

stability and oscillation, 146–48, 213–15; of strategies, 252–53

statistical character of evolutionary theory, 143, 216–19; *see also* thermodynamics

statistical methods, 184, 192

statistical propensity, 161–63

Stebbins, G. L., 122, 125, 132, 133, 145, 148, 149, 151, 152, 153

Steele, E., 134

strategy and tactics, 26–28, 65, 227, 228

Strickberger, M., 161

structural genes, 102–4, 134

struggle for survival, 123

Stryer, L., 225

subclan, 139, 140–50, 151

subcland, 142–46, 147–48, 150, 151, 179, 216

supervenience, 112–17, 164–69; defined, 113; of evolutionary theory, 166, 224; example, 113–14; of fitness, 164–66; formal treatment, 113–16; in reduction of Mendelian genetics, 111–17, 223

Suppe, F., 153

surface-to-volume ratios, 215–16

survival of the fittest, 125, 143

synthetic theory of natural selection, 31, 82, 130–31, 147, 153, 175–77, 191

systematics, 130, 179, 181; as a theoretical undertaking, 186–87; *see also* taxonomy and systematics

tactics, *see* strategy and tactics

Tatum, L., 100

tautology, theory of natural selection as, 125, 126–29, 141, 154, 158–59

taxonomy and systematics, 35; explanations of, 181, 202; *see also* natural kinds; speciation; species

Taylor, C., 62–63, 68

teleological explanations and descriptions, 33, 43, 44; compatible with mechanical causation,

45; indispensible, 49, 64; weakness of, 46, 52, 65, 235

teleological laws, 30, 32, 35, 52, 60, 105; autonomy of, 49, 59–61; schematic form (statement T), 49, 52, 53, 58, 59, 61

teleology, 9, 10, 34, 35, 70; as metaphysical overlay, 44, 45, 47, 48; physical explanation of, 53–57, 112

temperature, 155, 156, 174, 258

terrestrial evolution, 124–26, 134, 135, 136, 150–51, 169, 209, 212, 224; *see also* boundary conditions

tertiary structure, 76–77, 83, 89, 100, 118

testing of theory of natural selection, 124, 138–39, 146, 151, 169–79

theorem, 131, 132, 137; of theory of natural selection, 144–46; *see also* fundamental theorem

theory, 14, 30, 31, 64, 90, 108, 112, 136, 138, 148, 151, 158, 179, 219; defined, 121, 126

theory of evolution, 30, 31, 119; statistical character, 143, 216–19; *see also* theory of natural selection

theory of natural selection, 1, 34, 47; and essentialism, 189–90; ever disputed, 174; as extremal theory, 238–40; informal statement of, 123, 140, 180; and physical science, 152; and population genetics, 133; as sole biological theory, 219, 220–22, 242; *see also* Darwinian theory; evolution; natural selection; theory of evolution

thermodynamics, 16, 27, 30, 70, 93, 140, 143, 152, 156, 173, 217–18, 220, 238, 242, 258

thermometer, 156, 174, 258–59

Thompson, P., 153

tit-for-tat, 253–54

transfer RNA (tRNA), 84, 101, 247, 255, 260–61

triviality, of theory of natural selection, 126–30; *see also* tautology

tryptophan, 260

typology, theory of evolution as, 128

underlying mechanisms, 121; and essentialism, 188

unfalsifiability, 6–8, 10, 141, 159, 218

units of selection, 139, 145, 205

unity of science, 16, 17, 70, 72–73, 94, 107, 124

universality, of theory of natural selection, 152, 211, 212, 219

uracil–DNA glycosidase, 40, 59, 61, 168

Ursus ursus, 208

useful knowledge, 65; *see also* practical autonomy

vacuousness, of theory of natural selection, 124, 126–29, 158–59, 263
Valentine, J. W., 152
Van Valen, L., 198–200, 225
variation, 31, 123, 130, 133, 150, 233, 238, 240–41, 242, 253; and essentialism, 190–91; neutral, 149–50; *see also* punctuated equilibrium
virus, 230, 232
vitalism, 8, 9, 10, 20, 21, 22, 24, 36, 45, 63, 73, 90, 226
vocabulary of theory, 137
Volpe, E. P., 133
Vrba, E., 243–44

Waddington, C. H., 126
Watson, J., 69, 73, 77, 88, 94
wildlife cycles, 146–47
Williams, M. B., 136–46, 153, 162; *see also* axioms, of theory of natural selection
Wilson, E. O., 127
Wimsatt, W., 119
wolves, 148, 214
Woodfield, A., 68
Woodger, J. H., 11
Wright, L., 50, 68
Wright, S., 239

x-ray crystallography, 75, 81, 175, 176